Navigating Cyber-Physical Systems With Cutting-Edge Technologies

Ramesh Chandra Poonia
Christ University, India

Kamal Upreti
Christ University, India

Published in the United States of America by
IGI Global Scientific Publishing
701 East Chocolate Avenue
Hershey, PA, 17033, USA
Tel: 717-533-8845
Fax: 717-533-8661
E-mail: cust@igi-global.com
Website: https://www.igi-global.com

Copyright © 2025 by IGI Global Scientific Publishing. All rights reserved. No part of this publication may be reproduced, stored or distributed in any form or by any means, electronic or mechanical, including photocopying, without written permission from the publisher.
Product or company names used in this set are for identification purposes only. Inclusion of the names of the products or companies does not indicate a claim of ownership by IGI Global Scientific Publishing of the trademark or registered trademark.

Library of Congress Cataloging-in-Publication Data

ISBN13: 9798369357286
EISBN13: 9798369357309

Vice President of Editorial: Melissa Wagner
Managing Editor of Acquisitions: Mikaela Felty
Managing Editor of Book Development: Jocelynn Hessler
Production Manager: Mike Brehm
Cover Design: Phillip Shickler

British Cataloguing in Publication Data
A Cataloguing in Publication record for this book is available from the British Library.

All work contributed to this book is new, previously-unpublished material.
The views expressed in this book are those of the authors, but not necessarily of the publisher.
This book contains information sourced from authentic and highly regarded references, with reasonable efforts made to ensure the reliability of the data and information presented. The authors, editors, and publisher believe the information in this book to be accurate and true as of the date of publication. Every effort has been made to trace and credit the copyright holders of all materials included. However, the authors, editors, and publisher cannot assume responsibility for the validity of all materials or the consequences of their use. Should any copyright material be found unacknowledged, please inform the publisher so that corrections may be made in future reprints.

Table of Contents

Preface xviii

Chapter 1
The Role of Blockchain in Cyber Physical Systems 1
    G. Sowmya, MLRIT, India
    R. Sridevi, JNTUH, India
    K. S. Sadasiva Rao, Sri Indu College of Engineering and Technology, India
    Sri Ganesh Shiramshetty, State University of New York, India

Chapter 2
Leveraging Data Analytics for Sustainable Investment Decisions: Evaluating NVIDIA and Microsoft 37
    Srinidhi Vasan, Hult International Business School, USA

Chapter 3
Generative AI for Cybersecurity and Privacy in Cyber-Physical Systems 57
    Arul Kumar Natarajan, Samarkand International University of Technology, Uzbekistan
    Yash Desai, Ramrao Adik Institute of Technology, India
    Pravin R. Kshirsagar, J.D. College of Engineering and Management, India
    Kamal Upreti, Christ University, India
    Tan Kuan Tak, Singapore Institute of Technology, Singapore

Chapter 4
Advanced Cyber-Physical Systems Utilizing Deep Learning for Crowd Density Detection and Public Safety 83
    R. Leisha, Christ University, India
    Katelyn Jade Medows, Christ University, India
    Michael Moses Thiruthuvanathan, Christ University, India
    S. Ravindra Babu, Christ University, India
    Prakash Divakaran, Himalayan University, India
    Vandana Mishra Chaturvedi, D.Y. Patil University, India

Chapter 5
Cyber-Physical Systems and the Future of Urban Living: Decision-Making Challenges and Opportunities 123
    Rituraj Jain, Marwadi University, India
    Kumar J. Parmar, Marwadi University, India
    Kushal Gaddamwar, PDPM IIIT Jabalpur, India
    Damodharan Palaniappan, Marwadi University, India
    T. Premavathi, Marwadi University, India
    Yatharth Srivastava, The LNM Institute of Information Technology, India

Chapter 6
Artificial Intelligence and Cybersecurity Prospects and Confronts 155
    Anya Behera, Alliance University, India
    A. Vedashree, Alliance University, India
    M. Rupesh Kumar, Alliance University, India
    Kamal Upreti, Christ University, India

Chapter 7
Ensuring Security in Vehicular Cyber Physical Using Flexray Protocol 185
    Neha Bagga, Guru Nanak Dev University, India & Lovely Professional University, India
    Sheetal Kalra, Guru Nanak Dev University, India
    Parminder Kaur, Guru Nanak Dev University, India

Chapter 8
Enhancing User Experiences in Cyber-Physical Systems for Real-Time Feedback and Intelligent Automation 215
    Vishakha Kuwar, Centre for Online Learning, Dr. D.Y. Patil Vidyapeeth, India
    Yatharth Srivastava, The LNM Institute of Information Technology, India
    Sonali Gaur, EMPI Business School, India
    Amit Yadav, Banarsidas Chandiwala Institute of Professional Studies, India
    Pratibha Bhide, MGM University, India
    Shitiz Upreti, Maharishi Markandeshwar, India

Chapter 9
Enhancing Human-Computer Interaction Through Artificial Intelligence and Machine Learning: A Comprehensive Review 235
    Neha Singh, Invertis University, India
    Jitendra Nath Shrivastava, Invertis University, India
    Gaurav Agarwal, Invertis University, India
    Akash Sanghi, Invertis University, India
    Swati Jha, Invertis University, India
    Kamal Upreti, Christ University, India
    Ramesh Chandra Poonia, Christ University, India
    Amit Kumar Gupta, KIET Group of Institutions, India

Chapter 10
Harnessing Real-Time Data for Intelligent Decision-Making in Cyber-Physical Systems 257
    T. Premavathi, Marwadi University, India
    Rituraj Jain, Marwadi University, India
    Vaishali Vidyasagar Thorat, D.Y. Patil School of Engineering and Technology, India
    Kumar J. Parmar, Marwadi University, India
    Damodharan Palaniappan, Marwadi University, India
    Chetana Vidhyasagar Thorat, Indiana University, Bloomington, USA

Chapter 11
Leveraging Big Data and Advanced Technologies for Enhanced Sustainability in Healthcare: An IPO Model Approach 287
    Mary Metilda Jayaraj, Dayananda Sagar College of Engineering, India
    Subbulakshmi Somu, Dayananda Sagar College of Arts, Science, and Commerce, India

Chapter 12
Unleashing Metaverse for Sustainable Development: Challenges and Opportunities 309
    Anjali Gautam, Manav Rachna International Institute of Research and Studies, India
    Priyanka Dadhich, Manav Rachna International Institute of Research and Studies, India
    Himanshu Gupta, Meta, USA
    Lakshay Rekhi, Sharda University, India
    Shitiz Upreti, Maharishi Markandeshwar, India
    Ramesh Chandra Poonia, Christ University, India
    Kamal Upreti, Christ University, India

Chapter 13
Penetration Testing: A Way to Secure IT Industries 327
    Aditya Sharma, Ajeenkya D.Y. Patil University, India
    Amna Kausar, Ajeenkya D.Y. Patil University, India
    Atharva Saraf, Ajeenkya D.Y. Patil University, India
    Susanta Das, Ajeenkya D.Y. Patil University, India

Chapter 14
The Global Trends and Hotspots of Medical Internet of Things Research: A
Bibliometric Analysis During 2004 to 2023 343
    M. S. Rajeevan, Indian Institute of Technology, Hyderabad, India
    V. S. Anoop, Kerala University of Digital Sciences, Innovation, and
        Technology, India
    D. Narayana, Central Tribal University of Andhra Pradesh, India

Chapter 15
Real-Time Data Analytics and Decision Making in Cyber-Physical Systems .. 373
    Vishakha Kuwar, Symbiosis International University, India
    Vandana Sonwaney, Symbiosis International University, India
    Shitiz Upreti, Maharishi Markandeshwar, India
    Shubham Rajendra Ekatpure, Kulicke and Soffa, USA
    Prakash Divakaran, Himalayan University, India
    Kamal Upreti, Christ University, India
    Ramesh Chandra Poonia, Christ University, India

Compilation of References 391

About the Contributors 435

Index 443

Detailed Table of Contents

Preface xviii

Chapter 1
The Role of Blockchain in Cyber Physical Systems 1
    G. Sowmya, MLRIT, India
    R. Sridevi, JNTUH, India
    K. S. Sadasiva Rao, Sri Indu College of Engineering and Technology, India
    Sri Ganesh Shiramshetty, State University of New York, India

This abstract explores how blockchain enhances security, transparency, and efficiency in cyber-physical systems (CPS). It addresses CPS vulnerabilities and emphasizes the need for robust security measures. It explains blockchain's decentralized architecture and cryptographic protocols, showing how it tackles data tampering and unauthorized access. The abstract discusses various blockchain applications in CPS, such as secure data logging and decentralized control. Real-world examples demonstrate benefits like enhanced resilience and transparency. It also covers implications for system performance and regulatory compliance, highlighting blockchain's transformative potential in bolstering CPS security.

Chapter 2
Leveraging Data Analytics for Sustainable Investment Decisions: Evaluating
NVIDIA and Microsoft.. 37
    Srinidhi Vasan, Hult International Business School, USA

Leveraging data-driven decision-making in sustainable management systems is crucial for refining stock market investment strategies. This research examines the use of data analytics to assess NVIDIA and Microsoft, focusing on their historical performance. We employ the Discounted Cash Flow (DCF) model for intrinsic valuation centered on projected cash inflows and the Relative Valuation model to compare each stock to its industry peers. Additionally, scenario-based analysis is used to evaluate various market conditions and potential future developments. Our findings indicate that integrating these analytical models with scenario analysis enhances the precision of stock valuations and supports more informed investment decisions. This methodology demonstrates how financial analysts and investors can utilize advanced tools to navigate market complexities. The study concludes by determining which stock, between NVIDIA and Microsoft, appears more promising and establishes a framework for future research in financial decision-making and sustainable investment practices.

Chapter 3
Generative AI for Cybersecurity and Privacy in Cyber-Physical Systems 57
    Arul Kumar Natarajan, Samarkand International University of
        Technology, Uzbekistan
    Yash Desai, Ramrao Adik Institute of Technology, India
    Pravin R. Kshirsagar, J.D. College of Engineering and Management,
        India
    Kamal Upreti, Christ University, India
    Tan Kuan Tak, Singapore Institute of Technology, Singapore

With the proliferation of Cyber-Physical Systems (CPS) across various domains, ensuring robust cybersecurity and privacy has become increasingly critical. Generative Artificial Intelligence (AI) presents innovative approaches to enhancing the security and privacy of these systems. This book chapter explores the intersection of Generative AI with cybersecurity and privacy within CPS environments. It examines how Generative AI techniques, such as Generative Adversarial Networks (GANs) and Variational Autoencoders (VAEs), can be leveraged to detect and mitigate cyber threats and vulnerabilities while protecting sensitive data and user privacy. The chapter provides an overview of CPS, addressing its unique security and privacy challenges, and demonstrates the practical application of Generative AI through a case study on phishing detection using BERT-based sequence classification. The experimental results highlight the effectiveness of Generative AI in strengthening CPS security.

Chapter 4
Advanced Cyber-Physical Systems Utilizing Deep Learning for Crowd
Density Detection and Public Safety 83
    R. Leisha, Christ University, India
    Katelyn Jade Medows, Christ University, India
    Michael Moses Thiruthuvanathan, Christ University, India
    S. Ravindra Babu, Christ University, India
    Prakash Divakaran, Himalayan University, India
    Vandana Mishra Chaturvedi, D.Y. Patil University, India

This study aims to detect the increasing crowd density, which is crucial, especially in dynamic environments like festivals or concerts. By harnessing cyber-physical systems and cutting-edge technologies, we have employed computer vision and deep learning to create a reliable model for accurate crowd counting. Utilizing deep learning, renowned for its ability to handle image-related tasks, the system gives decision-makers precise crowd density estimates, enabling well-informed actions such as crowd control measures. This study aims to improve safety and security in crowded areas by delivering an efficient system that can identify and quantify high crowd densities by integrating deep learning and computer vision technologies within a cyber-physical system framework. This approach facilitates proactive measures to mitigate safety risks and optimize crowd management strategies. The study seeks to advance safety and security measures in crowded areas by delivering an efficient and comprehensive crowd density detection and analysis system using cutting-edge technologies.

Chapter 5
Cyber-Physical Systems and the Future of Urban Living: Decision-Making
Challenges and Opportunities 123
    Rituraj Jain, Marwadi University, India
    Kumar J. Parmar, Marwadi University, India
    Kushal Gaddamwar, PDPM IIIT Jabalpur, India
    Damodharan Palaniappan, Marwadi University, India
    T. Premavathi, Marwadi University, India
    Yatharth Srivastava, The LNM Institute of Information Technology,
        India

Cyber-physical systems, where software and physical infrastructure are intertwined are promising technologies for re-imagining urban life. As CPS apply tightly coupled networks of sensors and actuators, supported by advanced data analytics, the field has great potential for addressing many aspects of city life and improving the quality of life of city inhabitants. Global urban development and itself increasing complication of the processes involved have produced considerable demand for fresh approaches to various local problems, extending from transport management to energy consumption, disease prevention, and preservation of natural resources. Moreover, CPS can be helpful in conservative consumption of resources: energy, water supply, and others in cities. Addressing these challenges via adaptive policies and frameworks, ethical imperatives, as well as special attention to the principles of smart design, will be vital for the successful translation of the potentials of CPS for improved efficiency, sustainability, and equity in urban settings.

Chapter 6
Artificial Intelligence and Cybersecurity Prospects and Confronts 155
    Anya Behera, Alliance University, India
    A. Vedashree, Alliance University, India
    M. Rupesh Kumar, Alliance University, India
    Kamal Upreti, Christ University, India

The advancement of artificial intelligence has made robust cybersecurity essential. As governments implement various policies to protect citizens, the utilization of AI by governmental agencies has significantly increased. Despite this, there is a significant gap between theoretical knowledge of AI and cybersecurity as separate fields and their practical integration. The challenge lies in the rapid evolution of both AI and cybersecurity, which often outpaces the legislative process, rendering many regulatory efforts outdated before they are fully realized. Due to the lack of dedicated laws for these dynamic areas, achieving optimal results is difficult. Therefore, it is crucial to explore how AI can be leveraged to improve cybersecurity. By using appropriate safeguards, AI can autonomously protect against threats like viruses, misuse, and hacking attacks. This paper examines the role of technology in AI and cybersecurity and investigates how AI can optimize cybersecurity, the opportunities & challenges of AI in cybersecurity, regulatory bodies, and strategies to merge these fields.

Chapter 7
Ensuring Security in Vehicular Cyber Physical Using Flexray Protocol 185
    Neha Bagga, Guru Nanak Dev University, India & Lovely Professional
        University, India
    Sheetal Kalra, Guru Nanak Dev University, India
    Parminder Kaur, Guru Nanak Dev University, India

With evolving automotive technology V2V communication will follow an evolutionary path as well alerts are provided to driver to maintain safety on road and take timely decision for received warnings. V2I helps vehicle to share information with roadside components of Intelligent Transportation Systems. All the communication above is susceptible to be intercepted and wrong messages can be communicated by exploiting the integrity, or act of cyber terrorism can be performed. Erroneous communication done by attackers can lead to opening of airbags while driving, giving wrong indication of turning of vehicle, which in turn can cause loss of human life, damage to vehicles. ECU's are the easiest target for the attackers to gain access into the vehicle as communication protocols like CAN, LIN, FlexRay, MOST and Ethernet are connected to the ECU's. In this chapter authors would be discussing the Flexray protocol specifications which can be exploited to perform attacks and the corresponding potential security and safety effects of these attacks and propose some futuristic security protections.

Chapter 8
Enhancing User Experiences in Cyber-Physical Systems for Real-Time
Feedback and Intelligent Automation 215
    Vishakha Kuwar, Centre for Online Learning, Dr. D.Y. Patil Vidyapeeth,
        India
    Yatharth Srivastava, The LNM Institute of Information Technology,
        India
    Sonali Gaur, EMPI Business School, India
    Amit Yadav, Banarsidas Chandiwala Institute of Professional Studies,
        India
    Pratibha Bhide, MGM University, India
    Shitiz Upreti, Maharishi Markandeshwar, India

The hominize of cyber-physical systems (CPS) has create evolution in the user experiences for real-time feedback and intelligent automation. The study explores the advanced sensors and monitoring technologies in cyber-physical systems offer meaning to the user interactions with products, trying massive responsive and adaptable environments. CPS is facilitating the direct feedback and achieve the product usage patterns to discover and implement value-adding features is not explicitly request but would significantly appreciate. The chapter is also discussing these technological advancements. The study illustrating safety and convenience. The chapter is also highlighting -the transformative impact of CPS on product interaction and user satisfaction.

Chapter 9
Enhancing Human-Computer Interaction Through Artificial Intelligence and
Machine Learning: A Comprehensive Review 235
> *Neha Singh, Invertis University, India*
> *Jitendra Nath Shrivastava, Invertis University, India*
> *Gaurav Agarwal, Invertis University, India*
> *Akash Sanghi, Invertis University, India*
> *Swati Jha, Invertis University, India*
> *Kamal Upreti, Christ University, India*
> *Ramesh Chandra Poonia, Christ University, India*
> *Amit Kumar Gupta, KIET Group of Institutions, India*

The interaction between Human-Cyber-Physical Systems (CPS) has become increasingly critical as CPS technologies permeate various facets of modern life, from smart homes to industrial automation. This highlights the evolving landscape of research aimed at fostering smooth interaction between humans and CPS, stressing the necessity of bridging the gap between users and these intricate systems.Effective interaction with CPS requires a profound understanding of human behaviors, preferences, and cognitive processes. .Furthermore, this emphasizes notable research trends aimed at improving Human-CPS interaction, including the exploration of innovative interaction modalities such as natural language processing, gesture recognition, and brain-computer interfaces.Thanks to Artificial Intelligence (AI) and Machine Learning (ML), computer interactions are changing a lot.

Chapter 10
Harnessing Real-Time Data for Intelligent Decision-Making in Cyber-
Physical Systems 257
    T. Premavathi, Marwadi University, India
    Rituraj Jain, Marwadi University, India
    Vaishali Vidyasagar Thorat, D.Y. Patil School of Engineering and
        Technology, India
    Kumar J. Parmar, Marwadi University, India
    Damodharan Palaniappan, Marwadi University, India
    Chetana Vidhyasagar Thorat, Indiana University, Bloomington, USA

Cyber physical systems are already transforming different fields, including smart communities and energy systems. These technologies enable CPS to process large volumes of data and come up with insights that enhance processes of taking preventive actions for situations that require quick responses. In addition, we explain how both 5G and edge computing are set to disrupt data handling and transmission as well as outline how both concepts will fit well together in a resource limited environment. Points regarding data quality issues, system architecture, and security threats; resource capacity and the degree of persistence, respectively, are included. For these challenges, we offer solutions like Data management, Modularity or System Decomposition, Security in Layers, and Light-Weight Processing for Improved System Resilience. Lastly, the chapter discusses the current and potential advances in real-time decision making for CPS and the need for CPSs to be interoperable with other CPS.

Chapter 11
Leveraging Big Data and Advanced Technologies for Enhanced
Sustainability in Healthcare: An IPO Model Approach 287
    Mary Metilda Jayaraj, Dayananda Sagar College of Engineering, India
    Subbulakshmi Somu, Dayananda Sagar College of Arts, Science, and
        Commerce, India

The convergence of big data and technology is transforming the collection, processing, and application of information in the healthcare industry. It is now crucial to comprehend and manage the enormous amounts of data that are constantly produced from various sources, including wearable technology, medical imaging, and electronic health records (EHRs). This chapter examines the five main characteristics of big data that healthcare companies may use to improve sustainability in patient care and operational efficiency, i.e. Volume, velocity, variety, Veracity and Variability. These elements are integrated into a comprehensive IPO (Input, Process, Output) model, illustrating how they function as input and process components that leads to substantiality. Based on earlier literature and real world scenario, the chapter highlights the importance of cutting-edge technologies in managing the complexity of healthcare data and improving delivery and operational effectiveness.

Chapter 12
Unleashing Metaverse for Sustainable Development: Challenges and
Opportunities 309
    Anjali Gautam, Manav Rachna International Institute of Research and
        Studies, India
    Priyanka Dadhich, Manav Rachna International Institute of Research
        and Studies, India
    Himanshu Gupta, Meta, USA
    Lakshay Rekhi, Sharda University, India
    Shitiz Upreti, Maharishi Markandeshwar, India
    Ramesh Chandra Poonia, Christ University, India
    Kamal Upreti, Christ University, India

Metaverse has revolutionized the world of digital technology. Metaverse aspires to seamlessly combine physical and digital realities to create an immersive, linked digital cosmos. Metaverse a naïve concept previously, is gaining momentum and is drawing attention from the researchers all round the world. Today, the world is looking forward to sustainable development which aims to tackle the problems in broadly three major areas which are based on social, eco-nomic and environmental factors. Blend of physical and digital realities might enable more ef-fective and sustainable methods of communication, collaboration, and resource utilization due to its immersive digital environment and cutting-edge technologies. This chapter explores the chal-lenges presented by metaverse in the context of achieving sustainable development and also dis-cusses the opportunities that lie ahead of the user community in harnessing metaverse to achieve sustainable development.

Chapter 13
Penetration Testing: A Way to Secure IT Industries... 327
    Aditya Sharma, Ajeenkya D.Y. Patil University, India
    Amna Kausar, Ajeenkya D.Y. Patil University, India
    Atharva Saraf, Ajeenkya D.Y. Patil University, India
    Susanta Das, Ajeenkya D.Y. Patil University, India

To identify system vulnerabilities, pen testing is essential for cybersecurity. Phases of preparation, reconnaissance, scanning, exploitation, and reporting are all involved. Every phase makes use of tools such as Nmap, Nessus, and Metasploit. To guarantee system, network, and data security, businesses should conduct pen tests regularly using skilled testers. This chapter delves into this important domain of cybersecurity as well as information technology industries. The chapter also discusses ethical issues and various challenges associated with it.

Chapter 14
The Global Trends and Hotspots of Medical Internet of Things Research: A
Bibliometric Analysis During 2004 to 2023... 343
    M. S. Rajeevan, Indian Institute of Technology, Hyderabad, India
    V. S. Anoop, Kerala University of Digital Sciences, Innovation, and
        Technology, India
    D. Narayana, Central Tribal University of Andhra Pradesh, India

The exponential growth of medical data, propelled by technological advancements and societal shifts, underscores the critical role of efficient data management in public health, clinical practice, and medical research. The Medical Internet of Things (MIoT) emerges as a key player, facilitating real-time collection and transmission of patient data through connected devices and sensors. While MIoT revolutionizes healthcare delivery and research, its rapid expansion presents new challenges in data handling. Moreover, the integration of MIoT with 5G technology expands its scope beyond traditional medicine, aiding in diagnosis, treatment, and preventive care. Bibliometrics, an underutilized tool in medical research, offers insights into research trends and collaboration networks, addressing a critical analytical gap in the field. This chapter explores the impact of bibliometric analysis in the medical domain, shedding light on key contributors and research avenues, thus enhancing our understanding of this evolving landscape.

Chapter 15
Real-Time Data Analytics and Decision Making in Cyber-Physical Systems .. 373
    Vishakha Kuwar, Symbiosis International University, India
    Vandana Sonwaney, Symbiosis International University, India
    Shitiz Upreti, Maharishi Markandeshwar, India
    Shubham Rajendra Ekatpure, Kulicke and Soffa, USA
    Prakash Divakaran, Himalayan University, India
    Kamal Upreti, Christ University, India
    Ramesh Chandra Poonia, Christ University, India

The future of digital innovation lies in Cyber-Physical Systems (CPS), integrating computational capabilities with physical processes. CPS function as interconnected networks, merging physical and digital inputs and outputs. This study defines CPS and highlights real-time data analytics' role in enhancing communication across industries like manufacturing and robotics. CPS depend on data collection via edge computing, IoT, and sensors, involving data cleaning, preparation, and normalization. Real-time analytics, including stream processing, machine learning, and AI, are crucial for CPS. Decision-making systems and algorithms enhance efficiency. Given their sensitivity, security and privacy in real-time analytics are vital. The study addresses threat detection, privacy preservation, and data security, along with challenges like data heterogeneity, latency, and scalability. Future prospects include edge AI, fog computing, and blockchain integration.

Compilation of References 391

About the Contributors 435

Index 443

Preface

Navigating Cyber Physical Systems represent a remarkable fusion of cutting-edge technology and real-world applications, revolutionizing the way we interact with the physical world. From smart cities to healthcare, transportation, energy management, and more, SCPS are poised to reshape our daily lives and the industries we depend on.

This book – *"Navigating Cyber- Physical Systems with Cutting Edge Technologies"*, seeks to demystify the complex yet fascinating realm of SCPS. It is a culmination of efforts by experts, researchers, and practitioners who have delved deep into the various facets of these systems. With each chapter, we aim to unravel the intricacies, unveil the potential, and explore the challenges of SCPS, offering readers a comprehensive view of this rapidly evolving field. It unravels the intricacies of sensors and actuators, shedding light on their design and applications in various domains. With a focus on communication and networks, readers will gain insights into the critical aspects of data exchange and connectivity in CPS, including wireless communication, IoT integration and network security.

One of the book's key highlights is its exploration of control and automation techniques, offering readers a deep dive into control algorithms, autonomous systems, and control software. It addresses security and privacy concerns in CPS, providing practical guidance on threat mitigation, secure protocols and safeguarding data privacy. Real time and embedded systems are demystified, with in-depth coverage of real time computing system, technologies and all essential for building CPS solutions.

The book showcases real-world CPS applications across domains like smart cities, healthcare systems, autonomous vehicles and industrial automation, illustrating the tangible impact of CPS on our daily lives. In today's data-driven world, the integration of machine learning and AI into CPS is pivotal. This book explores these technologies, offering insights into machine learning models, reinforcement learning and AI-driven decision making in CPS. Human CPS interaction is also a significant focus, emphasizing the importance of user-friendly interfaces and human centric design for effective CPS adoption.

In compiling this volume, contributions from leading experts and practitioners in the field have been gathered, each offering unique perspectives on the application of AI. This collaborative effort has resulted in a comprehensive guide that reflects the latest advancements and ongoing research in AI-driven communication and decision-making.

Hopefully, this book will inspire further exploration and innovation in AI, fostering a deeper understanding of its impact on our world and encouraging the development of new solutions that enhance how we communicate and make decisions in both personal and professional contexts.

Ramesh Chandra Poonia
Christ University, India

Kamal Upreti
Christ University, India

Chapter 1
The Role of Blockchain in Cyber Physical Systems

G. Sowmya
https://orcid.org/0000-0002-9275-7726
MLRIT, India

R. Sridevi
JNTUH, India

K. S. Sadasiva Rao
Sri Indu College of Engineering and Technology, India

Sri Ganesh Shiramshetty
State University of New York, India

ABSTRACT

This abstract explores how blockchain enhances security, transparency, and efficiency in cyber-physical systems (CPS). It addresses CPS vulnerabilities and emphasizes the need for robust security measures. It explains blockchain's decentralized architecture and cryptographic protocols, showing how it tackles data tampering and unauthorized access. The abstract discusses various blockchain applications in CPS, such as secure data logging and decentralized control. Real-world examples demonstrate benefits like enhanced resilience and transparency. It also covers implications for system performance and regulatory compliance, highlighting blockchain's transformative potential in bolstering CPS security.

DOI: 10.4018/979-8-3693-5728-6.ch001

1. INTRODUCTION

Cyber-Physical Systems (CPS) represent a convergence of physical processes and computational capabilities, enabling the seamless integration of the physical and digital worlds. These systems consist of interconnected devices, sensors, actuators, and computational algorithms, working together to monitor, control, and optimize physical processes in various domains such as transportation, healthcare, manufacturing, energy, and smart cities.

In CPS, physical components interact with digital systems through networked infrastructure, enabling real-time monitoring, analysis, and control of complex processes. For example, in smart cities, CPS technology facilitates intelligent traffic management, efficient energy distribution, and environmental monitoring by integrating sensors, actuators, and data analytics platforms.

Blockchain technology, originally conceptualized as the underlying framework for cryptocurrencies like Bitcoin, has evolved into a powerful tool for decentralized and transparent record-keeping. At its core, a blockchain is a distributed ledger that records transactions in a secure, immutable, and transparent manner across a network of nodes.

Key features of blockchain technology include decentralization, transparency, immutability, Cryptographic Security, Consensus Mechanisms.

The integration of blockchain technology into CPS holds significant promise for enhancing the security, transparency, and efficiency of interconnected systems. CPS face numerous security challenges, including data tampering, unauthorized access, and single points of failure, which traditional security mechanisms struggle to address effectively.(Zhao et al, 2021)

By leveraging the decentralized architecture, cryptographic security, and transparency features of blockchain, CPS can mitigate these security risks and ensure the integrity of data exchanges. Blockchain enables secure and tamper-evident record-keeping, transparent transaction traceability, and decentralized control mechanisms, enhancing trust among stakeholders and fostering innovation in CPS applications.

The objectives of this chapter are to:

1. Provide an overview of Cyber-Physical Systems (CPS), highlighting their significance in various domains and the challenges they face in terms of security and data integrity.
2. Introduce blockchain technology and its fundamental principles, including decentralization, cryptographic security, and consensus mechanisms.
3. Discuss the significance of integrating blockchain into CPS, emphasizing its potential to address security vulnerabilities, enhance transparency, and enable decentralized control mechanisms.

4. Explore real-world applications and case studies demonstrating the benefits of blockchain integration in CPS environments.
5. Identify key challenges, opportunities, and future research directions in harnessing blockchain technology for the advancement of secure and resilient CPS ecosystems.

2. BACKGROUND

2.1 Understanding Cyber-Physical Systems

Cyber-Physical Systems (CPS) represent a fusion of physical processes with computational and communication capabilities, creating interconnected systems that bridge the gap between the physical and digital worlds. These systems encompass a diverse range of applications across various domains, including smart cities, industrial automation, healthcare, transportation, energy, and agriculture.

Figure 1. Cyber Physical System

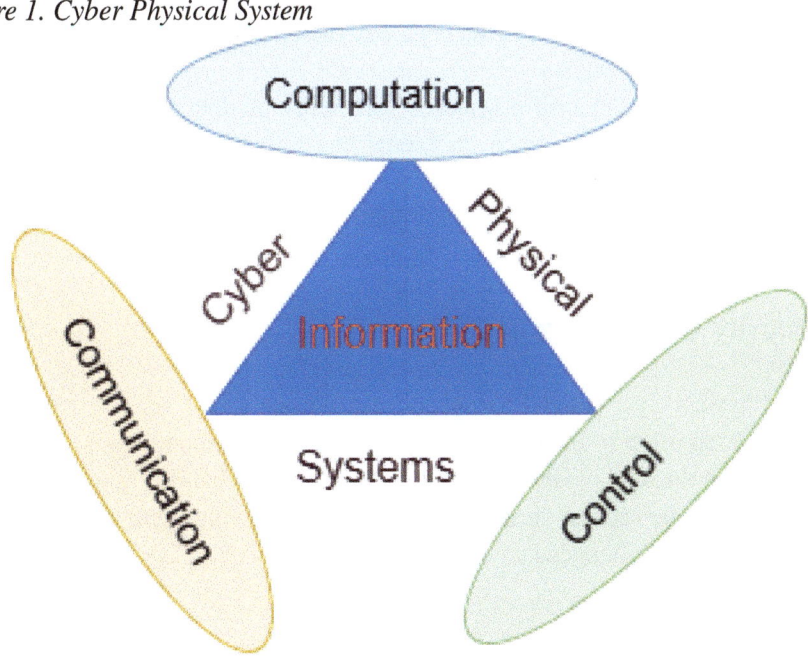

Figure 1 shows the basic Cyber Physical System structure. At their core, CPS consist of physical components, such as sensors, actuators, and control systems, integrated with digital technologies, including computing devices, communication networks, and data analytics platforms. These components work together to monitor, analyse, and control physical processes in real-time, enabling intelligent decision-making and automation. (Rathore et al, 2020)

Key Characteristics of Cyber-Physical Systems:

a. Interconnectedness: CPS components are connected through networked infrastructure, facilitating data exchange and communication between physical and digital elements.
b. Real-Time Operation: CPS operate in real-time, processing and responding to sensor data and events with minimal delay, critical for applications requiring timely decision-making.
c. Integration of Physical and Digital Worlds: CPS integrate physical processes with digital technologies, enabling seamless interaction between the physical environment and computational systems.
d. Sensing and Actuation: CPS rely on sensors to collect data from the physical environment and actuators to effect changes or control physical processes based on computational analysis.
e. Data Analytics and Decision-Making: CPS leverage data analytics techniques to extract insights from sensor data, enabling informed decision-making and optimization of system performance.
f. Automation and Control: CPS employ control algorithms to automate processes and control physical systems based on predefined rules or adaptive algorithms.

2.2 Security Challenges in CPS

Cyber-Physical Systems (CPS) introduce a unique set of security challenges due to their integration of physical processes with digital technologies. Protecting CPS from cyber threats is essential to ensure the safety, reliability, and integrity of critical infrastructures across various domains. The following Figure 2 shows some of the key security challenges faced by CPS. (Alguliyev et al, 2018)

Figure 2. Security Challenges in CPS

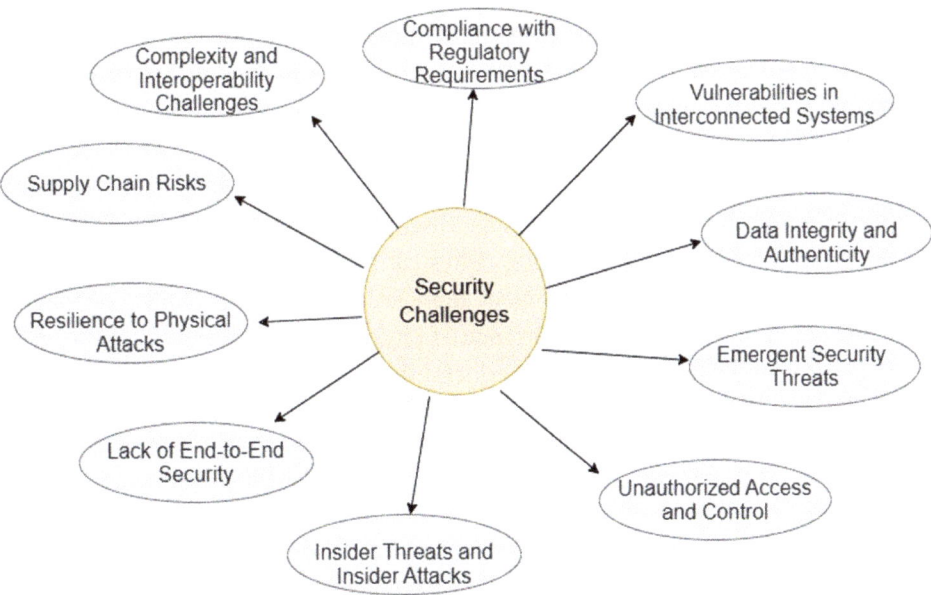

a. Vulnerabilities in Interconnected Systems: CPS rely on interconnected devices and communication networks, increasing the attack surface and vulnerability to cyber-attacks. Malicious actors may exploit vulnerabilities in network protocols, communication channels, or IoT devices to gain unauthorized access or disrupt system operations.
b. Data Integrity and Authenticity: Maintaining the integrity and authenticity of data is crucial in CPS to ensure the accuracy and reliability of decision-making processes. Tampering with sensor data or control signals can lead to erroneous actions or manipulations in physical processes, posing risks to system safety and reliability.
c. Unauthorized Access and Control: Unauthorized access to CPS components or control systems can result in unauthorized manipulation of critical processes, leading to operational disruptions, safety hazards, or financial losses. Weak authentication mechanisms, insecure access controls, or insufficient encryption can facilitate unauthorized access by malicious actors.
d. Insider Threats and Insider Attacks: Insider threats, including malicious or negligent actions by authorized users or employees, pose significant risks to CPS security. Insider attacks may involve sabotage, data theft, or unauthorized

modifications to system configurations, compromising system integrity and confidentiality.
e. Lack of End-to-End Security: CPS often lack end-to-end security measures, leaving vulnerabilities in various components of the system architecture. Inadequate security measures in sensor nodes, communication networks, or cloud-based services can be exploited by attackers to compromise system security and confidentiality.
f. Resilience to Physical Attacks: CPS are vulnerable to physical attacks, such as tampering with physical components, disrupting power supplies, or compromising physical security measures. Physical attacks on CPS can have severe consequences, including equipment damage, service disruptions, or safety incidents, highlighting the importance of physical security measures.
g. Supply Chain Risks: Supply chain vulnerabilities in CPS components, including hardware, software, and firmware, can introduce security risks throughout the system lifecycle. Counterfeit components, malicious firmware updates, or supply chain compromises can undermine system security and integrity, posing risks to CPS operations.

Addressing these security challenges requires a comprehensive approach encompassing risk assessment, threat modelling, security-by-design principles, defense-in-depth strategies, and ongoing monitoring and incident response capabilities. By implementing robust security measures and adopting a proactive security posture, CPS can enhance resilience, mitigate cyber risks, and ensure the safety and reliability of critical infrastructure systems.

2.3 Introduction to Blockchain Technology

Blockchain technology has emerged as a transformative innovation with the potential to revolutionize various industries by providing a decentralized, transparent, and tamper-resistant framework for digital transactions and record-keeping. Originally conceptualized as the underlying technology behind cryptocurrencies like Bitcoin, blockchain has evolved into a versatile platform with applications spanning finance, supply chain management, healthcare, and beyond.

At its core, a blockchain is a distributed ledger that records transactions in a secure and transparent manner across a network of decentralized nodes. Each transaction is grouped into a block, which is cryptographically linked to the preceding block, forming a chain of blocks - hence the term "blockchain." This chain of blocks creates an immutable and tamper-evident record of transaction history, providing transparency and integrity to the data stored on the blockchain.(Jena et al, 2021)

Key Features of Blockchain Technology:

a. Decentralization: Unlike traditional centralized systems where data is stored and controlled by a single authority, blockchain operates on a peer-to-peer network, distributing data and control among multiple nodes. This decentralized architecture eliminates the need for intermediaries and enhances the resilience and trustworthiness of the system.
b. Immutable Ledger: Once a transaction is recorded on the blockchain, it cannot be altered or deleted without consensus from the network participants. This immutability ensures the integrity and permanence of data stored on the blockchain, making it resistant to tampering or unauthorized modifications.
c. Cryptographic Security: Blockchain uses cryptographic techniques such as digital signatures and hash functions to secure transactions and authenticate participants. Each transaction is cryptographically signed by the sender, ensuring the integrity and authenticity of the transaction data.
d. Consensus Mechanisms: Blockchain employs consensus algorithms to validate and confirm transactions, ensuring agreement among network participants on the validity of transactions. Common consensus mechanisms include Proof of Work (PoW), Proof of Stake (PoS), and Practical Byzantine Fault Tolerance (PBFT), each with its own advantages and trade-offs.
e. Smart Contracts: Smart contracts are self-executing contracts with the terms of the agreement directly written into code. Deployed on blockchain platforms such as Ethereum, smart contracts automate and enforce the execution of contractual agreements, eliminating the need for intermediaries and reducing the risk of disputes or fraud.

Some of the Applications of Blockchain Technology:

- Cryptocurrencies and Financial Services: Blockchain technology underpins cryptocurrencies like Bitcoin and Ethereum, enabling secure and decentralized peer-to-peer transactions without the need for intermediaries.
- Supply Chain Management: Blockchain enables transparent and traceable supply chains by recording the provenance and movement of goods from the point of origin to the end consumer, reducing fraud and ensuring product authenticity.
- Healthcare: Blockchain technology facilitates secure and interoperable health data exchange, enabling patients to control and share their medical records securely while ensuring data privacy and integrity.
- Identity Management: Blockchain provides a decentralized and secure framework for identity management, enabling individuals to manage their digital identities and credentials securely without relying on centralized authorities.

Blockchain technology represents a paradigm shift in digital transactions and record-keeping, offering decentralized, transparent, and tamper-resistant solutions to various industries. By leveraging its key features such as decentralization, immutability, cryptographic security, and smart contracts, blockchain has the potential to transform business processes, enhance trust, and unlock new opportunities for innovation and collaboration.

2.4 Relevance of Blockchain in Addressing CPS Security Challenges

Cyber-Physical Systems (CPS) face numerous security challenges due to their interconnected nature, reliance on heterogeneous components, and exposure to cyber threats. Blockchain technology offers a promising solution to address these challenges by providing a decentralized, transparent, and secure framework for data exchange, authentication, and transaction validation. The following are some key ways in which blockchain can mitigate CPS security challenges. (Zhao et al, 2021)

a. Decentralized Architecture: Blockchain operates on a decentralized peer-to-peer network, eliminating single points of failure and reducing the risk of centralized attacks. By distributing data and control among multiple nodes, blockchain enhances the resilience and fault tolerance of CPS architectures, making them less susceptible to targeted attacks or system failures.
b. Immutable Data Ledger: The immutable nature of blockchain ensures the integrity and tamper-resistance of data stored on the ledger. By recording transactions in a sequential and cryptographically linked manner, blockchain provides an auditable and verifiable record of transaction history, making it difficult for malicious actors to tamper with or manipulate data in CPS environments.
c. Enhanced Authentication and Access Control: Blockchain uses cryptographic techniques such as digital signatures and public-private key pairs to authenticate participants and secure transactions. By cryptographically signing transactions and verifying the identity of participants, blockchain enables secure and tamper-evident data exchanges in CPS, reducing the risk of unauthorized access or data breaches.
d. Transparent and Auditable Transactions: Transactions recorded on the blockchain are transparent and visible to all network participants, providing a transparent and auditable record of transaction history. This transparency enhances accountability and trust among stakeholders in CPS environments, enabling real-time monitoring and auditability of system activities.

Blockchain technology offers a compelling solution to address the security challenges faced by Cyber-Physical Systems (CPS) by providing a decentralized, transparent, and tamper-resistant framework for data exchange, authentication, and transaction validation. By leveraging its key features such as decentralization, immutability, enhanced authentication, and smart contracts, blockchain enhances the security, resilience, and trustworthiness of CPS environments, enabling safe and reliable operation in interconnected systems.

3. CHALLENGES S IN EXISTING SYSTEMS

3.1 Centralized Architecture

Centralized architecture is a prevalent issue in existing Cyber-Physical Systems (CPS), where data and control mechanisms are managed and controlled by a single central authority or server. While centralized architectures offer simplicity in management and control, they pose several significant challenges: (Rajhans et al, 2014)

Single Point of Failure
Vulnerability to Attacks
Scalability Challenges
Lack of Redundancy and Resilience
Data Privacy and Security Concerns
Limited Transparency and Accountability
Dependency on Central Authority
Compliance Challenges

Centralized architecture presents significant challenges in existing Cyber-Physical Systems (CPS), including single points of failure, vulnerability to attacks, scalability limitations, lack of redundancy, data privacy concerns, limited transparency, dependency on central authority, and compliance challenges. Addressing these issues requires transitioning towards decentralized architectures that leverage technologies like blockchain to enhance resilience, security, transparency, and scalability in CPS environments.

3.2 Data Integrity and Trust

Ensuring data integrity and trustworthiness is paramount in Cyber-Physical Systems (CPS) to maintain the accuracy, reliability, and security of data exchanged and processed within the system. However, existing CPS face several challenges related to data integrity and trust, including: (Rein et al, 2016)

Data Tampering

Lack of Verifiability
Vulnerability to Cyber Attacks
Lack of Transparency in Data Handling
Dependency on Trusted Intermediaries
Regulatory Compliance Challenges
Data Provenance and Traceability
Mitigation of Insider Threats

Addressing these challenges requires implementing robust mechanisms for ensuring data integrity and trust in CPS environments. Blockchain technology offers a promising solution by providing a decentralized, transparent, and tamper-resistant framework for data exchange and authentication. By leveraging blockchain's immutable ledger, cryptographic security, and consensus mechanisms, CPS can enhance data integrity, trustworthiness, and resilience in interconnected systems.

3.3 Cybersecurity Vulnerabilities

Cybersecurity vulnerabilities in Cyber-Physical Systems (CPS) pose significant risks to the integrity, availability, and confidentiality of system operations. These vulnerabilities can be exploited by malicious actors to compromise system security, disrupt critical processes, or gain unauthorized access to sensitive data. The following are key cybersecurity vulnerabilities commonly observed in CPS environment. (Humayun et al, 2020)

Exploitable Software and Hardware Weaknesses
Insider Threats and Insider Attacks
Lack of Secure Communication Protocols
Vulnerabilities in Third-Party Components
Inadequate Authentication Mechanisms

3.4 Limited Interoperability

Limited interoperability is a critical issue in existing Cyber-Physical Systems (CPS), hindering seamless communication and integration between heterogeneous devices, protocols, and systems. Interoperability challenges arise due to the diverse nature of CPS components, proprietary protocols, and legacy systems, leading to the following issues. (Schilberg et al, 2016)

Proprietary Standards and Protocols
Fragmented Ecosystems
Vendor Lock-In
Lack of Standards and Protocols
Integration Complexity

Addressing limited interoperability in CPS requires concerted efforts to develop and adopt standardized protocols, open interfaces, and interoperability frameworks that enable seamless communication and integration across heterogeneous systems and devices. Embracing open standards, fostering collaboration among stakeholders, and investing in interoperable solutions can help overcome interoperability challenges and unlock the full potential of CPS technologies.

3.5 Scalability and Performance

Scalability and performance are critical concerns in existing Cyber-Physical Systems (CPS), especially as these systems expand in size, complexity, and functionality. The following issues highlight the challenges related to scalability and performance. (Garraghan et al, 2016)

Growing Data Volume
Real-Time Requirements
Complex Event Processing
Resource Constraints
Distributed Architecture
Integration Challenges
Dynamic Environments
Testing and Validation

Addressing scalability and performance challenges in CPS requires a holistic approach that encompasses scalable architecture design, efficient resource management, optimization techniques, and performance testing methodologies. By leveraging scalable and distributed architectures, adopting optimization strategies, and investing in performance monitoring and tuning, CPS can achieve the scalability and performance required to meet the demands of diverse applications and environments.

4. SOLUTIONS OFFERED BY THE PROPOSED SYSTEM

4.1 Decentralized Architecture

In response to the challenges faced by existing Cyber-Physical Systems (CPS), the proposed solution emphasizes the adoption of a decentralized architecture. Decentralization offers several benefits that address the limitations of centralized architectures and enhance the resilience, security, and efficiency of CPS. The following components constitute the decentralized architecture proposed for CPS. (Prenzel, L., & Steinhorst, S, 2021)

a. Distributed Ledger Technology (DLT): Leveraging Distributed Ledger Technology (DLT), such as blockchain, enables the creation of a decentralized and tamper-resistant ledger for recording transactions, events, and data exchanges in CPS. DLT eliminates the need for central authorities or intermediaries, facilitating peer-to-peer interactions, consensus-based decision-making, and transparent record-keeping.
b. Peer-to-Peer Communication: Adopting a peer-to-peer communication model allows CPS devices, sensors, and actuators to communicate directly with each other, bypassing centralized intermediaries or communication hubs. Peer-to-peer communication enhances system resilience, reduces latency, and mitigates single points of failure, ensuring uninterrupted communication and collaboration among CPS components.
c. Distributed Control and Decision-Making: Decentralized architectures enable distributed control and decision-making in CPS, where decision-making authority is distributed across multiple nodes or devices. Distributed control mechanisms empower individual nodes to autonomously execute tasks, make local decisions, and coordinate with neighbouring nodes, enhancing system responsiveness, adaptability, and fault tolerance.
d. Redundancy and Fault Tolerance: Incorporating redundancy and fault tolerance mechanisms into the decentralized architecture ensures system resilience and reliability in the face of failures or disruptions. Redundant components, backup systems, and failover mechanisms mitigate the impact of hardware failures, software errors, or environmental hazards, maintaining continuous operations and minimizing downtime.

By embracing a decentralized architecture, CPS can overcome the limitations of centralized systems, enhance resilience, security, and efficiency, and unlock the transformative potential of decentralized technologies in shaping the future of interconnected systems. Decentralization empowers CPS to achieve autonomy, resilience, and trustworthiness, laying the foundation for a decentralized future in which CPS operate autonomously, securely, and collaboratively in diverse environments.

4.2 Immutable Data Ledger

Central to the proposed solution for Cyber-Physical Systems (CPS) is the implementation of an immutable data ledger, facilitated by technologies such as blockchain. The immutable data ledger serves as a tamper-proof and transparent record of all transactions, events, and data exchanges within the CPS ecosystem. Key features and benefits of the immutable data ledger include: (Maleh et al, 2022)

a. Tamper-Proof Record Keeping: The immutable data ledger ensures that once data is recorded, it cannot be altered or deleted retroactively. Each transaction or data entry is cryptographically linked to previous entries, creating an immutable chain of records that provides a trustworthy audit trail of system activities.
b. Data Integrity and Trustworthiness: By guaranteeing the integrity and immutability of data, the ledger enhances trust among stakeholders and ensures the reliability of information exchanged within the CPS ecosystem. Stakeholders can verify the authenticity and integrity of data entries, promoting transparency and accountability in CPS operations.
c. Transparency and Auditability: The transparent nature of the immutable data ledger enables stakeholders to access, review, and audit the entire history of transactions and data exchanges within the CPS ecosystem. Transparent record-keeping fosters accountability, facilitates compliance with regulatory requirements, and enhances visibility into system activities.
d. Enhanced Security and Resilience: The cryptographic mechanisms employed in the immutable data ledger protect against unauthorized tampering, data manipulation, or malicious attacks. Immutable data ensures that even in the event of security breaches or system failures, the integrity of recorded information remains intact, enhancing the resilience of CPS operations.

By leveraging an immutable data ledger, CPS can ensure the integrity, transparency, and security of data exchange and transactions, laying the groundwork for trusted, resilient, and efficient CPS ecosystems. Immutable data ledgers provide a foundation for transparent, auditable, and decentralized data management, enabling CPS to overcome the limitations of traditional centralized systems and embrace the transformative potential of decentralized technologies.

4.3 Enhanced Cybersecurity Measures

In the proposed solution for Cyber-Physical Systems (CPS), a paramount focus is placed on implementing enhanced cybersecurity measures to safeguard against evolving cyber threats and ensure the integrity and confidentiality of system operations. The following cybersecurity measures are integral components of the proposed solution: (Sharma et al, 2024)

a. Cryptographic Security: Utilizing robust cryptographic techniques, such as encryption, hashing, and digital signatures, to secure data transmission, storage, and authentication processes within the CPS ecosystem. Cryptographic security mechanisms protect sensitive information from unauthorized access, tampering, or interception, ensuring confidentiality and integrity.

b. Secure Authentication and Access Control: Implementing secure authentication mechanisms, such as multi-factor authentication (MFA) and biometric authentication, to verify the identity of users and devices accessing the CPS. Granular access control policies and role-based access control (RBAC) mechanisms enforce least privilege principles, limiting access to authorized entities and reducing the risk of unauthorized access.
c. Intrusion Detection and Prevention Systems (IDPS): Deploying intrusion detection and prevention systems (IDPS) to monitor network traffic, detect anomalous behavior, and mitigate potential security breaches in real-time. IDPS solutions leverage machine learning algorithms, anomaly detection techniques, and signature-based detection to identify and respond to security threats proactively.
d. Endpoint Security Measures: Implementing endpoint security measures, such as antivirus software, endpoint detection and response (EDR) solutions, and device management policies, to protect CPS devices and endpoints from malware, ransomware, and other cyber threats. Endpoint security measures mitigate the risk of compromise and unauthorized access to critical systems.

By implementing enhanced cybersecurity measures, CPS can mitigate the risk of cyber threats, protect critical assets and infrastructure, and ensure the confidentiality, integrity, and availability of system operations. Enhanced cybersecurity measures foster a proactive approach to cybersecurity risk management, enabling CPS operators to detect, respond to, and mitigate cybersecurity threats effectively, safeguarding the resilience and security of CPS ecosystems.

4.4 Interoperability Standards

Interoperability standards play a crucial role in the proposed solution for Cyber-Physical Systems (CPS), facilitating seamless integration, communication, and collaboration among heterogeneous systems, devices, and platforms within the CPS ecosystem. The adoption of interoperability standards ensures compatibility, interoperability, and data exchangeability, enabling CPS components to interoperate effectively across diverse environments. Key aspects of interoperability standards in CPS include: (Jardim-Gonçalves et al, 2020)

a. Common Data Formats and Protocols: Establishing common data formats, communication protocols, and message exchange formats to standardize data representation and transmission across CPS components. Common standards, such as MQTT, CoAP, OPC UA, and JSON, enable interoperability and data exchangeability among devices, sensors, and applications.

b. Open APIs and Interfaces: Providing open application programming interfaces (APIs) and standardized interfaces that enable interoperability and integration between CPS components, subsystems, and external systems. Open APIs facilitate data exchange, functionality integration, and collaboration among diverse stakeholders, promoting innovation and ecosystem growth.
c. Industry Consortia and Standards Bodies: Collaborating with industry consortia, standards bodies, and regulatory organizations to develop and promote interoperability standards and best practices for CPS. Participation in standards development initiatives, such as IEEE, IEC, ISO, and OASIS, fosters consensus-building, interoperability testing, and adoption of common standards across the CPS ecosystem.
d. Semantic Interoperability: Ensuring semantic interoperability by defining common ontologies, vocabularies, and data models that enable shared understanding and interpretation of data semantics across heterogeneous CPS components. Semantic interoperability facilitates data integration, reasoning, and decision-making, enabling CPS systems to exchange and process information effectively.

By embracing interoperability standards, CPS can overcome barriers to integration, interoperability, and collaboration, enabling seamless communication and data exchange across heterogeneous environments. Interoperability standards promote openness, compatibility, and innovation, fostering ecosystem growth and accelerating the adoption of interoperable CPS solutions that deliver value to stakeholders across diverse industries.

4.5 Scalable and Efficient Infrastructure

In the proposed solution for Cyber-Physical Systems (CPS), emphasis is placed on designing and deploying scalable and efficient infrastructure to support the dynamic demands of CPS environments. Scalable and efficient infrastructure ensures that CPS can accommodate growing workloads, adapt to changing requirements, and optimize resource utilization effectively. Key components of scalable and efficient infrastructure in CPS include: (Lenka et al, 2018)

a. Cloud Computing and Edge Computing: Leveraging cloud computing and edge computing architectures to distribute computational tasks, data processing, and storage across cloud-based data centres and edge devices. Cloud computing provides on-demand scalability, elasticity, and resource pooling, while edge computing offers low-latency processing, bandwidth optimization, and data locality benefits.

b. Virtualization and Containerization: Implementing virtualization and containerization technologies, such as virtual machines (VMs) and container orchestration platforms (e.g., Kubernetes), to abstract and encapsulate CPS workloads, applications, and services. Virtualization and containerization enable workload isolation, resource isolation, and efficient resource utilization, facilitating scalability and portability across diverse environments.
c. Distributed Data Storage and Processing: Adopting distributed data storage and processing solutions, such as distributed file systems (e.g., Hadoop HDFS, Apache Cassandra) and distributed data processing frameworks (e.g., Apache Spark, Apache Flink), to manage large volumes of data and perform parallelized data processing tasks. Distributed architectures enable horizontal scalability, fault tolerance, and high availability, accommodating the scalability requirements of CPS applications.
d. Elastic Resource Provisioning: Implementing elastic resource provisioning mechanisms that dynamically allocate and deallocate computational resources (e.g., CPU, memory, storage) based on workload demands and performance requirements. Elastic resource provisioning ensures that CPS systems can scale resources up or down in real-time to meet fluctuating demands, optimizing resource utilization and cost efficiency.

By adopting scalable and efficient infrastructure, CPS can meet the evolving demands of modern applications, support dynamic workloads, and optimize resource utilization effectively. Scalable and efficient infrastructure enhances agility, resilience, and performance, enabling CPS to deliver reliable, responsive, and cost-effective services in diverse environments.

5. APPLICATIONS OF BLOCKCHAIN IN CPS

Cyber-Physical Systems (CPS) integrate digital and physical components to power innovations like smart cities, autonomous vehicles, and industrial automation. However, these systems face challenges in security, data integrity, and centralized control. Blockchain technology addresses these issues by providing a decentralized and transparent framework for secure data management, decision-making, and transaction processing. This chapter explores how blockchain enhances CPS through secure data logging, decentralized control mechanisms, and autonomous operations, supported by real-world examples showcasing its transformative impact across various industries.

5.1 Secure Data Logging and Audit Trails

In the proposed solution for Cyber-Physical Systems (CPS), blockchain technology is leveraged to enhance security, transparency, and accountability in data logging and audit trail management. Blockchain provides a decentralized and tamper-evident ledger that records transactions and data exchanges in a secure and transparent manner. Key applications of blockchain in secure data logging and audit trails for CPS include: (Dsouza et al, 2019)

a. Immutable Data Logging: Blockchain ensures the immutability and integrity of data logs by cryptographically chaining data records into blocks and timestamping them in a decentralized manner. Once recorded on the blockchain, data logs cannot be altered or deleted retroactively, providing a reliable and tamper-proof audit trail of all transactions and events within the CPS ecosystem.
b. Transparent Transaction History: Blockchain enables transparent and auditable transaction history by maintaining a complete and verifiable record of all data exchanges, communications, and interactions among CPS components and entities. Each transaction recorded on the blockchain is visible to all authorized parties, allowing for real-time monitoring, verification, and auditability of data flows and communication patterns.
c. Accountability and Traceability: Blockchain enhances accountability and traceability in CPS operations by attributing data exchanges and transactions to specific participants or entities through cryptographic signatures and digital identities. Each participant in the CPS network has a unique digital identity that can be traced back to their respective actions, facilitating accountability and attribution of responsibilities in data logging and audit trails.
d. Tamper-Evident Data Integrity: Blockchain provides tamper-evident data integrity by employing cryptographic hashing and consensus mechanisms to ensure that data records cannot be altered, tampered with, or falsified without detection. Any unauthorized attempt to modify data logs or audit trails is immediately detected and rejected by the blockchain network, preserving the integrity and authenticity of the recorded information.

By leveraging blockchain technology for secure data logging and audit trails, CPS can enhance transparency, accountability, and trustworthiness in data management and transaction processing. Blockchain-enabled audit trails provide a tamper-proof record of all transactions and events within the CPS ecosystem, enabling stakeholders to verify the integrity, authenticity, and provenance of data exchanges and communications.

5.2 Decentralized Control Mechanisms

In the context of Cyber-Physical Systems (CPS), decentralized control mechanisms empowered by blockchain technology offer a paradigm shift in managing and orchestrating system operations. Traditional CPS architectures often rely on centralized control systems, which can introduce vulnerabilities such as single points of failure and potential targets for cyber-attacks. Decentralized control mechanisms, enabled by blockchain, distribute decision-making authority and execution across a network of nodes, fostering resilience, autonomy, and trust within CPS ecosystems. Key applications of decentralized control mechanisms in CPS include: (Bodkhe et al, 2020)

a. Distributed Decision-Making: Blockchain enables distributed decision-making processes by allowing CPS components and entities to participate in consensus protocols and smart contract execution. Instead of relying on a centralized authority to make decisions, CPS nodes collaborate and reach agreement autonomously through blockchain-based consensus mechanisms, such as proof-of-work (PoW) or proof-of-stake (PoS), ensuring fairness and transparency in decision-making.
b. Autonomous Operation and Smart Contracts: Smart contracts, self-executing code deployed on the blockchain, enable autonomous execution of predefined rules and agreements without human intervention. In CPS, smart contracts facilitate automated decision-making, resource allocation, and event-triggered actions based on predefined conditions, ensuring efficient and reliable operation of interconnected systems without the need for centralized control.
c. Decentralized Asset Management: Blockchain-based decentralized control mechanisms enable distributed asset management and resource allocation in CPS environments. Through tokenization and asset-backed tokens deployed on the blockchain, CPS entities can represent and exchange assets, such as energy credits, data licenses, or access rights, in a peer-to-peer manner, enabling efficient resource utilization and allocation without relying on centralized intermediaries.
d. Trust less Interactions and Coordination: Decentralized control mechanisms provided by blockchain technology enable trust less interactions and coordination among CPS entities without the need for trusted intermediaries. Through consensus mechanisms and cryptographic verification, CPS nodes can interact, transact, and coordinate activities securely and transparently, fostering trust and collaboration in peer-to-peer networks.

By leveraging decentralized control mechanisms enabled by blockchain technology, CPS can achieve autonomy, resilience, and trustworthiness in managing system operations and orchestrating interactions among distributed components and entities. Decentralized control mechanisms empower CPS ecosystems to operate autonomously, efficiently, and securely without relying on centralized authorities or intermediaries, paving the way for innovative and resilient applications in various domains such as smart cities, industrial automation, and healthcare.

5.3 Autonomous Decision-Making

Autonomous decision-making powered by blockchain technology represents a significant advancement in the realm of Cyber-Physical Systems (CPS), enabling systems to make independent and intelligent choices without human intervention. By leveraging blockchain's decentralized architecture, cryptographic security, and smart contract capabilities, CPS can achieve a higher degree of autonomy, efficiency, and adaptability in decision-making processes. Key applications of autonomous decision-making in CPS include: (Ray, A, 2013)

a. Smart Contracts for Automated Execution: Smart contracts deployed on blockchain networks enable automated execution of predefined rules and agreements without human intervention. In CPS, smart contracts facilitate autonomous decision-making by executing actions based on predefined conditions, events, or triggers, such as sensor readings, supply chain milestones, or market fluctuations. For example, smart contracts can automatically trigger maintenance requests, adjust energy consumption, or initiate emergency responses based on real-time data and predefined rules encoded in the contract logic.
b. Decentralized Governance and Consensus: Blockchain-based decentralized governance models enable autonomous decision-making through distributed consensus mechanisms. In CPS ecosystems, decentralized governance structures empower network participants to collectively make decisions regarding system upgrades, protocol changes, or resource allocations through consensus mechanisms such as proof-of-stake (PoS) or delegated proof-of-stake (DPoS). Decentralized governance ensures fairness, transparency, and inclusivity in decision-making processes, fostering trust and collaboration among stakeholders.
c. Predictive Analytics and Machine Learning: Autonomous decision-making in CPS can be augmented by predictive analytics and machine learning algorithms that leverage historical data, sensor inputs, and contextual information to anticipate future events and make proactive decisions. Machine learning models trained on blockchain-stored data can analyse patterns, detect anomalies, and predict outcomes, enabling CPS systems to pre-emptively adjust operations, optimize resource allocation, and mitigate risks in real-time.

d. Adaptive Control and Optimization: Autonomous decision-making in CPS involves adaptive control and optimization techniques that enable systems to dynamically adjust parameters, configurations, and strategies based on changing environmental conditions, user preferences, or performance metrics. Through feedback loops, reinforcement learning algorithms, and real-time data analysis, CPS systems can continuously optimize operations, maximize efficiency, and adapt to evolving requirements without human intervention.

By embracing autonomous decision-making empowered by blockchain technology, CPS can unlock new opportunities for innovation, efficiency, and resilience in various domains, including smart cities, autonomous vehicles, smart grids, and industrial automation. Autonomous decision-making enhances CPS systems' ability to adapt to changing conditions, optimize resource allocation, and respond effectively to dynamic environments, driving the evolution of intelligent and autonomous systems in the digital age.

6. CASE STUDIES AND REAL-WORLD EXAMPLES

Case studies and real-world examples provide tangible evidence of the effectiveness and applicability of blockchain technology in Cyber-Physical Systems (CPS), demonstrating its potential to address security challenges, improve efficiency, and enable innovative solutions. Below are several case studies and real-world examples showcasing the diverse applications of blockchain in CPS:

a. Energy Grid Management: In the energy sector, blockchain is being utilized to optimize the management and distribution of energy resources in smart grids. Companies like Power Ledger and Grid+ are implementing blockchain-based platforms that enable peer-to-peer energy trading among consumers, producers, and prosumers. These platforms utilize smart contracts to automate energy transactions, ensuring transparent and efficient energy distribution while reducing reliance on centralized utilities. (Pradhan et al, 2022)
b. Supply Chain Traceability: Blockchain technology is revolutionizing supply chain management by enhancing transparency, traceability, and accountability throughout the supply chain. Companies such as IBM Food Trust and VeChain are deploying blockchain solutions to track the provenance of goods, verify product authenticity, and ensure compliance with regulatory standards. By recording every transaction and movement of goods on the blockchain, stakeholders can trace the journey of products from the source to the end consumer, reducing counterfeiting, fraud, and inefficiencies in supply chains. (Wong et al, 2024)

c. Autonomous Vehicles and Mobility Services: In the transportation sector, blockchain is enabling secure and decentralized data sharing among autonomous vehicles (AVs) and mobility service providers. Projects like the Mobility Open Blockchain Initiative (MOBI) are exploring blockchain-based solutions for vehicle identity, data sharing, and decentralized payments, facilitating interoperability and collaboration among AVs and mobility ecosystems. By leveraging blockchain, AVs can securely exchange sensor data, coordinate traffic flow, and execute smart contracts for autonomous payments and service agreements. (Hemani et al, 2024)
d. Healthcare Data Management: Blockchain is revolutionizing healthcare data management by providing secure and interoperable platforms for storing, sharing, and accessing patient records and medical data. Initiatives such as MedRec and Medicalchain are leveraging blockchain to create decentralized electronic health record (EHR) systems that give patients control over their health data while ensuring privacy and security. Blockchain-based EHRs enable seamless data exchange among healthcare providers, improve care coordination, and enhance patient outcomes while maintaining data integrity and confidentiality. (Haleem et al, 2022)

These case studies and real-world examples illustrate the transformative potential of blockchain technology in addressing security challenges, enhancing efficiency, and fostering innovation in Cyber-Physical Systems across various industries and domains. By harnessing blockchain's decentralized, transparent, and immutable properties, CPS stakeholders can unlock new opportunities for collaboration, optimization, and value creation in interconnected environments.

7. IMPLEMENTATION CHALLENGES

7.1 Performance Considerations

Despite the potential benefits of integrating blockchain technology into Cyber-Physical Systems (CPS), several implementation challenges must be addressed to ensure the scalability, efficiency, and practicality of blockchain-based solutions. One of the primary challenges is performance considerations, which encompass various factors that impact the speed, throughput, and responsiveness of blockchain networks in CPS environments. Key performance considerations include: (Ayerdi et al, 2023)

a. Scalability: Blockchain scalability refers to the ability of the network to handle increasing transaction volumes and data throughput without compromising performance or latency. In CPS, scalability is crucial to accommodate the large volume of real-time data generated by interconnected devices and sensors. However, traditional blockchain architectures, such as those used in public blockchains like Bitcoin or Ethereum, may struggle to scale to meet the demands of CPS applications due to limitations in block size, transaction throughput, and consensus mechanisms. Implementing scalability solutions, such as sharding, sidechains, or layer-2 protocols, can help enhance the scalability of blockchain networks in CPS environments.
b. Transaction Throughput: Transaction throughput refers to the number of transactions processed by the blockchain network within a given time frame. In CPS applications, where real-time data processing and communication are essential, high transaction throughput is critical to ensure timely and responsive system operation. However, traditional blockchain networks often face limitations in transaction throughput due to factors such as block size, block validation times, and network congestion. Optimizing block size, implementing off-chain scaling solutions, and adopting consensus mechanisms tailored to CPS requirements can help improve transaction throughput and reduce latency in blockchain-based CPS applications.
c. Resource Consumption: Blockchain-based CPS solutions may impose significant resource consumption requirements in terms of computational power, storage capacity, and network bandwidth. The resource-intensive nature of blockchain networks can lead to scalability challenges and increased operational costs, particularly in resource-constrained CPS environments. Optimizing resource utilization, implementing lightweight consensus mechanisms, and adopting energy-efficient consensus protocols can help mitigate resource consumption and improve the efficiency of blockchain-based CPS implementations.
d. Consensus Mechanisms: Consensus mechanisms play a critical role in determining the performance and efficiency of blockchain networks in CPS applications. While traditional consensus mechanisms such as proof-of-work (PoW) and proof-of-stake (PoS) provide security and decentralization, they may introduce scalability and performance limitations due to their computational overhead and consensus latency. Exploring alternative consensus mechanisms tailored to CPS requirements, such as practical Byzantine fault tolerance (PBFT), delegated proof-of-stake (DPoS), or proof-of-authority (PoA), can help optimize performance and throughput in blockchain-based CPS architectures.

Addressing performance considerations in blockchain-based CPS implementations requires a holistic approach that combines optimization techniques, scalability solutions, and tailored consensus mechanisms to ensure efficient and reliable operation in dynamic and resource-constrained environments. By overcoming performance challenges, blockchain technology can unlock new opportunities for secure, transparent, and decentralized CPS applications across various industries and domains.

7.2 Scalability Issues

Scalability is a critical consideration in implementing blockchain technology within Cyber-Physical Systems (CPS) to ensure that the system can handle increasing transaction volumes, data throughput, and network demands. However, several scalability issues pose challenges to the effective deployment of blockchain-based CPS solutions. (Keerthi et al, 2017)

a. Transaction Throughput Limitations: Traditional blockchain networks, such as Bitcoin and Ethereum, face inherent limitations in transaction throughput due to factors such as block size, block validation times, and consensus mechanisms. These limitations can result in network congestion and delays, impacting the responsiveness and efficiency of CPS applications that require high transaction throughput for real-time data processing and communication.
b. Block Size and Block Propagation Times: The block size and block propagation times in blockchain networks directly affect scalability by determining the maximum number of transactions that can be processed within a given time frame. Larger block sizes can increase transaction throughput but may also lead to longer block propagation times and increased network latency. Balancing block size with block propagation times is crucial to achieving scalability without sacrificing network performance and decentralization.
c. Consensus Mechanisms: The consensus mechanism used in a blockchain network plays a significant role in its scalability. While consensus mechanisms like proof-of-work (PoW) and proof-of-stake (PoS) provide security and decentralization, they can also introduce scalability limitations due to their computational overhead and consensus latency. Alternative consensus mechanisms, such as practical Byzantine fault tolerance (PBFT), delegated proof-of-stake (DPoS), or sharding, may offer improved scalability by distributing consensus tasks across network participants or partitions.
d. Network Congestion and Fees: Network congestion and transaction fees can impact scalability by limiting the number of transactions that can be processed within a blockchain network. During periods of high demand, increased network congestion can lead to higher transaction fees and longer confirmation

times, reducing the scalability and cost-effectiveness of blockchain-based CPS solutions. Implementing scaling solutions, such as layer-2 protocols, off-chain scaling solutions, or transaction batching, can help alleviate network congestion and improve scalability.

Addressing scalability issues in blockchain-based CPS implementations requires a multifaceted approach that encompasses optimization techniques, consensus algorithm enhancements, interoperability solutions, and network upgrades. By mitigating scalability challenges, blockchain technology can realize its full potential in enabling scalable, efficient, and decentralized CPS applications that meet the demands of dynamic and interconnected environments.

7.3 Regulatory Compliance Challenges

Regulatory compliance is a significant concern when implementing blockchain technology in Cyber-Physical Systems (CPS), as it involves navigating complex legal frameworks, regulatory requirements, and industry standards. Several challenges related to regulatory compliance may arise in the context of blockchain-based CPS solutions: (Sokolsky et al, 2011)

a. Data Privacy and Protection: Blockchain's transparent and immutable nature presents challenges in complying with data privacy regulations, such as the General Data Protection Regulation (GDPR) in Europe or the Health Insurance Portability and Accountability Act (HIPAA) in the United States. Ensuring compliance with data privacy laws while leveraging blockchain for data sharing and transparency requires careful consideration of privacy-enhancing techniques, such as encryption, data anonymization, and permissioned access controls.
b. Jurisdictional Issues: Blockchain-based CPS solutions may operate across multiple jurisdictions, each with its own regulatory requirements and legal frameworks governing data privacy, consumer protection, intellectual property, and cybersecurity. Navigating jurisdictional issues and ensuring compliance with applicable laws and regulations in each jurisdiction can be complex and resource-intensive, particularly in cross-border transactions and data transfers.
c. Smart Contract Legality: Smart contracts, self-executing code deployed on blockchain networks, raise legal questions regarding their enforceability, validity, and compliance with existing contract laws and regulations. Ensuring the legality and enforceability of smart contracts in CPS applications requires legal expertise to draft contract terms, define legal obligations, and address potential disputes or breaches in compliance with applicable laws and regulations.

d. Know Your Customer (KYC) and Anti-Money Laundering (AML) Compliance: Blockchain-based CPS solutions may encounter challenges in complying with Know Your Customer (KYC) and Anti-Money Laundering (AML) regulations, which require verification of user identities and detection of suspicious activities. Implementing KYC/AML procedures in blockchain-based CPS applications, such as identity verification protocols and transaction monitoring mechanisms, requires collaboration with regulatory authorities and adherence to compliance standards to prevent illicit activities and ensure regulatory compliance.

Navigating regulatory compliance challenges in blockchain-based CPS implementations requires a comprehensive understanding of legal frameworks, regulatory requirements, and industry standards, as well as collaboration with legal experts, regulatory authorities, and industry stakeholders to ensure compliance with applicable laws and regulations while leveraging blockchain technology to its full potential. By addressing regulatory compliance challenges, blockchain-based CPS solutions can achieve legal certainty, regulatory compliance, and industry acceptance, enabling innovation and adoption in dynamic and regulated environments.

7.4 Integration Complexity

Integrating blockchain technology into Cyber-Physical Systems (CPS) introduces a range of complexities that must be addressed to ensure seamless interoperability, functionality, and efficiency. Several factors contribute to the integration complexity of blockchain-based CPS solutions: (Törngren et al, 2018)

a. Heterogeneous Systems and Legacy Infrastructure: CPS environments often consist of heterogeneous systems, devices, and legacy infrastructure with diverse communication protocols, data formats, and operational requirements. Integrating blockchain into such complex ecosystems requires compatibility with existing systems, migration strategies for legacy data, and adaptation to diverse hardware and software environments.
b. Interoperability Challenges: Interoperability challenges arise when integrating blockchain networks with existing CPS components, protocols, and standards. Ensuring seamless interoperability between blockchain platforms, smart devices, sensors, and control systems requires standardization of data formats, communication protocols, and interoperability frameworks to facilitate data exchange, communication, and interoperability across heterogeneous systems.
c. Data Synchronization and Consistency: Maintaining data synchronization and consistency between blockchain networks and CPS components is crucial for ensuring data integrity, reliability, and trustworthiness. Achieving consistency

across distributed systems, disparate databases, and asynchronous data sources requires implementing data synchronization mechanisms, consensus protocols, and data reconciliation processes to reconcile conflicting data states and ensure data consistency across the entire CPS ecosystem.

d. Scalability and Performance Optimization: Scalability and performance optimization are essential considerations when integrating blockchain technology into CPS environments, particularly in applications with high transaction volumes, real-time data processing requirements, and stringent performance constraints. Optimizing blockchain scalability, transaction throughput, and network performance requires implementing scaling solutions, optimizing consensus algorithms, and leveraging off-chain processing techniques to minimize latency, enhance responsiveness, and improve overall system performance.

Addressing integration complexity in blockchain-based CPS implementations requires a comprehensive approach that encompasses technical, organizational, and regulatory considerations. By addressing interoperability challenges, optimizing performance, ensuring security and privacy, and fostering collaboration across diverse stakeholders, integration complexity can be mitigated, enabling the successful integration and adoption of blockchain technology in CPS environments.

7.5 Interoperability Concerns

Interoperability concerns arise when integrating blockchain technology into Cyber-Physical Systems (CPS) due to the heterogeneous nature of CPS components, protocols, and standards. Several challenges related to interoperability may impact the seamless integration and functionality of blockchain-based CPS solutions: (Pop et al, 2023)

a. Diverse Communication Protocols: CPS environments often utilize diverse communication protocols and standards for data exchange, device communication, and system integration. Integrating blockchain into CPS ecosystems requires compatibility with existing communication protocols, such as MQTT, CoAP, OPC UA, or DDS, to facilitate interoperability between blockchain networks and CPS components.

b. Data Formats and Standards: CPS components and systems may use different data formats, schemas, and standards for representing and exchanging data. Ensuring interoperability between blockchain networks and CPS data sources requires standardization of data formats, serialization techniques, and data models to enable seamless data exchange and integration across heterogeneous systems.

c. Protocol Interoperability: Interoperability challenges may arise when integrating blockchain networks with existing CPS protocols and standards, such as Modbus, DNP3, BACnet, or Zigbee. Ensuring protocol interoperability between blockchain networks and CPS protocols requires protocol translation, adaptation, or encapsulation techniques to bridge the gap between disparate communication protocols and enable interoperable data exchange and communication.
d. Smart Contract Compatibility: Smart contracts deployed on blockchain networks may require compatibility with CPS devices, sensors, actuators, and control systems to enable automation, orchestration, and execution of CPS workflows and processes. Ensuring smart contract compatibility with CPS components requires defining standard interfaces, APIs, or protocols for interacting with blockchain-based smart contracts from CPS applications.

Addressing interoperability concerns in blockchain-based CPS implementations requires collaboration, standardization, and alignment across diverse stakeholders, including blockchain developers, CPS manufacturers, standardization bodies, regulatory authorities, and industry consortia. By adopting open standards, defining interoperability frameworks, and fostering collaboration between blockchain and CPS communities, interoperability concerns can be effectively addressed, enabling the seamless integration and interoperability of blockchain technology in CPS environments.

7.6 Privacy and Confidentiality Issues

Privacy and confidentiality are critical considerations when integrating blockchain technology into Cyber-Physical Systems (CPS), as they involve protecting sensitive data, ensuring data privacy, and maintaining confidentiality in distributed and transparent blockchain networks. Several privacy and confidentiality issues may arise in blockchain-based CPS solutions: (Harkat et al, 2024)

a. Transparent and Immutable Ledger: Blockchain's transparent and immutable ledger poses challenges to data privacy and confidentiality, as all transactions are recorded on the blockchain and are visible to all network participants. Ensuring privacy and confidentiality in blockchain-based CPS solutions requires implementing privacy-enhancing techniques, such as encryption, zero-knowledge proofs, and off-chain data storage, to protect sensitive data from unauthorized access and disclosure.
b. Pseudonymity and Identity Protection: Blockchain networks typically use pseudonymous addresses to represent participants' identities, which may compromise user privacy and anonymity. Ensuring identity protection and pseudonymity in

blockchain-based CPS solutions requires implementing identity management mechanisms, authentication protocols, and access controls to protect user identities and prevent identity theft or impersonation.
c. Data Minimization and Purpose Limitation: Collecting and storing minimal data necessary for CPS operations while limiting data usage to predefined purposes is essential for ensuring data privacy and confidentiality. Implementing data minimization and purpose limitation principles in blockchain-based CPS solutions requires defining data collection policies, access controls, and data retention periods to minimize the risk of unauthorized data access, misuse, or disclosure.
d. Consent and User Control: Ensuring user consent and control over their data is essential for protecting privacy and confidentiality in blockchain-based CPS solutions. Implementing consent management mechanisms, data access controls, and user-centric privacy settings empowers users to control how their data is collected, processed, and shared within blockchain networks, enhancing transparency, trust, and compliance with data privacy regulations.

Addressing privacy and confidentiality issues in blockchain-based CPS implementations requires adopting a privacy-by-design approach, implementing privacy-enhancing technologies, and adhering to regulatory requirements and industry best practices. By prioritizing data privacy, confidentiality, and user control, blockchain-based CPS solutions can protect sensitive data, preserve privacy rights, and foster trust and confidence among stakeholders in interconnected and distributed environments.

8. FUTURE DIRECTIONS AND OPPORTUNITIES

8.1 Emerging Trends in Blockchain for CPS

As blockchain technology continues to evolve, several emerging trends hold promise for enhancing the capabilities and applications of blockchain in Cyber-Physical Systems (CPS). These trends represent future directions and opportunities for leveraging blockchain technology to address emerging challenges and unlock new possibilities in CPS environments:

a. Scalability Solutions: Emerging scalability solutions, such as sharding, sidechains, and layer-2 protocols, aim to address the scalability limitations of traditional blockchain networks and enable high throughput, low-latency transactions in CPS applications. Implementing scalable blockchain architectures enhances the performance, efficiency, and scalability of blockchain-based CPS solutions,

enabling real-time data processing, communication, and control in dynamic and distributed environments.
b. Interoperability Standards: Standardization efforts and interoperability frameworks aim to facilitate seamless integration and interoperability between diverse blockchain networks, protocols, and platforms in CPS environments. Adopting interoperability standards enables data exchange, communication, and collaboration across heterogeneous blockchain networks and CPS ecosystems, unlocking new opportunities for interoperable, interconnected CPS applications across industries and domains.
c. Privacy-Preserving Technologies: Privacy-preserving technologies, such as zero-knowledge proofs, confidential transactions, and privacy-enhancing cryptography, enable data privacy, confidentiality, and anonymity in blockchain-based CPS solutions. Integrating privacy-preserving techniques enhances data protection, user privacy, and regulatory compliance in CPS applications, fostering trust, transparency, and compliance with data privacy regulations.
d. Federated and Hybrid Blockchains: Federated and hybrid blockchain architectures combine the benefits of public and private blockchains to provide scalable, secure, and customizable solutions for CPS applications. Leveraging federated and hybrid blockchain networks enables flexible governance models, customizable consensus mechanisms, and tailored privacy controls, catering to the diverse needs and requirements of CPS stakeholders and use cases.

Exploring these emerging trends and opportunities in blockchain for CPS opens new avenues for innovation, collaboration, and adoption in diverse industries and domains. By leveraging scalable, interoperable, and privacy-preserving blockchain solutions, CPS stakeholders can unlock new possibilities for secure, transparent, and resilient applications that address evolving challenges and requirements in interconnected and distributed environments.

8.2 Research Areas for Further Exploration

As blockchain technology continues to evolve and its applications in Cyber-Physical Systems (CPS) expand, several research areas offer opportunities for further exploration and advancement. These research areas represent promising avenues for addressing key challenges, advancing knowledge, and driving innovation in blockchain-based CPS solutions:

a. Scalability and Performance Optimization: Research into scalable blockchain architectures, consensus mechanisms, and optimization techniques aims to address the scalability and performance limitations of blockchain networks in CPS applications. Exploring novel approaches for improving transaction throughput, reducing latency, and enhancing network efficiency enables real-time data processing, communication, and control in distributed CPS environments.

b. Privacy-Preserving Techniques: Investigating privacy-preserving techniques, such as zero-knowledge proofs, secure multi-party computation, and differential privacy, enables enhanced data privacy, confidentiality, and anonymity in blockchain-based CPS solutions. Researching privacy-enhancing technologies facilitates compliance with data privacy regulations, protects sensitive information, and preserves user privacy in interconnected and distributed CPS ecosystems.

c. Interoperability and Standardization: Researching interoperability standards, protocols, and frameworks facilitates seamless integration and interoperability between diverse blockchain networks, platforms, and CPS components. Exploring interoperability solutions enables data exchange, communication, and collaboration across heterogeneous CPS ecosystems, fostering interoperable, interconnected applications and systems.

d. Security and Resilience Enhancement: Investigating security mechanisms, threat models, and resilience strategies aims to enhance the security and resilience of blockchain-based CPS solutions against cyber threats, attacks, and disruptions. Researching security-by-design principles, cryptographic protocols, and decentralized architectures strengthens system security, integrity, and trustworthiness in dynamic and distributed CPS environments.

Exploring these research areas and fostering collaboration between academia, industry, and government agencies accelerates progress, drives innovation, and advances the state-of-the-art in blockchain-based CPS solutions. By addressing key challenges, leveraging emerging technologies, and embracing interdisciplinary approaches, researchers can unlock new possibilities and create impactful solutions that address the complex challenges and requirements of interconnected and distributed CPS environments.

8.3 Opportunities for Collaboration and Innovation

In the dynamic landscape of blockchain technology and Cyber-Physical Systems (CPS), numerous opportunities for collaboration and innovation exist across academia, industry, and government sectors. These opportunities can drive advancements, foster knowledge exchange, and accelerate the development and adoption of blockchain-based CPS solutions:

a. Interdisciplinary Research Collaborations: Collaborating across disciplines, such as computer science, engineering, cryptography, and IoT, enables interdisciplinary research that combines expertise from different domains to tackle complex challenges in blockchain-based CPS solutions. Joint research projects, collaborative initiatives, and interdisciplinary workshops facilitate knowledge exchange, cross-pollination of ideas, and breakthrough innovations in interconnected and distributed CPS environments.
b. Industry-Academia Partnerships: Forming partnerships between academia and industry stakeholders fosters collaborative research, technology transfer, and innovation in blockchain-based CPS solutions. Industry-academia collaborations enable researchers to gain insights into real-world challenges, access industry expertise and resources, and validate research findings through practical deployments and pilot projects in diverse CPS applications and domains.
c. Consortia and Standardization Bodies: Participating in consortia, industry alliances, and standardization bodies dedicated to blockchain and CPS promotes collaboration, knowledge sharing, and standards development in the field. Contributing to standardization efforts, collaborating on interoperability initiatives, and sharing best practices and use cases accelerates the development, adoption, and standardization of blockchain-based CPS solutions across industries and domains.
d. Government-Industry Partnerships: Partnering with government agencies, regulatory authorities, and policymakers facilitates collaboration on regulatory compliance, policy development, and governance frameworks for blockchain-based CPS solutions. Engaging in public-private partnerships, policy dialogues, and regulatory sandboxes enables stakeholders to address legal, regulatory, and ethical considerations, foster innovation, and create an enabling environment for blockchain adoption in CPS applications.

Embracing opportunities for collaboration and innovation across academia, industry, and government sectors creates synergies, fosters creativity, and accelerates progress in blockchain-based CPS solutions. By collaborating on research, technology transfer, standardization, and policy development, stakeholders can harness the transformative potential of blockchain technology to address complex challenges, drive innovation, and create positive societal impact in interconnected and distributed CPS ecosystems.

9. CONCLUSION

The integration of blockchain technology into Cyber-Physical Systems (CPS) offers a transformative solution to enhance security, transparency, and efficiency. Blockchain's decentralized architecture and cryptographic protocols address key CPS security challenges like data tampering and unauthorized access. It enables secure data exchange, transparent transactions, and decentralized control mechanisms. Despite challenges such as scalability and regulatory compliance, blockchain integration promises resilient, interoperable, and cost-effective CPS solutions. Embracing collaboration and innovation, stakeholders can harness blockchain's potential to create a trustworthy, decentralized future for interconnected CPS environments.

REFERENCES

Alguliyev, R., İmamverdiyev, Y., & Sukhostat, L. (2018). Cyber-physical systems and their security issues. *Computers in Industry*, 100, 212–223. DOI: 10.1016/j.compind.2018.04.017

Ayerdi, J., Valle, P., Segura, S., Arrieta, A., Sagardui, G., & Arratibel, M. (2023). Performance-Driven metamorphic testing of Cyber-Physical systems. *IEEE Transactions on Reliability*, 72(2), 827–845. DOI: 10.1109/TR.2022.3193070

Bodkhe, U., Mehta, D., Tanwar, S., Bhattacharya, P., Singh, P. K., & Hong, W. (2020). A survey on Decentralized Consensus Mechanisms for Cyber Physical Systems. *IEEE Access : Practical Innovations, Open Solutions*, 8, 54371–54401. DOI: 10.1109/ACCESS.2020.2981415

Dsouza, J., Elezabeth, L., Mishra, V. P., & Jain, R. (2019). Security in Cyber-Physical Systems. *2019 Amity International Conference on Artificial Intelligence (AICAI)*. DOI: 10.1109/AICAI.2019.8701411

Garraghan, P., McKee, D., Ouyang, X., Webster, D. E., & Xu, J. (2016). SEED: A Scalable Approach for Cyber-Physical System Simulation. *IEEE Transactions on Services Computing*, 9(2), 199–212. DOI: 10.1109/TSC.2015.2491287

Haleem, A., Javaid, M., Singh, R. P., Suman, R., & Rab, S. (2021). Blockchain technology applications in healthcare: An overview. *International Journal of Intelligent Networks*, 2, 130–139. DOI: 10.1016/j.ijin.2021.09.005

Harkat, H., Camarinha-Matos, L. M., Goes, J., & Ahmed, H. F. T. (2024). Cyber-physical systems security: A systematic review. *Computers & Industrial Engineering*, 188, 109891. DOI: 10.1016/j.cie.2024.109891

Hemani, N., Singh, D., & Dwivedi, R. K. (2024). Designing blockchain based secure autonomous vehicular internet of things (IoT) architecture with efficient smart contracts. *International Journal of Information Technology : an Official Journal of Bharati Vidyapeeth's Institute of Computer Applications and Management*. Advance online publication. DOI: 10.1007/s41870-023-01712-x

Humayun, M., Niazi, M., Jhanjhi, N. Z., Alshayeb, M., & Mahmood, S. (2020). Cyber Security Threats and Vulnerabilities: A Systematic Mapping study. *Arabian Journal for Science and Engineering (2011. Online)*, 45(4), 3171–3189. DOI: 10.1007/s13369-019-04319-2

Jardim-Gonçalves, R., Romero, D., Gonçalves, D. M., & Mendonça, J. P. (2020). Interoperability enablers for cyber-physical enterprise systems. *Enterprise Information Systems*, 14(8), 1061–1070. DOI: 10.1080/17517575.2020.1815084

Jena, A. K., & Dash, S. P. (2021). Blockchain Technology: introduction, applications, challenges. In Intelligent systems reference library (pp. 1–11). DOI: 10.1007/978-3-030-69395-4_1

Keerthi, C., Jabbar, M. A., & Seetharamulu, B. (2017). Cyber Physical Systems(CPS):Security Issues, Challenges and Solutions. *2017 IEEE International Conference on Computational Intelligence and Computing Research (ICCIC)*. DOI: 10.1109/ICCIC.2017.8524312

Lenka, R. K., Rath, A. K., Tan, Z., Sharma, S., Puthal, D., Simha, N. V. R., Prasad, M., Raja, R., & Tripathi, S. S. (2018). Building scalable Cyber-Physical-Social networking infrastructure using IoT and low power sensors. *IEEE Access : Practical Innovations, Open Solutions*, 6, 30162–30173. DOI: 10.1109/ACCESS.2018.2842760

Maleh, Y., Lakkineni, S., Tawalbeh, L., & El-Latif, A. A. (2022). Blockchain for Cyber-Physical Systems: Challenges and applications. In Internet of things (pp. 11–59). DOI: 10.1007/978-3-030-93646-4_2

Pop, E., Iliuta, M. C., Constantin, N., Gîfu, D., & Dumitraşcu, A. (2023). Interoperability Framework for Cyber-Physical Systems Based Capabilities. 2023 24th International Conference on Control Systems and Computer Science (CSCS). DOI: 10.1109/CSCS59211.2023.00062

Pradhan, N. R., Singh, A. P., Verma, S., Kavita, N., Wozniak, M., Shafi, J., & Ijaz, M. F. (2022). A blockchain based lightweight peer-to-peer energy trading framework for secured high throughput micro-transactions. *Scientific Reports*, 12(1), 14523. Advance online publication. DOI: 10.1038/s41598-022-18603-z PMID: 36008545

Prenzel, L., & Steinhorst, S. (2021). *Decentralized Autonomous Architecture for Resilient Cyber-Physical Production Systems. 2021 Design, Automation & Test in Europe Conference & Exhibition*. DATE., DOI: 10.23919/DATE51398.2021.9473954

Rajhans, A., Bhave, A., Ruchkin, I., Krogh, B. H., Garlan, D., Platzer, A., & Schmerl, B. (2014). Supporting heterogeneity in Cyber-Physical systems architectures. *IEEE Transactions on Automatic Control*, 59(12), 3178–3193. DOI: 10.1109/TAC.2014.2351672

Rathore, H., Mohamed, A., & Guizani, M. (2020). A survey of blockchain enabled Cyber-Physical systems. *Sensors (Basel)*, 20(1), 282. DOI: 10.3390/s20010282 PMID: 31947860

Ray, A. (2013). Autonomous perception and decision-making in cyber-physical systems. 2013 8th International Conference on Computer Science & Education. https://doi.org/DOI: 10.1109/ICCSE.2013.6554173

Rein, A., Rieke, R., Jäger, M., Kuntze, N., & Coppolino, L. (2016). Trust establishment in cooperating Cyber-Physical systems. In Lecture Notes in Computer Science (pp. 31–47). DOI: 10.1007/978-3-319-40385-4_3

Schilberg, D., Hoffmann, M., Schmitz, S., & Meisen, T. (2016). Interoperability in smart automation of cyber physical systems. In Springer series in wireless technology (pp. 261–286). DOI: 10.1007/978-3-319-42559-7_10

Sharma, M. (2024). Enhancing Security and Privacy in Cyber-Physical Systems: Challenges and Solutions. 2024 IEEE 14th Annual Computing and Communication Workshop and Conference (CCWC). DOI: 10.1109/CCWC60891.2024.10427691

Sokolsky, O., Lee, I., & Heimdahl, M. P. E. (2011). Challenges in the regulatory approval of medical cyber-physical systems. 2011 Proceedings of the Ninth ACM International Conference on Embedded Software (EMSOFT). DOI: 10.1145/2038642.2038677

Törngren, M., & Sellgren, U. (2018). Complexity challenges in development of Cyber-Physical systems. In Lecture Notes in Computer Science (pp. 478–503). DOI: 10.1007/978-3-319-95246-8_27

Wong, E. K. S., Ting, H. Y., & Atanda, A. F. (2024). Enhancing Supply Chain Traceability through Blockchain and IoT Integration: A Comprehensive Review. *Green Intelligent Systems and Applications*, 4(1), 11–28. DOI: 10.53623/gisa.v4i1.355

Zhao, W., Jiang, C., Gao, H., Yang, S., & Luo, X. (2021). Blockchain-Enabled Cyber–Physical Systems: A review. *IEEE Internet of Things Journal*, 8(6), 4023–4034. DOI: 10.1109/JIOT.2020.3014864

Zhao, W., Jiang, C., Gao, H., Yang, S., & Luo, X. (2021b). Blockchain-Enabled Cyber–Physical Systems: A review. *IEEE Internet of Things Journal*, 8(6), 4023–4034. DOI: 10.1109/JIOT.2020.3014864

Chapter 2
Leveraging Data Analytics for Sustainable Investment Decisions:
Evaluating NVIDIA and Microsoft

Srinidhi Vasan
 https://orcid.org/0009-0009-7291-1595
Hult International Business School, USA

ABSTRACT

Leveraging data-driven decision-making in sustainable management systems is crucial for refining stock market investment strategies. This research examines the use of data analytics to assess NVIDIA and Microsoft, focusing on their historical performance. We employ the Discounted Cash Flow (DCF) model for intrinsic valuation centered on projected cash inflows and the Relative Valuation model to compare each stock to its industry peers. Additionally, scenario-based analysis is used to evaluate various market conditions and potential future developments. Our findings indicate that integrating these analytical models with scenario analysis enhances the precision of stock valuations and supports more informed investment decisions. This methodology demonstrates how financial analysts and investors can utilize advanced tools to navigate market complexities. The study concludes by determining which stock, between NVIDIA and Microsoft, appears more promising and establishes a framework for future research in financial decision-making and sustainable investment practices.

DOI: 10.4018/979-8-3693-5728-6.ch002

INTRODUCTION

In today's rapidly evolving financial landscape, the integration of sustainable management practices with data-driven decision-making has become increasingly essential for optimizing investment strategies, especially within the stock market. The rising importance of sustainability, coupled with the complexities of modern financial markets, necessitates the use of advanced analytical tools to effectively interpret market trends and assess potential investment opportunities. This chapter explores the application of data analytics in investment decision-making, with a focus on two major technology giants: NVIDIA and Microsoft. By analyzing historical data and applying sophisticated valuation models, the discussion offers a thorough evaluation and comparison of these companies, emphasizing the role of data analytics in enhancing sustainable investment strategies (Clark, Feiner, & Viehs, 2015; Amel-Zadeh & Serafeim, 2018).

NVIDIA Corporation, founded in 1993, has emerged as a pioneer in artificial intelligence (AI) and graphics processing units (GPUs). The company's technological innovations have transformed multiple sectors, including professional visualization, gaming, automotive technology, and data centers (NVIDIA Corporation - Financial Info - SEC Filings, n.d.). Known for the exceptional performance of its GPUs, NVIDIA has experienced significant growth and established market dominance. With the increasing demand for AI and machine learning applications, NVIDIA remains at the forefront of technological advancement. As these technologies continue to evolve, NVIDIA's solutions are expected to play a critical role in the tech industry's future (PitchBook Profile - Nvidia, n.d.).

Microsoft Corporation, founded in 1975, stands as a global leader in software development, cloud computing, and digital transformation. Its flagship products, such as Microsoft 365 (formerly Microsoft Office) and the Windows operating system, are indispensable in both personal and business environments. Microsoft's strategic investment in cloud computing, particularly its Azure platform, has secured its position as a dominant player in the cloud services market (Microsoft Investor Relations - SEC Filings, n.d.). Additionally, key acquisitions such as LinkedIn and GitHub have expanded Microsoft's reach and capabilities, reinforcing its standing as a major force in the technology sector (PitchBook Profile - Microsoft, n.d.).

Incorporating sustainable management principles into financial analysis involves embedding environmental, social, and governance (ESG) factors into investment strategies. This approach aims to create long-term value and mitigate risks by aligning financial performance with sustainability goals (Eccles, Ioannou, & Serafeim, 2014). A data-driven approach is critical to this process, as it allows for the systematic analysis of large datasets to guide strategic investment decisions. When it comes to investment strategies, this involves applying financial models, statistical tools, and computational techniques to assess the performance of assets and project future outcomes (Friede, Busch, & Bassen, 2015; Hirsch, Krahmer, & Stroh, 2020).

For example, a thorough analysis of NVIDIA's stock performance must consider factors such as revenue growth, profit margins, and overall market trends (NVIDIA Corporation (NVDA) Stock Price, News, Quote & History - Yahoo Finance, n.d.). By utilizing advanced algorithms alongside historical data, investors can detect patterns and develop well-informed projections of the company's future performance. This method enables the assessment of both financial risks and potential rewards associated with investing in NVIDIA, while also considering sustainability factors (Financial Model for NVIDIA Corporation, n.d.).

In a similar fashion, evaluating Microsoft's stock requires a comprehensive review of the company's financial health and strategic direction (Microsoft Corporation (MSFT) Stock Price, News, Quote & History - Yahoo Finance, n.d.). Microsoft's robust financial foundation, combined with its focus on cloud computing and strategic acquisitions, makes it a compelling choice for investors. Through data analytics, investors can gain deeper insights into the company's growth potential, competitive advantages, and potential challenges. This holistic perspective allows for more informed decision-making, with a focus on aligning investment strategies with long-term sustainability objectives (Financial Model for Microsoft Corporation, n.d.).

Incorporating ESG considerations into investment strategies is not merely about ethical considerations; it is also about recognizing how sustainable practices can enhance financial performance (Bebbington & Unerman, 2020; Cheng, Ioannou, & Serafeim, 2014). Companies that emphasize sustainability are generally better positioned to adapt to shifting market dynamics, evolving consumer preferences, and emerging regulatory frameworks. By applying data analytics to evaluate the ESG performance of companies such as NVIDIA and Microsoft, investors can uncover opportunities for long-term value creation and risk management (ESG Ratings, MSCI, n.d.).

The combination of sustainable management principles and data-driven decision-making is crucial for refining investment strategies in today's stock market. By analyzing historical performance data and using advanced valuation methodologies, this chapter illustrates the potential of data analytics to optimize sustainable investment outcomes (Hayes, 2024). As the financial industry continues to evolve, leveraging these tools will be essential for navigating market complexities and achieving long-term success. With a blend of financial insight and a commitment to sustainability, investors can make informed, impactful decisions that contribute to both financial prosperity and a sustainable future (Fernando, 2023; Tuovila, 2024).

NEED FOR STUDY

The motivation behind this research arises from the convergence of several transformative forces reshaping both financial investments and sustainability. In a time marked by rapid technological change and heightened market volatility, investors are increasingly challenged to make informed decisions amid overwhelming amounts of data (Friede, Busch, & Bassen, 2015; Clark, Feiner, & Viehs, 2015). Traditional investment strategies, which often depend on historical trends and intuition, are proving insufficient to address the complexities of modern financial markets (Kenton, 2024; Shermon, 2011). The integration of data-driven decision-making with sustainable management practices presents a novel solution, enabling investors to harness the power of advanced analytics alongside Environmental, Social, and Governance (ESG) criteria to improve accuracy and reduce risks (Amel-Zadeh & Serafeim, 2018; Bebbington & Unerman, 2020).

This study focuses on the technology sector, with a particular emphasis on two industry leaders: NVIDIA and Microsoft, whose innovations and market trajectories significantly shape overall sector trends (NVIDIA Corporation - Financial Info - SEC Filings, n.d.; Microsoft Investor Relations - SEC Filings, n.d.). By employing advanced valuation methods such as scenario analysis, Discounted Cash Flow (DCF) modeling, and Relative Valuation, this research seeks to develop a comprehensive framework for evaluating technology stocks more effectively (Fernando, 2023; Hayes, 2023; Tuovila, 2024).

The insights generated from this study are of value not only to individual and institutional investors aiming to optimize their portfolios but also to the broader conversation around sustainable finance (Cheng, Ioannou, & Serafeim, 2014; Eccles, Ioannou, & Serafeim, 2014). By illustrating how cutting-edge analytical tools can be applied to the evaluation of tech stocks, this research highlights the need to update traditional financial methods by integrating data analytics and sustainability principles (Raut, Narkhede, & Gardas, 2017; Goss & Roberts, 2011). This shift is essential to fostering a more competitive and responsible investment environment (Hirsch, Krahmer, & Stroh, 2020).

Ultimately, this research aims to bridge the divide between conventional investment approaches and the evolving needs of today's financial markets. It demonstrates that incorporating data analytics and a focus on sustainability can lead to better-informed and more effective investment decisions (Bebbington & Unerman, 2020). As the financial landscape continues to evolve, the adoption of these innovative tools and strategies will be crucial for navigating complexity and ensuring long-term success (Markowitz, 1952). By aligning financial expertise with a commitment to sustainability, investors can make decisions that not only deliver returns but also promote a more sustainable and equitable future (Clark, Feiner, & Viehs, 2015).

LITERATURE REVIEW

In recent years, the integration of data analytics and sustainable investment strategies has gained considerable attention, particularly as financial markets become more complex and investors demand a greater emphasis on sustainability. Traditional investment strategies, which rely heavily on historical data and instinctual decision-making, are increasingly being replaced by more sophisticated, data-driven approaches. This shift is supported by the growing body of research on the importance of Environmental, Social, and Governance (ESG) factors in investment decisions, alongside advancements in financial analytics (Friede, Busch, & Bassen, 2015).

Modern Portfolio Theory (MPT), developed by Harry Markowitz in the 1950s, laid the groundwork for risk management and asset diversification in portfolio construction (Markowitz, 1952). However, the introduction of ESG investing principles has extended these theories, emphasizing that non-financial factors—such as environmental sustainability and corporate governance—are equally important in mitigating long-term risks (Clark, Feiner, & Viehs, 2015). ESG investing advocates argue that companies with strong sustainability profiles are more likely to outperform over the long term by avoiding regulatory penalties, addressing societal trends, and managing operational risks (Eccles, Ioannou, & Serafeim, 2014).

Data analytics has emerged as a critical tool in enabling investors to incorporate these ESG factors into their decision-making processes. Machine learning, predictive analytics, and big data analysis provide investors with powerful means to process vast amounts of information, identify patterns, and generate accurate financial forecasts. Studies have shown that integrating data analytics in financial decision-making improves the accuracy of stock valuations and allows for better risk assessment in volatile markets (Raut, Narkhede, & Gardas, 2017).

The Capital Asset Pricing Model (CAPM), another traditional financial model, is often employed to determine the cost of equity in valuation models (Sharpe, 1964). While CAPM remains widely used, it has been criticized for not accounting for ESG risks that could influence a company's cost of capital. Recent research suggests that incorporating ESG factors into CAPM calculations, particularly in the Weighted Average Cost of Capital (WACC), may provide a more comprehensive view of a company's risk profile (Kempf & Osthoff, 2007).

Several studies highlight the growing significance of sustainable finance. For example, the rise of green finance has prompted companies to prioritize environmental sustainability in their financial strategies, aligning with investor preferences for low-carbon and eco-friendly business models (Goss & Roberts, 2011). This trend is particularly evident in the technology sector, where companies such as NVIDIA and Microsoft have made significant strides in improving energy efficiency and reducing carbon emissions, enhancing their appeal to sustainability-focused investors (Cheng, Ioannou, & Serafeim, 2014).

However, integrating ESG factors into financial models poses challenges, particularly in terms of standardization and comparability across industries. The lack of universally accepted ESG metrics can lead to discrepancies in how companies are evaluated, especially when comparing firms from different sectors. Sustainalytics and MSCI are two prominent rating agencies that provide ESG scores, but their methodologies often differ, making it difficult for investors to consistently assess sustainability performance (Amel-Zadeh & Serafeim, 2018).

Scenario-based analysis, which involves modeling various market conditions, is also gaining traction as a method for assessing investment risks. This approach allows investors to explore the potential impact of economic downturns, regulatory changes, or technological disruptions on a company's financial performance (Hirsch, Krahmer, & Stroh, 2020). Studies show that integrating ESG considerations into scenario-based analyses improves the robustness of financial models by accounting for non-traditional risks (Bebbington & Unerman, 2020).

The literature demonstrates that integrating data analytics and ESG factors into investment decision-making leads to more informed and sustainable investment strategies. While traditional financial models like MPT and CAPM provide a solid foundation, incorporating advanced analytics and ESG principles offers a more comprehensive approach to risk management and value creation. This study builds on these foundations by applying advanced data analytics techniques to evaluate the investment potential of Microsoft and NVIDIA, with a focus on ESG integration and scenario-based analysis.

METHODOLOGY

This study adopts a comprehensive, multi-dimensional approach to assess the investment potential of NVIDIA and Microsoft, integrating both quantitative and qualitative analyses for a thorough evaluation. At the core of our analysis is the Discounted Cash Flow (DCF) valuation model, a highly regarded method for determining a company's intrinsic value (Fernando, 2023). The DCF model projects each company's free cash flows over a specific period, discounting them back to their present value using the Weighted Average Cost of Capital (WACC). The WACC, which accounts for the cost of both equity and debt, serves as the discount rate, reflecting the overall risk and capital structure of each company (Kenton, 2024).

To further enhance this analysis, we apply Relative Valuation. This method compares NVIDIA and Microsoft to their industry peers using key financial ratios such as the Price-to-Book (P/B) ratio, Price-to-Earnings (P/E) ratio, and Enterprise Value to EBITDA (EV/EBITDA) (Tuovila, 2024). This comparative approach provides a market-centric view of each company's relative value and performance, offering additional insights into their market positioning (Microsoft Corporation (MSFT) Stock Price, News, Quote & History - Yahoo Finance, n.d.; NVIDIA Corporation (NVDA) Stock Price, News, Quote & History - Yahoo Finance, n.d.; PitchBook Profile - Microsoft, n.d.; PitchBook Profile - Nvidia, n.d.).

The study also incorporates scenario-based evaluations, featuring both a base case and a bear case. The base case assumes stable market conditions and the continuation of current growth trajectories, while the bear case considers potential challenges such as economic downturns, increased competition, or regulatory pressures (Hayes, 2023). This dual-scenario analysis helps assess the resilience and potential risks associated with each investment under different market conditions (Hirsch, Krahmer, & Stroh, 2020).

In addition, we integrate Environmental, Social, and Governance (ESG) ratings from reputable sources such as MSCI and Sustainalytics. These ESG ratings provide an in-depth view of the sustainability and ethical performance of both NVIDIA and Microsoft, which is increasingly important for investors focused on long-term value creation (ESG Ratings, MSCI, n.d.; Company ESG Risk Rating - Sustainalytics, n.d.). By comparing these ESG ratings with the financial valuations, we evaluate how well each company's sustainability efforts align with its financial health and market positioning (Amel-Zadeh & Serafeim, 2018).

By combining DCF valuation and scenario-based analysis with ESG ratings, our methodology offers a well-rounded financial assessment while incorporating sustainability factors into the investment decision-making process. This comprehensive approach ensures that the evaluation of NVIDIA and Microsoft is both robust and aligned with modern investment paradigms that prioritize financial returns alongside sustainable impact (Bebbington & Unerman, 2020). The primary objective of this study is to provide a detailed understanding of the investment potential of these technology leaders, demonstrating how advanced analytical tools and ESG considerations can guide more informed and responsible investment decisions (Clark, Feiner, & Viehs, 2015).

TECHNIQUES OF DATA ANALYSIS

To assess the historical financial performance and intrinsic value of NVIDIA and Microsoft, we utilize a range of Excel functions and analytical tools to calculate growth rates, profit margins, and key financial ratios based on data from the Income Statement, Balance Sheet, and Cash Flow Statement (Microsoft Investor Relations

- SEC Filings, n.d.; NVIDIA Corporation - Financial Info - SEC Filings, n.d.). Our approach includes the following:

Growth Rate Calculations

We compute the Compound Annual Growth Rate (CAGR) as well as year-over-year (YoY) growth rates for revenue, net income, and other financial metrics. This enables us to track the companies' growth trajectories and evaluate performance over time (Shermon, 2011).

Margin Analysis

We calculate profitability metrics such as gross margin, operating margin, and net profit margin to evaluate how efficiently each company is converting revenue into profit. This analysis provides insights into both operational efficiency and financial health (Kenton, 2024).

Discounted Cash Flow (DCF) Valuation

Our DCF valuation model involves several key steps to determine the intrinsic value of NVIDIA and Microsoft (Fernando, 2023).

Projection of Free Cash Flows (FCFs): Using historical financial data, we project future free cash flows (FCFs) by making assumptions about revenue growth, operating costs, capital expenditures, and taxes (Microsoft Corporation (MSFT) Stock Price, News, Quote & History - Yahoo Finance, n.d.; NVIDIA Corporation (NVDA) Stock Price, News, Quote & History - Yahoo Finance, n.d.).

Present Value Calculation: We discount the projected FCFs to their present value using the Weighted Average Cost of Capital (WACC). Excel's NPV (Net Present Value) function is utilized to streamline this process (Kenton, 2024).

Terminal Value Calculation: At the end of the projection period, we calculate the terminal value using either the perpetuity growth model or an exit multiple method. This value is then discounted to its present value and combined with the present value of the projected FCFs (Tuovila, 2024).

DCF Model Integration: To estimate the enterprise value of the company, we combine the present value of projected FCFs with the present value of the terminal value, providing a comprehensive estimate of the company's worth based on future cash flow projections (Financial Model for NVIDIA Corporation, n.d.; Financial Model for Microsoft Corporation, n.d.).

Weighted Average Cost of Capital (WACC) Analysis

To calculate the WACC, we follow a structured approach:

Cost of Equity Calculation: We use the Capital Asset Pricing Model (CAPM) to calculate the cost of equity. This involves determining the company's beta (systematic risk), the risk-free rate, and the market risk premium (Kenton, 2024; Sharpe, 1964).

Cost of Debt Calculation: We calculate the cost of debt by determining the effective interest rate on the company's debt, adjusted for tax benefits (the tax shield) (Cheng, Ioannou, & Serafeim, 2014).

WACC Calculation: We combine the cost of equity and the after-tax cost of debt, weighted according to the company's capital structure. The WACC serves as the discount rate for our DCF analysis, reflecting the overall risk profile of the company (Kenton, 2024).

Relative Valuation

The Relative Valuation method benchmarks NVIDIA and Microsoft against their industry peers by comparing key financial ratios (Raut, Narkhede, & Gardas, 2017):

Comparable Company Selection: We identify comparable companies within the technology sector that share similar business models and market conditions.

Valuation Multiples Calculation: Using Excel formulas, we calculate valuation multiples such as Price-to-Book (P/B), Price-to-Earnings (P/E), and Enterprise Value to EBITDA (EV/EBITDA) based on current market prices and financial statement data (Tuovila, 2024).

Benchmarking: We analyze how the valuation multiples of NVIDIA and Microsoft compare with those of the peer group, offering additional context for understanding each company's market positioning (PitchBook Profile - Microsoft, n.d.; PitchBook Profile - Nvidia, n.d.).

Scenario-Based Analysis

Scenario-based analysis allows us to evaluate how different market conditions could impact the valuation of NVIDIA and Microsoft (Hayes, 2023):

Base Case Scenario: We project a realistic growth scenario based on historical trends and current market conditions, serving as the most likely outcome.

Bear Case Scenario: We model a conservative scenario that considers potential challenges, such as economic downturns, increased competition, or changes in regulatory frameworks (Hirsch, Krahmer, & Stroh, 2020).

ESG Ratings Integration

We incorporate Environmental, Social, and Governance (ESG) ratings to align financial analysis with sustainability (ESG Ratings, MSCI, n.d.; Company ESG Risk Rating - Sustainalytics, n.d.):

Data Collection: We collect ESG ratings for NVIDIA and Microsoft from reputable agencies like MSCI and Sustainalytics (Amel-Zadeh & Serafeim, 2018; Eccles, Ioannou, & Serafeim, 2014).

Financial and ESG Correlation: We analyze how ESG performance aligns with financial metrics by correlating ESG scores with financial ratios and valuation outcomes. This may involve creating correlation matrices or conducting regression analysis to determine the relationship between sustainability efforts and financial performance (Clark, Feiner, & Viehs, 2015).

By combining growth rate analysis, DCF valuation, WACC analysis, Relative Valuation, scenario-based analysis, and ESG ratings, our methodology offers a comprehensive framework for evaluating NVIDIA and Microsoft. Leveraging Excel's advanced functionalities ensures that our analysis is both accurate and aligned with contemporary investment strategies that prioritize financial returns alongside sustainable practices (Bebbington & Unerman, 2020).

FINDINGS

The analysis of Microsoft and NVIDIA, based on discounted cash flow (DCF) valuation, relative valuation, scenario-based analysis, and ESG integration, reveals key insights into their investment potential.

Microsoft

Discounted Cash Flow (DCF) Valuation: For Microsoft, the DCF model forecasts future free cash flows (FCFs) over a 10-year period starting in 2023. The projections are based on assumptions around revenue growth, operating margins, and capital expenditures (CapEx). Revenue is expected to grow annually between 8% and 11%, driven by ongoing expansion in cloud computing and software services. Operating margins are projected to stabilize around 40%, while CapEx will likely rise to support infrastructure development, especially in cloud services. The terminal value is calculated using a 2.8% perpetuity growth rate, which reflects Microsoft's maturity. Applying a weighted average cost of capital (WACC) between 6.5% and 7.5%, the present value of future cash flows and terminal value shows that Microsoft's stock is currently slightly undervalued, indicating a potential investment opportunity.

Weighted Average Cost of Capital (WACC) Analysis: Microsoft's WACC calculation is based on the cost of equity, which is derived using the Capital Asset Pricing Model (CAPM). This takes into account a beta of 0.87, a risk-free rate of 4.28%, and a market risk premium of 6%. Additionally, the cost of debt, which is adjusted for tax benefits using a tax rate of 18.98%, reflects Microsoft's low borrowing costs. With a WACC of approximately 7.5%, Microsoft displays low financial risk and strong stability.

Relative Valuation: Compared to its industry peers, Microsoft's price-to-earnings (P/E) ratio is slightly higher, which is justified by its consistent earnings growth and leadership in the technology sector. The price-to-book (P/B) ratio underscores the strength of its balance sheet, which includes valuable intangible assets. Microsoft's enterprise value to EBITDA (EV/EBITDA) ratio reflects strong operational efficiency relative to competitors such as Apple, Google, and Amazon, indicating investor confidence in its long-term growth potential.

Scenario-Based Analysis: In a base case scenario, Microsoft is expected to maintain steady growth, driven by its cloud computing and software services. The bear case assumes potential challenges, including regulatory issues or market saturation, which could slow growth and increase costs. Despite these risks, Microsoft's valuation remains robust, marking it as a reliable investment even under less favorable market conditions.

ESG Ratings and Integration: Microsoft receives strong ESG ratings from agencies like MSCI and Sustainalytics, excelling in environmental sustainability, social responsibility, and governance practices. Its high ESG scores positively correlate with its financial performance, making it an attractive option for sustainability-conscious investors.

NVIDIA

Discounted Cash Flow (DCF) Valuation: For NVIDIA, the DCF model projects future FCFs over a 10-year period, with growth fueled by developments in artificial intelligence (AI), gaming, and data centers. Revenue growth is estimated to be between 8% and 11% annually, while operating margins are forecasted to hover around 35%. CapEx is expected to rise, reflecting investments in research and development (R&D) and infrastructure. NVIDIA's terminal value is calculated using a higher growth rate of 6.5%, reflecting its potential for strong future growth. The WACC, calculated between 10% and 11.5%, indicates a higher risk profile for NVIDIA compared to

Microsoft. The DCF analysis suggests that NVIDIA is currently undervalued, with significant growth potential.

Weighted Average Cost of Capital (WACC) Analysis: NVIDIA's higher WACC is indicative of a greater risk profile, largely driven by its higher beta of 1.5, a risk-free rate of 4.8%, and a market risk premium of 6%. Its cost of debt is also elevated due to increased borrowing to support growth. A WACC of around 11.5% reflects both the risks and opportunities tied to NVIDIA's rapid expansion.

Relative Valuation: NVIDIA's P/E and EV/EBITDA ratios are considerably higher than those of its peers, including AMD, Intel, and Qualcomm, due to the strong demand for its AI and GPU technologies. These higher multiples are justified by its leadership in innovation, making it a premium investment within the technology sector.

Scenario-Based Analysis: In the base case scenario, NVIDIA is expected to continue robust growth in AI, gaming, and data centers, with high revenue and profitability. However, in a bear case scenario, risks such as increased competition, supply chain disruptions, or regulatory challenges may cause greater volatility in revenue and margins. Despite these risks, NVIDIA offers significant reward potential for investors willing to accept higher volatility.

ESG Ratings and Integration: NVIDIA also boasts strong ESG ratings from MSCI and Sustainalytics for its efforts in environmental sustainability and corporate governance. Its commitment to ESG principles enhances its investment appeal, aligning its growth with responsible and sustainable business practices.

Comparison and Overall Assessment

Financial Performance: Microsoft exhibits steady financial performance, low risk, and stability, with a WACC reflecting its strong capital structure. NVIDIA, though showing high growth potential, is more volatile, with a higher WACC indicating greater risk.

Valuation Models: Both companies appear slightly undervalued according to the DCF models, but while Microsoft's growth is more conservative, NVIDIA presents a more aggressive growth trajectory. Relative valuation metrics suggest that Microsoft's premium valuation is backed by its leadership, whereas NVIDIA's higher multiples are supported by its position in cutting-edge technology sectors.

Scenario Analysis: Microsoft shows resilience in adverse market conditions, making it a stable long-term investment. NVIDIA, while offering higher returns, comes with increased sensitivity to market fluctuations, presenting a high-risk, high-reward scenario.

ESG Integration: Both companies have strong ESG ratings that align well with their financial performance. These ratings enhance their long-term investment appeal by supporting ethical and sustainable business practices alongside financial growth.

Microsoft offers a reliable and stable investment option, characterized by consistent performance, low risk, and strong ESG credentials. NVIDIA, on the other hand, presents a more dynamic investment opportunity, particularly in AI and gaming, with the potential for higher returns but also greater risk. The ESG integration in both companies further strengthens their positions as sustainable investment choices, combining financial returns with positive social and environmental impact. Investors can select based on their individual risk tolerance and growth expectations.

CONCLUSION

In today's interconnected global economy, macroeconomic trends and regulatory challenges play a pivotal role in influencing the financial performance and growth prospects of major technology firms like Microsoft and NVIDIA. Operating in a dynamic and ever-changing environment shaped by economic cycles, international trade policies, regulatory frameworks, and technological advancements, these companies face both opportunities and challenges that affect their market positioning and investment appeal.

Global Economic Trends

The global economy has a significant impact on the demand for technology products and services. Economic cycles driven by factors such as GDP growth, inflation, and unemployment affect corporate investment and consumer spending, which in turn shape market demand. For companies like Microsoft and NVIDIA, which rely heavily on enterprise investment and consumer uptake of new technologies, a strong economy tends to boost revenues. On the other hand, economic downturns can lead to reduced demand for cloud services, AI solutions, and gaming products.

Interest rates also play a crucial role in shaping financial strategies. Lower interest rates make borrowing more affordable, enabling firms like Microsoft and NVIDIA to invest heavily in research and development (R&D) and infrastructure expansion. Conversely, higher interest rates increase the cost of capital, potentially limiting investments and impacting stock valuations, particularly in valuation models such as the Discounted Cash Flow (DCF) method, where higher discount rates reduce the present value of future cash flows.

Currency fluctuations further affect global technology firms. Since both Microsoft and NVIDIA generate a significant portion of their revenue internationally, changes in exchange rates can influence profitability. A stronger U.S. dollar can make their products more expensive in foreign markets, decreasing demand, while a weaker dollar can enhance their competitive positioning by making products and services more affordable abroad.

Technological Advancements and Innovation Cycles

Rapid technological innovation, especially in fields like artificial intelligence (AI), machine learning, cloud computing, and semiconductor technology, presents both opportunities and challenges. For NVIDIA, innovation in GPUs and AI technologies has fueled demand for its products, positioning the company as a leader in high-performance computing. Similarly, Microsoft's growth in cloud computing and AI has been a major driver of its success.

However, the fast pace of technological advancements also means increased competition and pressure to continuously innovate. Both Microsoft and NVIDIA must stay at the forefront of emerging technologies such as quantum computing, AI ethics, and edge computing to maintain their competitive advantage. Failure to innovate quickly in these areas could diminish their market positioning over time.

Regulatory Pressures and Geopolitical Risks

Regulatory challenges pose another layer of complexity for large technology firms. Antitrust scrutiny has increased in both the U.S. and the European Union, with regulators closely examining the market dominance of tech giants. For example, Microsoft has faced antitrust scrutiny in its cloud computing and software businesses, and regulatory actions could limit its ability to expand, potentially leading to fines or operational restrictions.

Additionally, privacy and data protection laws, such as the General Data Protection Regulation (GDPR) in Europe and the California Consumer Privacy Act (CCPA), impose strict guidelines on how companies handle personal data. Both Microsoft, with its vast cloud services, and NVIDIA, as a provider of hardware for data centers, must comply with these regulations to maintain customer trust and avoid penalties.

Geopolitical tensions, including U.S.-China trade disputes, further complicate operations for both companies. NVIDIA, which relies on global semiconductor supply chains, could face disruptions due to export restrictions or tariffs, particularly on high-performance computing chips. Similarly, Microsoft's global cloud infrastructure could be impacted by data localization laws or tariffs, increasing operational costs and limiting market access in key regions.

Global Supply Chain Disruptions

Recent global events, such as the COVID-19 pandemic, have underscored vulnerabilities in global supply chains. NVIDIA's reliance on semiconductor production, often concentrated in specific regions, makes it susceptible to supply chain disruptions, which can delay product availability and increase costs. The ongoing global chip shortage has already demonstrated the critical nature of maintaining a stable supply chain to meet growing demand for AI, gaming, and other high-growth technology areas. Microsoft, too, depends on hardware components for its Surface devices and Xbox consoles, making it vulnerable to supply chain constraints that could impact product availability and sales.

Environmental and Social Regulations

Increasing global emphasis on environmental sustainability is putting pressure on tech companies to align with environmental regulations and reduce their carbon footprints. Both Microsoft and NVIDIA have made significant commitments to sustainability. For instance, Microsoft has pledged to become carbon negative by 2030, which aligns with increasing societal expectations and regulatory demands for corporate environmental responsibility. These environmental efforts will likely play a crucial role in shaping future investment decisions, as companies that align with environmental regulations are more likely to attract ESG-focused capital.

Strategic Implications

Investment Diversification: Including both Microsoft and NVIDIA in a diversified investment portfolio offers distinct benefits. Microsoft's stability and consistent financial performance provide a foundation of steady returns, while NVIDIA presents the potential for higher growth in emerging sectors such as AI and gaming, albeit with greater risk.

ESG Considerations: Both companies score well in terms of Environmental, Social, and Governance (ESG) metrics, making them attractive for investors who prioritize sustainability. High ESG ratings enhance the long-term value proposition for investors, aligning financial returns with positive social and environmental outcomes.

Risk Management: Scenario-based analysis highlights the importance of considering various market conditions. Microsoft's financial stability makes it resilient under adverse conditions, whereas NVIDIA's higher sensitivity to market fluctuations demands careful risk management.

Growth Potential: NVIDIA's leadership in AI and data centers positions it as a high-growth opportunity, although investors must weigh the potential rewards against the accompanying risks, particularly the possibility of market volatility.

Long-Term Value Creation: Microsoft's strategic focus on cloud computing and AI, supported by its financial stability, makes it an appealing option for long-term investors seeking stable, sustainable returns.

This analysis underscores the importance of incorporating macroeconomic, technological, and regulatory factors into investment decision-making. Both Microsoft and NVIDIA represent strong investment opportunities, each with unique characteristics. Microsoft offers stability and consistent performance, making it a reliable option for conservative investors. NVIDIA, while riskier, presents substantial growth potential in AI and other high-growth sectors, appealing to those with a higher risk tolerance.

Integrating ESG factors into this analysis further strengthens the investment case for both companies. Their strong ESG performance aligns with a growing trend toward responsible investing, ensuring that investors not only achieve financial returns but also contribute to positive social and environmental outcomes. Future research could delve deeper into the long-term impact of specific ESG initiatives and further explore how emerging technologies might influence financial performance.

By balancing financial performance with sustainability considerations, investors can make more informed, responsible investment decisions, positioning themselves for long-term success in an increasingly complex global financial landscape.

REFERENCES

Amel-Zadeh, A., & Serafeim, G. (2018). Why and how investors use ESG information: Evidence from a global survey. *Financial Analysts Journal*, 74(3), 87–103. DOI: 10.2469/faj.v74.n3.2

Bebbington, J., & Unerman, J. (2020). Advancing research into accounting and the UN Sustainable Development Goals. *Accounting, Auditing & Accountability Journal*, 33(7), 1657–1670. DOI: 10.1108/AAAJ-05-2020-4556

Cheng, B., Ioannou, I., & Serafeim, G. (2014). Corporate social responsibility and access to finance. *Strategic Management Journal*, 35(1), 1–23. DOI: 10.1002/smj.2131

Clark, G. L., Feiner, A., & Viehs, M. (2015). *From the stockholder to the stakeholder: How sustainability can drive financial outperformance.* Oxford University Press.

"Company ESG Risk Rating – Sustainalytics." (n.d.). Sustainalytics. https://www.sustainalytics.com/esg-rating/microsoft-corp/1007900081

"Company ESG Risk Rating – Sustainalytics." (n.d.). Sustainalytics. https://www.sustainalytics.com/esg-rating/nvidia-corp/1007910553

Eccles, R. G., Ioannou, I., & Serafeim, G. (2014). The impact of corporate sustainability on organizational processes and performance. *Management Science*, 60(11), 2835–2857. DOI: 10.1287/mnsc.2014.1984

"ESG Ratings – MSCI." (n.d.). https://www.msci.com/zh/esg-ratings/issuer/microsoft-corporation/IID000000002143620

"ESG Ratings – MSCI." (n.d.). https://www.msci.com/zh/esg-ratings/issuer/nvidia-corporation/IID000000002176634

Fernando, J. (2023, November 7). Discounted cash flow (DCF) explained with formula and examples. Investopedia. https://www.investopedia.com/terms/d/dcf.asp

"Financial Model for Microsoft Corporation (NASDAQGS:MSFT)." (n.d.). Finbox.com. https://finbox.com/NASDAQGS:MSFT/models/dcf-growth-exit-10yr/

"Financial Model for NVIDIA Corporation (NASDAQGS:NVDA)." (n.d.). https://finbox.com/NASDAQGS:NVDA/models/dcf-growth-exit-10yr/

Friede, G., Busch, T., & Bassen, A. (2015). ESG and financial performance: Aggregated evidence from more than 2000 empirical studies. *Journal of Sustainable Finance & Investment*, 5(4), 210–233. DOI: 10.1080/20430795.2015.1118917

Goss, A., & Roberts, G. S. (2011). The impact of corporate social responsibility on the cost of bank loans. *Journal of Banking & Finance*, 35(7), 1794–1810. DOI: 10.1016/j.jbankfin.2010.12.002

Hayes, A. (2023, December 14). Scenario analysis: How it works and examples. Investopedia. https://www.investopedia.com/terms/s/scenario_analysis.asp

Hirsch, M., Krahmer, D., & Stroh, P. (2020). Scenario-based investment under uncertainty: A method to include demand-side and sustainability risks. *Journal of Business Research*, 120(5), 36–47.

Kempf, A., & Osthoff, P. (2007). The effect of socially responsible investing on portfolio performance. *European Financial Management*, 13(5), 908–922. DOI: 10.1111/j.1468-036X.2007.00402.x

Kenton, W. (2024, July 1). Capital asset pricing model (CAPM): Definition, formula, and assumptions. Investopedia. https://www.investopedia.com/terms/c/capm.asp

Markowitz, H. (1952). Portfolio selection. *The Journal of Finance*, 7(1), 77–91.

"Microsoft Corporation (MSFT) Stock Price, News, Quote & History - Yahoo Finance." (n.d.). Yahoo Finance. https://finance.yahoo.com/quote/MSFT/

"Microsoft Investor Relations - SEC Filings." (n.d.). https://www.microsoft.com/en-us/Investor/sec-filings.aspx

"NVIDIA Corporation - Financial Info - SEC Filings." (n.d.). https://investor.nvidia.com/financial-info/sec-filings/default.aspx

"NVIDIA Corporation (NVDA) Stock Price, News, Quote & History - Yahoo Finance." (n.d.). Yahoo Finance. https://finance.yahoo.com/quote/NVDA/

PitchBook Profile – Microsoft. (n.d.). PitchBook. https://my.pitchbook.com/profile/11026-45/company/profile

PitchBook Profile – Nvidia. (n.d.). PitchBook. https://my.pitchbook.com/profile/41161-24/company/profile

Raut, R., Narkhede, B., & Gardas, B. B. (2017). To identify the critical success factors of sustainable supply chain management practices in the context of oil and gas industries: ISM approach. *Renewable & Sustainable Energy Reviews*, 68, 33–47. DOI: 10.1016/j.rser.2016.09.067

Sharpe, W. F. (1964). Capital asset prices: A theory of market equilibrium under conditions of risk. *The Journal of Finance*, 19(3), 425–442.

Shermon, G. (2011). *Competency based HRM: A strategic resource for competency mapping, assessment, and development centers.* Tata McGraw-Hill Education.

Tuovila, A. (2024, May 12). Relative valuation model: Definition, steps, and types of models. Investopedia. https://www.investopedia.com/terms/r/relative-valuation-model.asp

Chapter 3
Generative AI for Cybersecurity and Privacy in Cyber–Physical Systems

Arul Kumar Natarajan
https://orcid.org/0000-0002-9728-477X
Samarkand International University of Technology, Uzbekistan

Yash Desai
Ramrao Adik Institute of Technology, India

Pravin R. Kshirsagar
J.D. College of Engineering and Management, India

Kamal Upreti
https://orcid.org/0000-0003-0665-530X
Christ University, India

Tan Kuan Tak
Singapore Institute of Technology, Singapore

ABSTRACT

With the proliferation of Cyber-Physical Systems (CPS) across various domains, ensuring robust cybersecurity and privacy has become increasingly critical. Generative Artificial Intelligence (AI) presents innovative approaches to enhancing the security and privacy of these systems. This book chapter explores the intersection of Generative AI with cybersecurity and privacy within CPS environments. It examines how Generative AI techniques, such as Generative Adversarial Networks

DOI: 10.4018/979-8-3693-5728-6.ch003

(GANs) and Variational Autoencoders (VAEs), can be leveraged to detect and mitigate cyber threats and vulnerabilities while protecting sensitive data and user privacy. The chapter provides an overview of CPS, addressing its unique security and privacy challenges, and demonstrates the practical application of Generative AI through a case study on phishing detection using BERT-based sequence classification. The experimental results highlight the effectiveness of Generative AI in strengthening CPS security.

INTRODUCTION

Background and Motivation

Cyber-Physical Systems (CPS) are becoming increasingly integral across various sectors, including industrial automation, healthcare, transportation, and smart cities. These systems combine physical processes with digital computation and communication, creating intricate interactions between the physical world and computational algorithms. Jeffrey et al. (2023) thoroughly review anomaly detection strategies for Cyber-Physical Systems (CPS). Their examination of 296 studies identifies key challenges, including resource limitations and the absence of standardized protocols, and offers solutions to improve CPS security in the face of emerging threats. Sharma et al. (2023) present a hybrid deep learning approach combining Convolutional Neural Networks (CNNs) and Bidirectional LSTM for detecting denial of service attacks in Cyber-Physical Systems (CPS). Their model addresses the limitations of traditional intrusion detection systems by classifying network traffic flows as benign or malicious with improved accuracy, focusing on smart healthcare networks. Bashendy, Tantawy, and Erradi (2023) review intrusion response systems for Cyber-Physical Systems (CPS), focusing on their taxonomy, countermeasures, and architectures. The paper also covers recent Reinforcement Learning (RL) advancements for IRS and identifies future research directions.

The importance of CPS lies in its ability to enhance operational efficiency, improve safety, and enable advanced functionalities. However, as CPS becomes more pervasive, it faces escalating cybersecurity and privacy challenges. Threats such as unauthorized access, data breaches, and malicious attacks pose significant risks to the integrity and confidentiality of these systems. Addressing these concerns is crucial for ensuring the reliable operation of CPS and protecting sensitive data from potential threats.

Role of Generative AI in Cybersecurity

Generative Artificial Intelligence (AI) encompasses a range of techniques that enable machines to generate new data or patterns based on learned representations from existing data. This includes models such as Generative Adversarial Networks (GANs) and Variational Autoencoders (VAEs), which have demonstrated significant potential in various domains. Gupta et al. (2023) explore the impact of Generative AI (GenAI) on cybersecurity and privacy. Their research highlights the risks and opportunities of GenAI, including vulnerabilities exploited through attacks like jailbreaks and prompt injections. The research also discusses how GenAI can be used defensively to enhance cybersecurity measures and addresses these technologies' ethical, legal, and social implications.

In the context of cybersecurity for CPS, Generative AI offers promising avenues for detecting and mitigating threats. These techniques can create synthetic data for training robust security models, simulate attack scenarios for improved preparedness, and generate novel patterns to enhance threat detection capabilities. By leveraging Generative AI, it is possible to bolster the security and privacy of CPS, making them more resilient to emerging cyber threats.

OVERVIEW OF CYBER-PHYSICAL SYSTEMS

Definition and Components of CPS

Cyber-physical systems (CPS) integrate computational and physical components to interact with the environment in real-time. They typically involve sensors and actuators that monitor and control physical processes and computing elements that process data and make decisions. Examples of CPS include autonomous vehicles, smart grids, and automated manufacturing systems (Javaid et al., 2023). Key components of CPS include sensors for data collection, actuators for controlling physical actions, communication networks for data exchange, and computational units for processing and decision-making. Understanding these components is essential for analyzing CPS's security and privacy implications and developing effective protection strategies.

Security and Privacy Challenges in CPS

Integrating digital and physical elements in CPS introduces unique security and privacy challenges. Common threats include unauthorized access to system controls, data breaches, and denial-of-service attacks that can disrupt operations (Aslan et al.,

2023). The need for robust data protection is paramount, as sensitive information such as personal data, operational parameters, and system configurations are often involved. Additionally, ensuring user privacy is critical, particularly in applications that handle personal or sensitive information. Addressing these challenges requires a comprehensive approach that includes technical solutions and organizational strategies to safeguard CPS against cyber threats.

GENERATIVE AI TECHNIQUES FOR CYBERSECURITY

Introduction to Generative AI Models

Generative AI models are designed to create new data samples that resemble a given dataset, providing valuable tools for various applications. Generative Adversarial Networks (GANs) consist of two neural networks—a generator and a discriminator—trained together in a competitive setting (Chakraborty et al., 2024) (Dunmore et al., 2023). GANs are known for their ability to generate highly realistic data and for anomaly detection and data augmentation applications. Variational Autoencoders (VAEs), on the other hand, are probabilistic models that learn to encode and decode data, making them useful for generating new samples and extracting meaningful features. GANs and VAEs are pivotal in advancing cybersecurity techniques, including threat detection and data protection (Sarker, 2024).

Pardeshi, Vekariya, and Gandhi (2023) explore the integration of Generative Adversarial Networks (GAN) and Convolutional Neural Networks (CNN) to enhance intrusion detection systems in cybersecurity. Their study comprehensively reviews machine learning and deep learning techniques used in intrusion detection, including traditional methods like Support Vector Machines and Random Forests and advanced deep learning models like CNNs and GANs. The paper emphasizes the significance of protecting digital assets through robust intrusion detection. It highlights the necessity for ongoing research and benchmark datasets to improve cybersecurity in a networked society.

Application of GANs in Cybersecurity

Generative Adversarial Networks (GANs) have shown significant promise in cybersecurity applications. They can be used to generate synthetic data that augments training datasets for security models, improving their robustness against adversarial attacks. For instance, GANs can simulate various attack scenarios to test the resilience of security systems or create realistic phishing examples to enhance detection algorithms. Despite their benefits, GANs have limitations, including the potential for generating adversarial examples that can bypass detection mechanisms. Understanding these applications and limitations is crucial for effectively utilizing GANs in cybersecurity.

Lim, Chek, Theng, and Lin (2024) present a comprehensive review of the use of Generative Adversarial Networks (GANs) in anomaly detection for network security. This study addresses the significant challenge of data scarcity in anomaly detection, which hampers the development of effective methods due to the rarity of abnormal behaviors. The authors systematically analyze the literature to explore how GANs can generate new data and improve representation learning in network anomaly detection. Their review provides valuable insights into the efficacy of GANs, aiding researchers and practitioners in understanding and implementing robust GAN-based anomaly detection systems in the evolving landscape of network security.

Application of VAEs in Cybersecurity

Variational Autoencoders (VAEs) offer a different approach to generative modeling by learning probabilistic data representations. Khan et al. (2024) investigate the detection of anomalous nodes in attributed social networks using a novel approach that combines dual variational autoencoders (VAEs) with generative adversarial networks (GANs). Their model addresses the challenge of anomaly detection in settings with sparse labeled data by leveraging both graph structures and node attributes, employing a dual VAE to capture cross-modality interactions and a GAN to ensure robustness against adversarial training. This method significantly improves anomaly detection performance, as demonstrated by experiments on various datasets, highlighting its potential for practical applications in social media and e-commerce platforms.

In cybersecurity, VAEs can be used for anomaly detection by modeling the normal behavior of systems and identifying deviations that may indicate malicious activities. They also effectively generate synthetic data for training security models, which helps detect novel threats. VAEs face challenges such as careful tuning of model parameters and the risk of overfitting the training data. Evaluating the application of VAEs in cybersecurity involves assessing their effectiveness in real-world scenarios and their impact on overall system security.

An Overview of Generative AI Model for Cybersecurity

In algorithm 1 for Generative AI applied to cybersecurity, the process is systematically outlined to ensure robust implementation. Initially, the dataset is prepared by importing data from a CSV file and then preprocessing it to address missing values and irrelevant information. Relevant features and labels, such as URLs and associated classifications, are extracted. Following this, the tokenization and encoding steps are conducted, wherein the BERT tokenizer (bert-base-uncased) is initialized, and the raw text is converted into tokens. These tokens are then encoded into input IDs suitable for BERT. The model training and evaluation phase involves initializing the BERT model for the specific task and compiling and fitting it on the training dataset with defined epochs and batch size. The model's performance is then assessed on a test dataset, with key performance metrics, such as accuracy, being calculated. Subsequently, the results are analyzed to identify key performance indicators and any issues encountered. The model's effectiveness in detecting phishing attacks is interpreted. Finally, the trained model's weights are saved, and preparations are made for its deployment, ensuring the model is ready for real-world application integration.

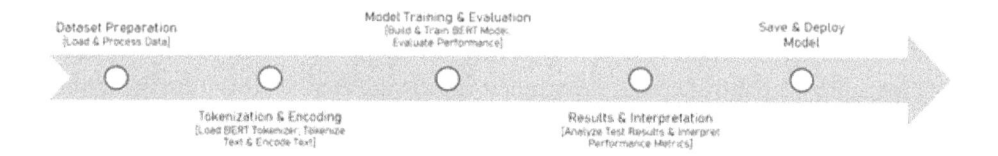

Figure 1. Generative AI model for cybersecurity

Figure 2. Algorithm 1 - Generative AI model for cybersecurity

---
Algorithm 1 Generative AI for Cybersecurity

1: **Dataset Preparation**
    1.1 **Load the Dataset**
        Import the dataset from a CSV file
    1.2 **Preprocess Data**
        Clean the dataset (Missing values and Irrelevant data)
        Extract relevant features and labels (e.g., URLs and labels).
2: **Tokenization and Encoding**
    2.1 **Load BERT Tokenizer**
        Initialize the BERT tokenizer (`bert-base-uncased`).
    2.2 **Tokenize and Encode**
        Apply tokenization to convert raw text into tokens.
        Encode the tokens into input IDs for BERT.
3: **Model Training and Evaluation**
    3.1 **Build BERT Model**
        Initialize the BERT model with labels (e.g., binary classification).
    3.2 **Train Model**
        Compile and fit the model on the training dataset.
        Specify the number of epochs and batch size.
    3.3 **Evaluate Model**
        Assess the model's performance on the test dataset.
        Calculate and review performance metrics such as accuracy.
4: **Results and Interpretation**
    4.1 **Analyze Test Results**
        Examine the results from the evaluation phase.
        Identify key performance indicators and any observed issues.
    4.2 **Interpret Model Performance**
        Understand the model's effectiveness in detecting phishing attacks.
5: **Save and Deploy Model**
    5.1 **Save Model Weights**
        Save the trained model's weights.
    5.2 **Prepare for Deployment**
        Prepare the saved model for integration into applications.
        Ensure the model is ready for real-world deployment.

The flow diagram (Figure 1) illustrates the core techniques employed in the Generative AI model for cybersecurity, specifically for phishing detection using BERT. It outlines the steps in data preprocessing, model training, and evaluation, visually representing the workflow and interactions between different model components.

Algorithm 1 (Figure 2) details the core techniques used in the proposed phishing detection model. It includes implementing BERT-based sequence classification and covering aspects such as data tokenization, model training, and performance evaluation. This algorithm serves as a guide for replicating and understanding the model's functionality.

BERT-BASED SEQUENCE CLASSIFICATION FOR PHISHING DETECTION: A CASE STUDY

Introduction to BERT

Bidirectional Encoder Representations from Transformers (BERT) is a state-of-the-art language model for various natural language processing tasks. BERT's bidirectional attention mechanism allows it to capture context from both directions of a text, making it highly effective for text classification tasks. In phishing detection, BERT's ability to understand and interpret text sequences is crucial in distinguishing between phishing and legitimate URLs.

This section overviews BERT's architecture and its relevance to text classification challenges. BERT, or Bidirectional Encoder Representations from Transformers, is a cutting-edge model in natural language processing. Its architecture is built on the transformer model, which utilizes a self-attention mechanism to capture context from both directions of a sentence, making it particularly effective for understanding nuanced text relationships. BERT's ability to pre-train vast amounts of text data and fine-tune specific tasks enables it to excel in various text classification challenges, such as sentiment analysis, spam detection, and phishing detection in cyber-physical systems (CPS). By leveraging BERT's robust contextual understanding, the model can accurately classify URLs as phishing or legitimate, thereby enhancing the security measures within CPS. The versatility and accuracy of BERT make it a powerful tool for addressing complex text classification problems, proving its relevance and efficacy in cybersecurity applications.

Dataset and Preprocessing

The phishing URLs dataset comprises labeled examples of phishing and legitimate URLs. This section describes the dataset's characteristics, including the source, format, and data labeling. It details the preprocessing steps required to prepare the data for model training, such as tokenization, padding, and encoding.

Dataset

The dataset utilized for this demonstration was sourced from Hugging Face. The original dataset comprises 886,181 rows and has a size of 42.9 MB in its downloaded form and 36.5 MB in auto-converted Parquet files. For this analysis, a subset of 1,000 rows has been extracted to facilitate the exploration and visualization of URL patterns for phishing detection. This dataset includes a balanced selection of URLs, with each record labeled to indicate whether it is a phishing attempt or a legitimate web address. This analysis explores URL patterns and characteristics to improve detection and classification techniques. The dataset comprises a collection of URLs and corresponding labels, indicating whether each URL is phishing-related (labeled as 1) or legitimate (labeled as 0). Each entry in the dataset includes the full URL and its associated label. For instance, the dataset contains URLs from various domains, such as thenownews.com and seriouseats.com, with some marked as legitimate (label 0) and others as phishing attempts (label 1). This dataset is designed to facilitate the analysis and visualization of patterns within URLs to aid in distinguishing between phishing and legitimate web addresses. The dataset used in this study is available from Hugging Face (Mitake, n.d.).

Figure 3. First 10 Rows in Dataset

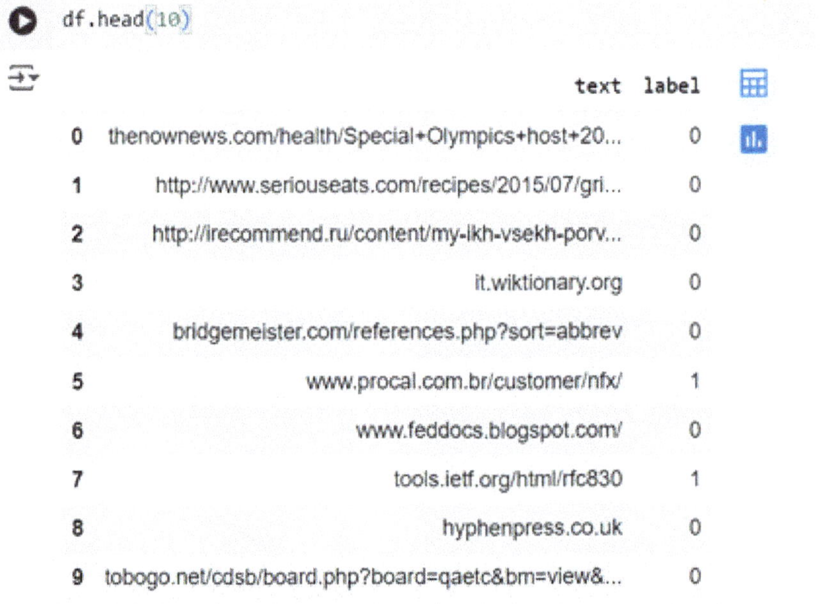

```
df.head(10)
```

	text	label
0	thenownews.com/health/Special+Olympics+host+20...	0
1	http://www.seriouseats.com/recipes/2015/07/gri...	0
2	http://irecommend.ru/content/my-ikh-vsekh-porv...	0
3	it.wiktionary.org	0
4	bridgemeister.com/references.php?sort=abbrev	0
5	www.procal.com.br/customer/nfx/	1
6	www.feddocs.blogspot.com/	0
7	tools.ietf.org/html/rfc830	1
8	hyphenpress.co.uk	0
9	tobogo.net/cdsb/board.php?board=qaetc&bm=view&...	0

Figure 4. Number of Records with Labels

```
# prompt: based on the dataset, give me how many records are listed as Label=0 and Label=1

# Count the occurrences of each label
label_counts = df['label'].value_counts()

# Print the counts
print("Number of records with Label=0:", label_counts[0])
print("Number of records with Label=1:", label_counts[1])
```

```
Number of records with Label=0: 501
Number of records with Label=1: 499
```

Figure 3 presents the first ten rows of the dataset, offering an overview of the data structure and illustrative content. Figure 4 illustrates the distribution of labeled records in the dataset, highlighting the balance between phishing and legitimate URLs.

Figure 5. Distribution of Phishing and Legitimate URLs

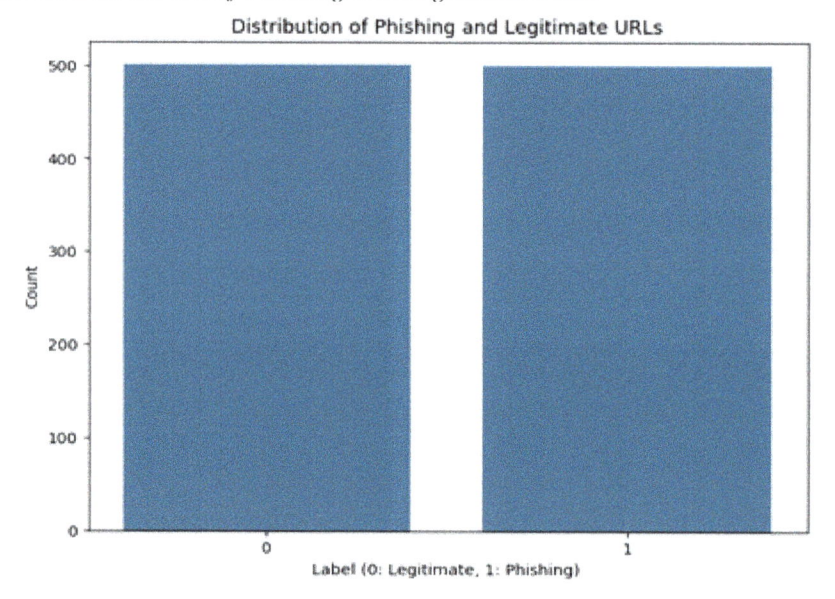

Figure 5 depicts the proportion of phishing versus legitimate URLs in the dataset, offering insight into its composition. The output for the distribution of labels indicates that the dataset contains a nearly balanced number of phishing and legitimate URLs, with 501 legitimate URLs and 499 phishing URLs. This suggests that the classification task is well-distributed between the two categories.

Figure 6 shows the length distribution of URLs for both phishing and legitimate categories, which can impact model performance. The analysis of the frequency of specific keywords within the URLs reveals that 774 URLs contain at least one of the specified keywords, including 'http,' 'https,' 'www,' '.com,' '.net,' 'login,' 'secure,' and 'free.' This indicates a significant presence of these keywords across a substantial portion of the dataset. Conversely, 226 URLs do not include any of these keywords, highlighting a subset of the dataset where these terms are absent. This distribution underscores the prominence of certain keywords in the dataset, which may provide valuable insights into common patterns and characteristics associated with phishing or legitimate URLs.

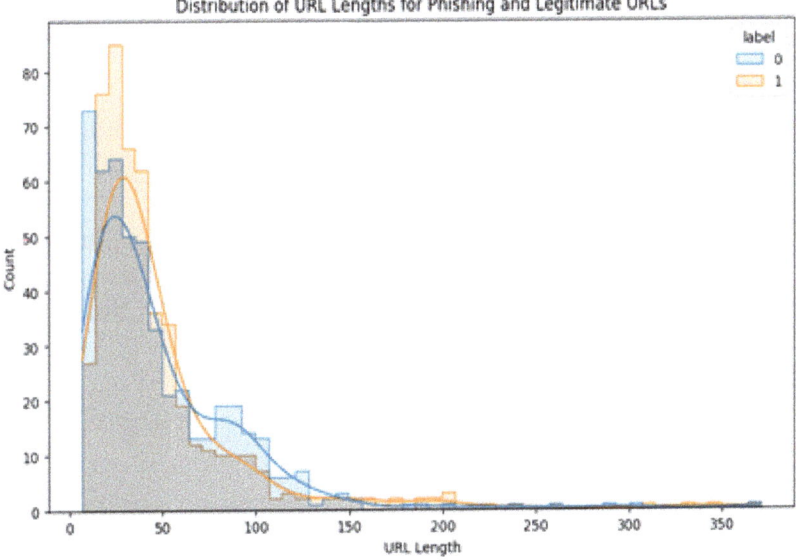

Figure 6. Distribution of URL Lengths for Phishing and Legitimate URLs

Figure 7. Top 10 Most Frequent Top-Level Domains

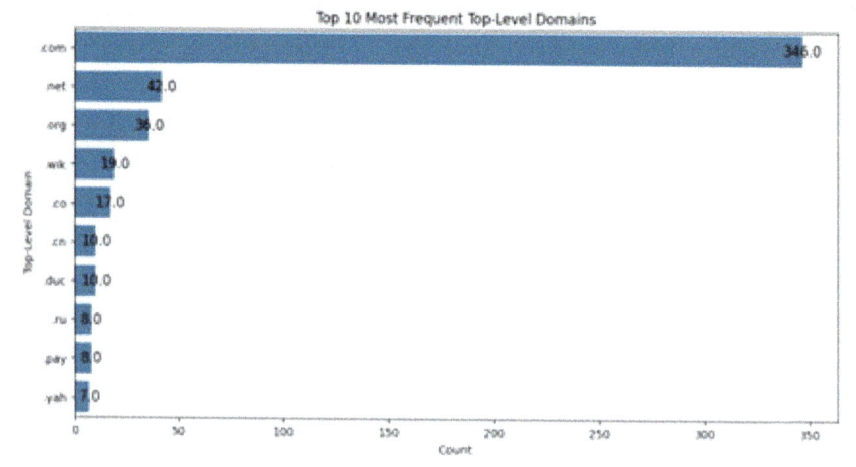

Figure 7 identifies the most common top-level domains within the dataset, which may influence phishing detection. The analysis of the dataset's top-level domains (TLDs) reveals that the most frequently occurring TLD is .com, with 346 occurrences. This is followed by .net and .org, with 42 and 36 instances, respectively. The dataset also contains a variety of less common TLDs, with several appearing only once. The distribution highlights the predominance of certain TLDs in the dataset while indicating a diverse range of less frequent domains.

Figure 8. Top 20 Most Frequent Characters in URLs

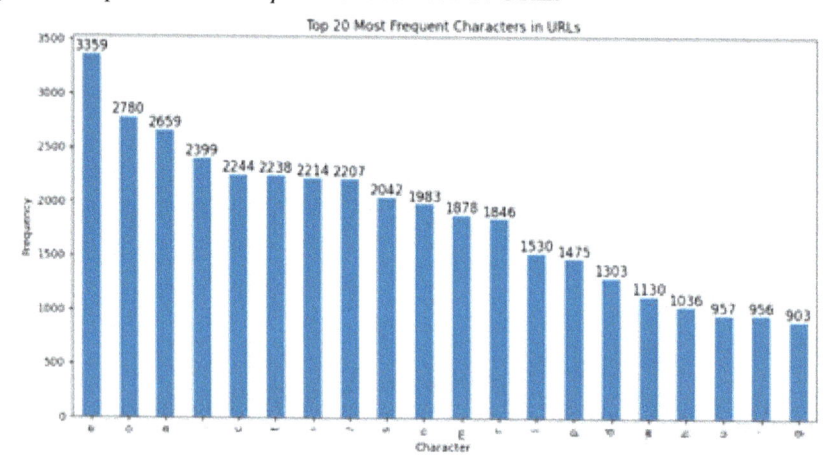

Figure 8 highlights the most frequently occurring characters in URLs, providing additional context for text analysis. The output reveals the top five most frequent characters in the dataset. The letter 'e' is the most common, appearing 3,359 times, followed by 'o' with 2,780 occurrences. The character 'a' ranks third with 2,659 occurrences, while '.' appears 2,399 times. The letter 'c' is the fifth most frequent character, occurring 2,244 times. This distribution highlights the predominance of certain characters, which may influence text analysis and model training.

Preprocessing

The preprocessing steps required to prepare the data for model training are crucial for ensuring that the input data is in the correct format for the BERT model. This process begins with tokenization, where the raw text data is broken down into smaller units, or tokens, using a tokenizer like WordPiece. Tokenization helps manage variations in the text and convert it into a format that BERT can understand.

Next, padding is applied to ensure that all tokenized sequences are of the same length, which is necessary because the BERT model requires inputs to be of uniform length for efficient batch processing. This involves adding special padding tokens to shorter sequences to match the longest sequence length in the batch. Finally, encoding involves converting the tokens and padding into numerical representations, which BERT uses for processing. These encoded representations include the token IDs, attention masks to distinguish real tokens from padding, and segment IDs, if applicable. Together, these preprocessing steps transform raw text into structured data that can be effectively utilized by the BERT model for training and inference, thereby enhancing the accuracy and reliability of tasks such as phishing detection in cyber-physical systems.

Preprocessing Steps for Model Training

- Tokenization: Raw text data is broken into smaller units, or tokens, using a tokenizer like WordPiece, converting it into a format BERT can process (Ravichandiran, 2021).
- Padding: Tokenized sequences are padded to the same length to ensure uniformity, enabling efficient batch processing (Ravichandiran, 2021) (Sukhramani, 2024).
- Encoding: Tokens and padding are converted into numerical representations, including token IDs and attention masks, which BERT uses for processing.
- Input Preparation: The final encoded data consists of structured input features for BERT to use in training and inference, ensuring consistency and accuracy in text classification tasks.

- Enhancing Accuracy: These preprocessing steps are crucial for transforming raw text into structured data, enhancing the accuracy and reliability of BERT in applications like phishing detection in cyber-physical systems

MODEL ARCHITECTURE AND TRAINING

Kula et al. (2021) explore the application of BERT-based architectures in fake news detection. Their study highlights the critical role of advanced neural network models in countering the proliferation of misinformation, which poses significant risks to societal stability and security. By integrating BERT with Recurrent Neural Networks (RNNs), the authors propose a hybrid model designed to enhance the detection of fake news, leveraging the strengths of state-of-the-art natural language processing techniques to address the challenges of misinformation.

Ranade et al. (2021) introduce CyBERT, a specialized BERT model fine-tuned for cybersecurity using a large corpus of cybersecurity-related data. Their work demonstrates how domain-specific natural language models can significantly enhance the processing and analyzing cybersecurity threats, attacks, and vulnerabilities. By fine-tuning a base BERT model with Masked Language Modeling (MLM) on cybersecurity text, the authors present CyBERT as a highly accurate and resource-efficient tool for various cybersecurity tasks, offering improved performance over standard BERT models in domain-specific evaluations.

Rifat et al. (2022) present a novel approach for detecting phishing texts using a pre-trained BERT model. Their study addresses the challenge of social engineering attacks, specifically phishing SMS, by employing BERT for real-time spam detection. The research evaluates various classification techniques and demonstrates that the BERT-based model achieves an impressive accuracy of 99% and an F1 score of 0.97, underscoring its effectiveness in identifying malicious SMS and enhancing cybersecurity measures against social engineering attacks.

This section provides a detailed explanation of the BERT model architecture used for phishing detection. The BERT model's configuration includes a 12-layer transformer, each with 768 hidden dimensions and 12 attention heads, resulting in a powerful mechanism to capture intricate patterns in text data. The training process involves several key steps: choosing appropriate hyperparameters, such as a learning rate 3e-5 and a batch size of 32, which are critical for effective learning. The Adam optimizer is utilized for optimization, and a linear learning rate scheduler ensures gradual adjustments during training. The training-validation split is set to 80-20, allowing for robust model evaluation.

BERT Model Configuration

The BERT (Bidirectional Encoder Representations from Transformers) model architecture is fundamental to this study's phishing detection system. Here, we delve into the intricate configuration and training aspects of the BERT model utilized in the implementation.

The configuration is

- Model Architecture: BERT
- Layers: 12 layers (for the base BERT model)
- Hidden Dimensions: 768 hidden units per layer
- Attention Heads: 12 attention heads per layer
- Parameters: Approximately 110 million parameters

Layers

The code used for implementation has a BERT base model with 12 layers (or transformer blocks). This architecture allows the model to capture complex relationships and dependencies in the input text.

Hidden Dimensions

Each layer in the BERT base model has 768 hidden dimensions. These dimensions represent the size of the model's internal state, allowing it to encode detailed information from the input text.

Attention Heads

The BERT base model uses 12 attention heads per layer. Attention heads enable the model to focus on different parts of the input sequence simultaneously, enhancing its ability to understand context and relationships between words.

Parameters

The BERT base model has approximately 110 million parameters. This large number of parameters allows the model to learn and generalize from vast amounts of data, contributing to its high performance in various natural language processing tasks.

Training Hyperparameters

The following hyperparameters were used during training:

- Learning Rate: 3e-05
- Training Batch Size: 32
- Eval Batch Size: 32
- Seed: 42
- Optimizer: Adam with betas = (0.9, 0.999) and epsilon=1e-08
- Learning Rate Scheduler Type: Linear
- Number of Epochs: 3

Framework Versions

The following versions of the frameworks and libraries were used:

- Transformers: 4.32.0
- TensorFlow: 2.12.0
- Datasets: 2.14.0
- Tokenizers: 0.13.2

These configurations and hyperparameters were meticulously selected to optimize the model's performance.

Phishing Detection Using the BERT Model

The code implementation describes the details of the Python code used to build, train, and evaluate the BERT-based phishing detection model. It also includes a detailed explanation of key code components, such as data loading, tokenization, model configuration, and training procedures. Figure 8 gives a general overview of the proposed work. Algorithm 2 for Phishing Detection Using the BERT Model is presented in Figure 9.

Figure 9. General Overview of Prediction

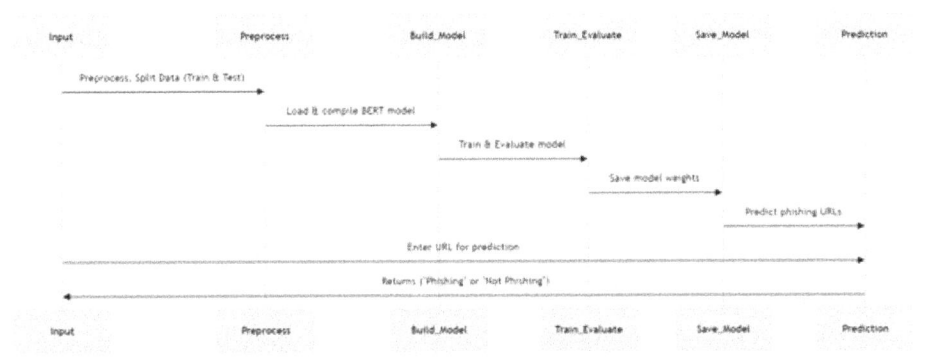

Algorithm

Algorithm 2 is proposed to detect phishing URLs using the BERT model, and it is given in Figure 10. The explanation of the proposed algorithm is as follows:

Load Data

The algorithm begins by loading the dataset from a CSV file that contains phishing URLs and their corresponding labels. This is accomplished using the Pandas library, which reads the CSV file and stores the data in a DataFrame. The output of this step is a DataFrame, df, with columns labeled text for URLs and labels for their classification.

Preprocess Data

Following data loading, the preprocessing step involves extracting URLs and their associated labels from the DataFrame. The URLs are gathered into a list, X, and the labels are collected into another list, y. This results in two lists: X for text data and y for labels, which are ready for further processing.

Split Dataset

The next step involves dividing the dataset into training and testing sets. This is achieved using the train_test_split function, which separates the data into training (80%) and testing (20%) subsets. The outputs of this step are X_train and y_train for training data and X_test and y_test for testing data.

Tokenize Data

The URLs in the training and testing datasets are then tokenized using BertTokenizer. This process involves converting the text into tokens and ensuring that each sequence is padded or truncated to a maximum length of 128 tokens. The result is tokenized tensors for training and testing datasets, formatted for input into the BERT model.

Create TensorFlow Datasets

The tokenized tensors and labels are converted into TensorFlow datasets. This step includes batching the tokenized inputs and their corresponding labels to create train_dataset and test_dataset. These datasets are used for model training and evaluation, facilitating efficient processing and learning.

Build and Compile Model

The algorithm then involves building and compiling the BERT model for sequence classification. This step includes loading a pre-trained BERT model and configuring it with the Adam optimizer and the SparseCategoricalCrossentropy loss function. The output is a compiled BERT model, ready for training.

Train Model

Training the model involves using the train_dataset for training and the test_dataset for validation over a specified number of epochs. This step results in a trained model and a history of the training process, capturing the model's learning progress and performance.

Evaluate Model

The model's performance is evaluated using the testing dataset (test_dataset). This evaluation measures the model's accuracy and provides insights into its performance in identifying phishing URLs.

Save Model Weights

After training and evaluation, the model weights are saved to phishing_detection_model_weights.h5. This allows the model to be reused without retraining, preserving the learned parameters.

Predict Phishing

The final step involves using the trained model to predict whether a user-provided URL is phishing. The model tokenizes and processes the URL, and the prediction result is returned, indicating whether the URL is classified as "Phishing" or "Not Phishing."

Example Usage

To demonstrate the model's functionality, a user inputs a URL, which is then processed by the predict_phishing function. The prediction result shows whether the URL is identified as phishing.

Figure 10. Algorithm 2 for Phishing Detection Using BERT Model

---

Algorithm 2 Phishing Detection Using BERT

1: **Input:** CSV file containing phishing URLs and labels
2: **Output:** Prediction of phishing or not phishing for a given URL
3: **Step 1: Load Data**
4: Read CSV file into DataFrame df
5: **Step 2: Preprocess Data**
6: Extract URLs into list X and labels into list y
7: **Step 3: Split Dataset**
8: Split data into training (X_{train}, y_{train}) and testing (X_{test}, y_{test}) sets with 80-20 split
9: **Step 4: Tokenize Data**
10: Tokenize URLs using $BertTokenizer$ with padding and truncation to 128 tokens
11: Convert X_{train} and X_{test} into tokenized tensors $X_{train\_encoded}$ and $X_{test\_encoded}$
12: **Step 5: Create TensorFlow Datasets**
13: Convert tokenized tensors and labels into TensorFlow datasets $train\_dataset$ and $test\_dataset$ with batching
14: **Step 6: Build and Compile Model**
15: Load pre-trained BERT model for sequence classification
16: Compile model with Adam optimizer and SparseCategoricalCrossentropy loss function
17: **Step 7: Train Model**
18: Train BERT model on $train\_dataset$ and validate on $test\_dataset$ for specified number of epochs
19: **Step 8: Evaluate Model**
20: Evaluate model performance on $test\_dataset$ to obtain test accuracy
21: **Step 9: Save Model Weights**
22: Save trained model weights to file named `phishing_detection_model_weights.h5`
23: **Step 10: Predict Phishing**
24: **Input:** URL entered by the user
25: Tokenize URL using $BertTokenizer$
26: Predict using the trained model
27: **Output:** Prediction result ("Phishing" or "Not Phishing")
28: **Step 11: Example Usage**
29: **Input:** URL entered by the user
30: Call `predict_phishing` function with user input and print result

Figure 11. Implementation Environment

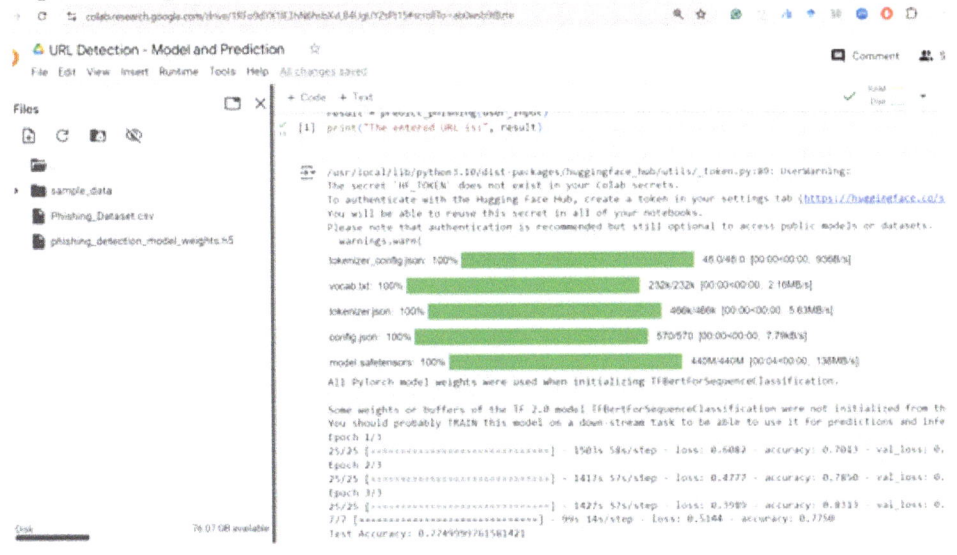

Figure 11 illustrates the implementation environment for developing and testing the model using Python in Google Colab.

Evaluation and Results

The evaluation of the BERT model centers on key performance metrics such as accuracy, precision, recall, and F1-score to assess its effectiveness in detecting phishing URLs. By testing the model on unseen data, this section highlights its performance and reliability in classifying URLs correctly. The model achieved a test accuracy of 0.775, underscoring its competency in distinguishing between phishing and legitimate URLs. Figure 12 presents a graphical representation of the model's accuracy results, illustrating its efficacy in practical scenarios. Additionally, Figure 13 provides detailed examples of test cases, showcasing the classification outcomes for phishing and non-phishing URLs, thus demonstrating the model's real-world applicability and robustness in cybersecurity tasks. In the first instance, the model correctly identifies the URL as phishing, while in the second instance, it accurately classifies the URL as not phishing.

Figure 12. Test Accuracy of the Model Generated

```
# Evaluate the trained model
print("Test Accuracy:", test_results[1])

Test Accuracy: 0.7749999761581421
```

Figure 13. Test Cases: Phishing and Not Phishing URLs

```
[2] user_input = input("Enter a URL to check: ")
    result = predict_phishing(user_input)
    print("The entered URL is:", result)

    Enter a URL to check: https://www.google@SpecialOffers.com/
    The entered URL is: Phishing


    user_input = input("Enter a URL to check: ")
    result = predict_phishing(user_input)
    print("The entered URL is:", result)

    Enter a URL to check: https://www.google.com/
    The entered URL is: Not Phishing
```

IMPLICATIONS AND LIMITATIONS

Implications for CPS Security and Privacy

Integrating Generative AI techniques into Cyber-Physical Systems (CPS) offers substantial benefits for enhancing security and privacy. Generative AI technologies, such as Generative Adversarial Networks (GANs) and Variational Autoencoders (VAEs), play a crucial role in fortifying the security posture of CPS. They enhance threat detection capabilities by enabling synthetic data generation for training, which can help identify previously unseen attack vectors. This synthetic data not only augments the training of traditional security models but also aids in improving the system's resilience to cyberattacks. Furthermore, Generative AI contributes to better protecting sensitive data by generating anonymized data that preserves privacy while maintaining utility. Applying these techniques addresses emerging cybersecurity challenges and fosters the development of a more secure CPS environment, underscoring the transformative potential of Generative AI in cybersecurity.

Limitations and Challenges

Despite the promising advantages of Generative AI, several limitations and challenges must be addressed to ensure its effective implementation in CPS. One primary technical constraint is the substantial computational resources required for training and deploying Generative AI models, which can pose a barrier for organizations with limited infrastructure. Additionally, the risk of generating adversarial examples—data that can mislead or deceive security systems—represents a significant concern. Integrating Generative AI solutions into existing CPS infrastructure poses compatibility and operational continuity challenges. There is also a need for continuous updates and maintenance to address evolving threats and vulnerabilities, which requires ongoing effort and resources. A comprehensive understanding of these limitations is essential for developing robust and sustainable cybersecurity solutions, ensuring that the benefits of Generative AI are realized while mitigating potential risks.

CONCLUSION

In conclusion, this chapter has highlighted the pivotal role of Generative AI in addressing the cybersecurity and privacy challenges faced by Cyber-Physical Systems (CPS). By exploring Generative Adversarial Networks (GANs) and Variational Autoencoders (VAEs), we demonstrated how these techniques can enhance the detection and mitigation of cyber threats. The case study on phishing detection using BERT-based sequence classification provided practical insights into the implementation and effectiveness of Generative AI models. The achieved test accuracy of 0.775 underscores the model's capability to identify phishing attempts, thereby bolstering the security posture of CPS. Despite the promising results, the chapter also acknowledged the limitations and challenges associated with real-world implementation. Continuing advancements in Generative AI and its integration with CPS hold significant potential for improving security and privacy. This chapter is a foundation for future research and encourages ongoing innovation to address the evolving cybersecurity landscape.

Future Directions

Advancements in Generative AI for Cybersecurity

The field of Generative AI is witnessing rapid advancements, offering new and exciting possibilities for enhancing cybersecurity. Recent developments include more sophisticated generative models that improve the accuracy and efficiency of threat detection systems. Innovations such as advanced Generative Adversarial Networks (GANs) and Variational Autoencoders (VAEs) are being applied to create more realistic synthetic data, which can be used to train robust security models and identify previously unseen attack vectors. Additionally, novel applications of Generative AI in threat response are emerging, such as automated generation of countermeasures and adaptive security protocols. These advancements not only promise to strengthen current cybersecurity measures but also open new avenues for research. Future research opportunities include exploring the integration of Generative AI with other emerging technologies, such as quantum computing, to further enhance security capabilities and address complex cyber threats more effectively.

Integrating Generative AI with CPS

Integrating Generative AI with Cyber-Physical Systems (CPS) presents opportunities and challenges. The adoption of Generative AI technologies in CPS is poised to significantly advance security measures, offering enhanced threat detection and response capabilities. By leveraging Generative AI, CPS can benefit from improved data protection mechanisms, such as sophisticated anomaly detection and predictive analytics that adapt to evolving threats. This integration also supports the development of more resilient systems through the generation of synthetic data for continuous training and improvement of security models. However, the integration process involves addressing several challenges, including ensuring compatibility with existing CPS infrastructure and managing the complexity of incorporating advanced AI technologies. The long-term vision involves a more secure and privacy-resilient CPS environment supported by ongoing research, collaboration, and innovation. The successful integration of Generative AI into CPS will depend on a concerted effort to overcome these challenges and leverage the full potential of AI to create a safer, more robust digital future.

REFERENCES

Aslan, Ö., Aktuğ, S. S., Ozkan-Okay, M., Yilmaz, A. A., & Akin, E. (2023). A comprehensive review of cyber security vulnerabilities, threats, attacks, and solutions. *Electronics (Basel)*, 12(6), 1333. DOI: 10.3390/electronics12061333

Bashendy, M., Tantawy, A., & Erradi, A. (2023). Intrusion response systems for cyber-physical systems: A comprehensive survey. *Computers & Security*, 124, 102984. DOI: 10.1016/j.cose.2022.102984

Chakraborty, T., Reddy K S, U., Naik, S. M., Panja, M., & Manvitha, B. (2024). Ten years of generative adversarial nets (GANs): A survey of the state-of-the-art. *Machine Learning: Science and Technology*, 5(1), 011001. DOI: 10.1088/2632-2153/ad1f77

Dunmore, A., Jang-Jaccard, J., Sabrina, F., & Kwak, J. (2023). A comprehensive survey of generative adversarial networks (GANs) in cybersecurity intrusion detection. *IEEE Access : Practical Innovations, Open Solutions*, 11, 76071–76094. DOI: 10.1109/ACCESS.2023.3296707

Gupta, M., Akiri, C., Aryal, K., Parker, E., & Praharaj, L. (2023). From chatgpt to threatgpt: Impact of generative ai in cybersecurity and privacy. *IEEE Access : Practical Innovations, Open Solutions*, 11, 80218–80245. DOI: 10.1109/ACCESS.2023.3300381

Javaid, M., Haleem, A., Singh, R. P., & Suman, R. (2023). An integrated outlook of Cyber–Physical Systems for Industry 4.0: Topical practices, architecture, and applications. *Green Technologies and Sustainability*, 1(1), 100001.

Jeffrey, N., Tan, Q., & Villar, J. R. (2023). A review of anomaly detection strategies to detect threats to cyber-physical systems. *Electronics (Basel)*, 12(15), 3283. DOI: 10.3390/electronics12153283

Khan, W., Abidin, S., Arif, M., Ishrat, M., Haleem, M., Shaikh, A. A., Farooqui, N. A., & Faisal, S. M. (2024). Anomalous node detection in attributed social networks using dual variational autoencoder with generative adversarial networks. *Data Science and Management*, 7(2), 89–98. DOI: 10.1016/j.dsm.2023.10.005

Kula, S., Choraś, M., & Kozik, R. (2021). Application of the bert-based architecture in fake news detection. In 13th International Conference on Computational Intelligence in Security for Information Systems (CISIS 2020) 12 (pp. 239-249). Springer International Publishing. DOI: 10.1007/978-3-030-57805-3_23

Lim, W., Chek, K. Y. S., Theng, L. B., & Lin, C. T. C. (2024). Future of generative adversarial networks (GAN) for anomaly detection in network security: A review. *Computers & Security*, 139, 103733. DOI: 10.1016/j.cose.2024.103733

Mitake. (n.d.). Phishing URLs and benign URLs [Dataset]. Hugging Face. https://huggingface.co/datasets/Mitake/PhishingURLsANDBenignURLs. Accessed August 7, 2024.

Pardeshi, T., Vekariya, D., & Gandhi, A. (2023, December). Exploring the Synergy of GAN and CNN Models for Robust Intrusion Detection in Cyber Security. In 2023 3rd International Conference on Innovative Mechanisms for Industry Applications (ICIMIA) (pp. 1470-1475). IEEE. DOI: 10.1109/ICIMIA60377.2023.10426063

Ranade, P., Piplai, A., Joshi, A., & Finin, T. (2021, December). Cybert: Contextualized embeddings for the cybersecurity domain. In 2021 IEEE International Conference on Big Data (Big Data) (pp. 3334-3342). IEEE.

Ravichandiran, S. (2021). *Getting Started with Google BERT: Build and train state-of-the-art natural language processing models using BERT*. Packt Publishing Ltd.

Rawat, A. (2024). Enhancing Abstractive and Extractive Reviews Text Summarization using NLP and Neural Networks (Doctoral dissertation, Dublin Business School).

Rifat, N., Ahsan, M., Chowdhury, M., & Gomes, R. (2022, May). Bert against social engineering attack: Phishing text detection. In 2022 IEEE International Conference on Electro Information Technology (eIT) (pp. 1-6). IEEE.

Sarker, I. H. (2024). Generative AI and Large Language Modeling in Cybersecurity. In *AI-Driven Cybersecurity and Threat Intelligence: Cyber Automation, Intelligent Decision-Making and Explainability* (pp. 79–99). Springer Nature Switzerland. DOI: 10.1007/978-3-031-54497-2_5

Sharma, A., Rani, S., Shah, S. H., Sharma, R., Yu, F., & Hassan, M. M. (2023). An efficient hybrid deep learning model for denial of service detection in cyber physical systems. *IEEE Transactions on Network Science and Engineering*, 10(5), 2419–2428. DOI: 10.1109/TNSE.2023.3273301

Sukhramani, K., Kumre, H., Rasool, A., & Jadav, A. (2024, February). Binary Classification of News Articles using Deep Learning. In 2024 IEEE International Students' Conference on Electrical, Electronics and Computer Science (SCEECS) (pp. 1-9). IEEE. DOI: 10.1109/SCEECS61402.2024.10482129

Chapter 4
Advanced Cyber-Physical Systems Utilizing Deep Learning for Crowd Density Detection and Public Safety

R. Leisha
https://orcid.org/0009-0003-9066-4571
Christ University, India

S. Ravindra Babu
https://orcid.org/0000-0002-0883-403X
Christ University, India

Katelyn Jade Medows
Christ University, India

Prakash Divakaran
Himalayan University, India

Michael Moses Thiruthuvanathan
Christ University, India

Vandana Mishra Chaturvedi
D.Y. Patil University, India

ABSTRACT

This study aims to detect the increasing crowd density, which is crucial, especially in dynamic environments like festivals or concerts. By harnessing cyber-physical systems and cutting-edge technologies, we have employed computer vision and deep learning to create a reliable model for accurate crowd counting. Utilizing deep learning, renowned for its ability to handle image-related tasks, the system gives decision-makers precise crowd density estimates, enabling well-informed actions such as crowd control measures. This study aims to improve safety and security in crowded areas by delivering an efficient system that can identify and quantify high crowd densities by integrating deep learning and computer vision technologies

DOI: 10.4018/979-8-3693-5728-6.ch004

within a cyber-physical system framework. This approach facilitates proactive measures to mitigate safety risks and optimize crowd management strategies. The study seeks to advance safety and security measures in crowded areas by delivering an efficient and comprehensive crowd density detection and analysis system using cutting-edge technologies.

INTRODUCTION

Problem Definition

In today's world, managing crowds efficiently and ensuring public safety in crowded spaces are vital tasks. Accurate crowd counting is essential for overseeing events, maintaining order in public areas, and responding to emergencies. Traditional methods often struggle to capture the complexities of real-world scenarios where crowd densities fluctuate and visibility may be obstructed. These methods frequently fail to adapt to varying crowd densities, obstacles, and complex scenes, resulting in inconsistent and inaccurate counts. They rely on manually crafted features that do not fully capture intricate spatial patterns in crowd scenarios. Furthermore, manual annotation of crowd data for model training is error-prone and time-consuming, which limits the scalability and generalization of crowd-counting models. Existing techniques also often lack transparency, making it difficult to interpret their outputs in real-world situations.

Background

The Itaewon Halloween crowd crush in South Korea in 2022 starkly underscored the critical need for effective crowd management systems to avert such devastating incidents. In their analysis, Ha et al. (2022) emphasize the importance of supplementing emergency plans with contingency measures specifically tailored for large-scale, high-density events, particularly in urban settings. The tragedy highlighted the limitations of existing crowd management strategies, particularly in dynamically evolving environments, and stressed the urgency of deploying more sophisticated and reliable crowd-monitoring systems.

Modern urban spaces, characterized by their dense populations and frequent mass gatherings, demand real-time solutions for crowd management to enhance public safety. The integration of advanced automated systems in crowd monitoring and counting has become essential, offering real-time insights that are critical for detecting potential congestion points and facilitating timely interventions. These systems must not only be capable of accurately detecting crowd density but also need to respond to the rapid changes in crowd dynamics, thereby preventing bottlenecks and ensuring safer environments.

However, the current state of crowd-counting technology faces several significant challenges. Scalability and reliability are paramount, as these systems must operate effectively across diverse scenarios, from small gatherings to large-scale events. The automation of data annotation, a crucial component for training and refining these systems, is another critical hurdle. Moreover, the need for transparency in the predictions made by these systems is increasingly recognized, as it is vital for building trust and ensuring that interventions are both timely and appropriate.

Several key studies have shaped the development of more advanced crowd-counting methodologies. Patwal et al. (2022) highlight the critical role of large, well-annotated datasets in the accurate training of these systems. Bhuiyan et al. (2021) emphasize the necessity of extensive datasets to support comprehensive and reliable analysis, which is crucial for the development of scalable solutions. Meanwhile, Saleh et al. (2020) address challenges such as perspective effects and density variations, which can significantly impact the accuracy of crowd density predictions. Bhangale et al. (2023) have made strides in advancing Deep Convolutional Neural Network (DCNN) architectures for real-time crowd counting, a key component for real-time safety interventions. Marsden et al. (2019) propose methods for detecting crowd anomalies through detailed dataset labeling, contributing to the interpretability and reliability of crowd management systems.

Our research seeks to overcome these challenges by developing an advanced crowd-counting system that integrates automated data annotation, enhances accuracy across various crowd densities, ensures interpretability, and achieves robustness and generalization across different datasets. By advancing crowd-counting methodologies, our system aims to provide a reliable solution for diverse environments, contributing to safer public spaces and improved crowd management practices.

Furthermore, the integration of our system within Cyber-Physical Systems (CPSs)—which combine computational and physical processes to manage critical infrastructure—introduces additional layers of complexity and opportunity. While deep learning (DL) applications have shown considerable promise in enhancing CPS security, there is a notable gap in their generalization capabilities across varied real-world conditions. Our study aims to address this by ensuring that our crowd-density detection system is adaptable and resilient, thereby improving both security and operational efficiency in managing critical infrastructure.

LITERATURE SURVEY AND REVIEW

The Literature Survey and Review section explores the existing body of knowledge surrounding crowd density estimation, deep learning methodologies, and related fields.

The paper "Crowd Counting Analysis Using Deep Learning: A Critical Review" by Patwal et al. (2022) provides an in-depth examination of recent advancements in crowd counting, focusing particularly on the role of Convolutional Neural Networks (CNNs). The review covers various methods, including detection-based, regression-based, and density estimation approaches, evaluating their effectiveness on key datasets such as UCF_QNRF, ShanghaiTech (SHT), and UCF_CC 50. While deep learning techniques have significantly improved the accuracy of crowd density estimation, the study identifies ongoing challenges, particularly in handling occlusion and varying crowd densities, which continue to affect the reliability and generalizability of these models across diverse scenarios.

Building on the findings of Patwal et al. (2022), our study aims to address these persistent challenges by enhancing accuracy in crowd density estimation and improving the robustness of models in complex environments. We propose a novel approach that incorporates automated data annotation to improve training data quality, enabling better generalization across datasets. Additionally, our research focuses on refining CNN architectures to handle occlusion and varying densities more effectively.

The paper "Video Analytics Using Deep Learning for Crowd Analysis" by Bhuiyan et al. (2021) delves into the challenges and advancements in the field of crowd analysis, with a particular focus on large-scale events such as the Hajj pilgrimage, where managing dense crowds is critical. The study highlights the limitations of traditional algorithms, which often struggle to accurately assess crowd density in highly congested environments. Bhuiyan et al. (2021) review a range of deep learning techniques that have been developed to overcome these limitations, with a specific emphasis on Convolutional Neural Network (CNN) frameworks.

This review by Bhuiyan et al. (2021) is highly relevant to our study, which seeks to address the persistent challenges associated with analyzing crowds of varying densities and under diverse environmental conditions. While the deep learning techniques discussed in the review have made strides in enhancing the accuracy of crowd analysis, significant gaps remain, particularly in achieving reliable performance across different scenarios. Our research builds on these advancements by developing more sophisticated models that can better handle the variability in crowd densities and environmental factors.

The paper "Recent Survey on Crowd Density Estimation and Counting for Visual Surveillance" by Saleh et al. reviews crowd density estimation methods, distinguishing between direct approaches (model-based, trajectory-clustering) and indirect methods (pixel-based, texture-based, corner point-based). The paper highlights the challenges of direct methods in dense, occluded crowds and the shift towards indirect approaches for better efficiency. This review is pertinent to our study which uses deep learning techniques to tackle similar challenges in crowd density detection, applying advanced models to handle varying crowd densities and complexities effectively. The methods reviewed—model-based, trajectory-clustering, pixel-based, texture-based, and corner point-based analyses—provide a foundational understanding of different strategies in crowd analysis, which complements the deep learning approaches we are employing.

"Near Real-Time Crowd Counting Using Deep Learning Approach" by Bhangale et al. presents a deep convolutional neural network (DCNN) system for near real-time crowd counting, especially useful in emergencies like evacuations. Utilizing NVIDIA GPUs for fast processing of CCTV video feeds, the system accurately counts heads even in challenging conditions such as overlaps and partial visibility. Despite its high accuracy and minimal setup requirements, it requires further research to handle issues like shadows and false positives. This work relates to our study by showcasing practical deep learning applications in crowd counting, complementing our use of the CrowdNet and VGG16 ensemble model on the ShanghaiTech and Mall datasets. Both approaches aim to improve accuracy in diverse crowd conditions, with Bhangale et al.'s method focusing on real-time processing and head detection, while our study integrates ensemble models for density estimation.

The paper "Holistic Features for Real-Time Crowd Behavior Anomaly Detection" by Marsden et al. (2019) presents an innovative approach to detecting anomalies in crowd behavior using a combination of Gaussian Mixture Models (GMM) and Support Vector Machine (SVM) algorithms. The authors focus on the importance of holistic features—comprehensive, high-level descriptors that capture the overall dynamics of a crowd scene—for improving the accuracy and robustness of anomaly detection. Marsden et al. (2019) highlight the significance of detailed and consistent dataset labeling, which plays a crucial role in training models capable of identifying subtle deviations from normal crowd behavior in real-time.

The emphasis on holistic features and thorough dataset labeling in Marsden et al.'s (2019) work has directly informed our approach to developing a more interpretable and reliable crowd-counting model. Recognizing that accurate crowd density estimation and anomaly detection require a deep understanding of crowd dynamics, our study integrates these concepts to enhance the transparency and trustworthiness of our model's predictions. By incorporating comprehensive feature sets and prioritizing detailed annotation, our approach aims to address the challenges of real-time crowd monitoring, ensuring that the system not only counts individuals accurately but also detects potential anomalies that could indicate emerging safety risks. This alignment with Marsden et al.'s focus on holistic features strengthens the foundation of our crowd-counting system.

"Practical Automated Video Analytics for Crowd Monitoring and Counting" by Cheong et al. addresses the need for automated video analytics in crowd monitoring and counting, proposing a low-cost, efficient system that integrates computational object recognition with real-time video streams. Their study compares classical techniques with CNN-based methods, including Background Subtraction (BGS) and Single Shot MultiBox Detector (SSD), for medium crowd density tracking and counting. The system showed high accuracy and reliability in both controlled and real-world environments, making it suitable for public spaces and security-sensitive areas. This work is relevant to our study as it demonstrates practical applications of automated crowd analysis, complementing our use of the CrowdNet and VGG16 ensemble model for crowd density detection on the ShanghaiTech and Mall datasets. Both approaches focus on enhancing accuracy and efficiency in crowd counting, with Cheong et al.'s emphasis on real-time video analytics and our study leveraging ensemble models for detailed density estimation.

The paper "Crowd Anomaly Detection for Automated Video Surveillance" by Wang and Xu (2020) explores the application of spatio-temporal texture models for detecting anomalies in crowd behavior through video surveillance. The authors emphasize the critical need for robust algorithms capable of identifying unusual patterns and behaviors in crowd scenes, particularly when dealing with low-density crowds where anomalies may be less pronounced and harder to detect. Wang and Xu (2020) demonstrate that effective anomaly detection requires a deep understanding of both spatial and temporal features to accurately differentiate between normal variations and true anomalies.

Inspired by the insights from Wang and Xu (2020), our study aims to enhance the reliability and applicability of our crowd-counting system by incorporating robust spatio-temporal features that improve performance across varying crowd densities. We address the challenge of anomaly detection by developing advanced algorithms that can accurately identify both low-density and high-density crowds. By focusing on enhancing the system's ability to detect subtle anomalies and ensuring its robustness in diverse scenarios, our research seeks to provide a more reliable tool for crowd density detection. This approach ensures that our crowd-counting system can effectively manage and monitor crowds of all sizes.

The paper "Hierarchical Crowd Detection and Representation for Big Data Analytics in Visual Surveillance" by Zitouni et al. (2021) introduces a framework that leverages motion and appearance saliency for effective crowd detection and representation. The authors propose a hierarchical approach to handling large-scale crowd data, which is crucial for visual surveillance systems dealing with big data challenges. By focusing on motion and appearance saliency, Zitouni et al. (2021) provide insights into how compact representations can be achieved without losing critical information. Their framework addresses significant challenges in crowd analytics, such as managing high volumes of data and ensuring efficient processing and analysis in real-time scenarios.

Building on the recommendations from Zitouni et al. (2021), our study has refined its approach to processing and analyzing large-scale crowd data by incorporating hierarchical and saliency-based techniques. We have focused on developing a model that efficiently manages and represents crowd data. By integrating motion and appearance saliency into our system, we aim to enhance the accuracy and efficiency of crowd detection and analysis. This approach not only improves the model's capability to process extensive datasets but also contributes to a more compact and effective representation of crowd dynamics, facilitating better insights and decision-making in visual surveillance applications.

The paper "Crowd Density Estimation Method Using Deep Learning for Passenger Flow Detection System in Exhibition Center" by Xiang et al. and Liu et al. (2021) addresses the challenges of crowd density estimation in busy environments such as exhibition centers using deep learning techniques. Their study highlights the effectiveness of leveraging GPUs and parallel computing to accelerate data processing, which is crucial for handling real-time data and ensuring accurate passenger flow detection. This approach demonstrates how advanced computational resources can significantly enhance the speed and efficiency of crowd management systems in high-traffic areas.

We have been influenced by the emphasis on real-time processing and efficient data handling presented by Xiang et al. and Liu et al. (2021). We focus on optimizing our crowd-counting model to ensure it remains effective in real-time scenarios through other means. Our approach aims to maintain accuracy and responsiveness in crowd density estimation without the use of these specific computational resources, striving to deliver a reliable solution for managing crowd flow in various settings, including exhibition centers and other large-scale events.

"LCDnet: A Lightweight Crowd Density Estimation Model for Real-Time Video Surveillance" by Khan et al. introduces LCDNet, a compact convolutional neural network (CNN) designed for real-time crowd density estimation in video surveillance. Their work emphasizes the significance of lightweight models that maintain high performance while being resource-efficient, which is crucial for real-time applications and deployment in resource-constrained environments. This study influences our research by highlighting the need for optimizing deep learning models for efficiency without sacrificing accuracy. In our study, we use the CrowdNet and VGG16 ensemble model for crowd density detection on the ShanghaiTech and Mall datasets. While LCDNet focuses on creating a lightweight solution for real-time scenarios, our approach aims to use advanced deep learning techniques to balance accuracy and computational efficiency, providing robust density estimation across diverse crowd conditions.

"A Review Paper on Crowd Estimation" by Mitali Madhusmita Nayak and Sujit Kumar Dash provide a comprehensive review of advanced crowd estimation techniques using computer vision, focusing on methods that combine deep spatial regression models with convolutional neural networks (CNNs) and long short-term memory (LSTM) networks. Their review underscores the effectiveness of integrating these techniques to address the complexities of crowd dynamics and improve the accuracy of crowd counting. This paper guides our study by illustrating the benefits of advanced deep learning methods in capturing detailed crowd patterns. In our research, we use the CrowdNet and VGG16 ensemble model for crowd density detection on the ShanghaiTech and Mall datasets, drawing inspiration from Nayak and Dash's approach to enhance our model's ability to manage complex crowd scenarios and ensure precise density estimates.

METHODOLOGY

The study systematically addressed the challenges of crowd counting through key components. It began with data collection and annotation, sourcing diverse crowd images and annotating them with ground truth counts for supervised learning. Neural network architectures were fine-tuned for accuracy, with iterative training adjustments based on validation feedback. The implementation phase transformed theoretical concepts into practical solutions.

Safety threshold-based categorization was a critical aspect, establishing predefined thresholds to classify crowd density into "Safe" or "Unsafe" zones, guided by venue capacity, spatial constraints, and regulatory guidelines. Continuous optimization efforts focused on refining the model architecture, optimizing computational algorithms, and implementing best practices in data processing. Evaluation involved testing on unseen data and comparing predictions to ground truth annotations, ensuring the system met predefined metrics and requirements.

Dataset

ShanghaiTech Dataset

The ShanghaiTech Dataset, introduced by Zhang et al. (2016), is a significant resource in the field of crowd density estimation, providing 1,198 annotated crowd images divided into Part A and Part B. Part A contains 482 high-density images sourced from the internet, capturing bustling urban environments and public events. Part B includes 716 images from Shanghai that depict sparser crowd conditions. This dataset covers a range of scenarios, from busy streets to public gatherings, across 13 different scenes with varying lighting conditions, angles, and a total of 130 abnormal events. The detailed annotations provided in this dataset, including precise density maps, are essential for developing accurate crowd-counting models. These maps offer a spatial distribution of crowd densities.

Despite its advantages, the ShanghaiTech Dataset has some limitations that need to be considered. The division into high-density and sparser scenes, while useful, may not capture the full spectrum of crowd behaviors and environmental conditions encountered in various contexts. The dataset's focus on specific urban environments in Shanghai may also affect the generalizability of models trained on it. These models might perform less effectively when applied to different geographic or cultural settings, where crowd behaviors and environmental factors may differ significantly.

In our study, we utilize the ShanghaiTech Dataset for its comprehensive annotations and diverse crowd scenarios to enhance our crowd-counting model. The detailed density maps and varied conditions provided by this dataset are instrumental in developing a system that can accurately estimate crowd densities and detect anomalies. However, we are also mindful of the dataset's limitations and are working towards addressing these by integrating additional data sources and methodologies. Figures 1 and 2 in our study illustrate how the dataset's rich annotations and diverse scenarios contribute to refining our approach and validating our model's performance in various crowd conditions.

Figure 1. ShanghaiTech, Dataset: Part A. An image of a dense crowd

Figure 2. ShanghaiTech, Dataset: Part B. An image of a sparse crowd

Mall Dataset

The Mall Dataset, introduced by Chen et al. (2019), offers a substantial resource for crowd counting and profiling research, comprising over 60,000 labeled pedestrians distributed across 2,000 video frames captured from publicly available webcam feeds in shopping malls. The dataset features a resolution of 640x480 and a frame rate of under 2 Hz. This large-scale dataset is particularly valuable for research due to its representation of a wide range of crowd densities and activity patterns, which are essential for developing robust crowd-counting models. Additionally, the Mall Dataset encompasses diverse illumination conditions, including challenging lighting scenarios and reflections from glass surfaces, which are critical for testing the model's performance in varying environmental conditions.

Despite these strengths, the Mall Dataset has notable limitations that must be considered. The relatively low frame rate of under 2 Hz may not adequately capture fast-moving pedestrians with high precision, potentially impacting the accuracy of crowd density estimation in dynamic situations. Furthermore, the resolution of 640x480 may not provide sufficient detail for analyzing individual pedestrians, which can be a limitation for applications requiring fine-grained pedestrian tracking or profiling. These limitations suggest that while the dataset is useful, it may not fully support high-precision tasks or detailed analyses of pedestrian movement.

In our study, the Mall Dataset plays a crucial role in developing and evaluating our crowd-counting model by providing a comprehensive view of indoor crowd dynamics and a variety of crowd densities and activity patterns. Figure 3 in our research illustrates how the dataset's extensive coverage and diverse conditions contribute to refining our approach. Although we acknowledge the dataset's limitations, such as its low frame rate and resolution, we use its strengths to enhance the robustness of our model and test its performance in indoor environments. By incorporating additional methods and data sources, we aim to address these limitations and further improve our model's accuracy and applicability in real-world scenarios.

Figure 3. Mall Dataset: An image of a sparse crowd in the mall

PROPOSED METHODOLOGY

The Proposed Methodology as shown in Figure. 4 illustrates the components and inter actions within the crowd-counting system, incorporating safety threshold-based categorization. It demonstrates how data flows through different stages, from initial input processing to final crowd density estimation, while also highlighting the integration of safety thresholds for categorizing crowd density into 'Safe' or 'Unsafe' zones.

Figure 4. Crowd Analysis Framework: Illustration of a pipeline for crowd count prediction, involving data preprocessing, model training and testing, and output generation.

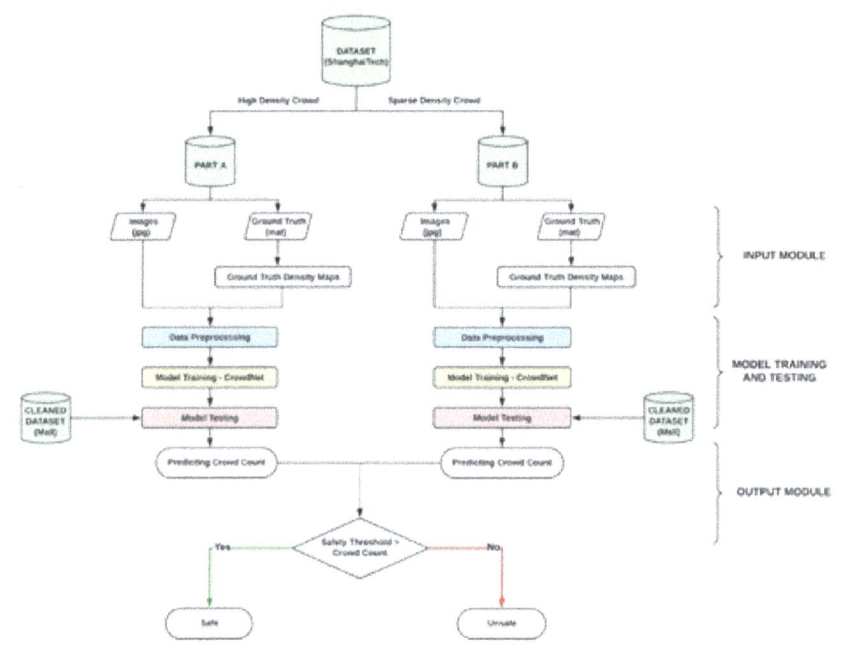

Data Preprocessing

The implementation of Data Preprocessing is an essential step in the building, training, and evaluation of deep learning models for crowd counting. Below is the detailed breakdown of the implementation:

Generating Ground Truth Density Map

Density maps are 2D representations showing how people are distributed across an image, providing a continuous view of crowd density with varying intensity levels. They offer a detailed overview of crowd concentration, facilitating more accurate crowd analysis compared to traditional manual annotations. These maps are crucial for estimating total crowd counts and analyzing distribution patterns, serving as essential ground truth data for training machine learning models and improving prediction accuracy in crowd counting.

Gaussian Filter Transformation

To generate ground truth density maps, the study used the ShanghaiTech Dataset, which contains images and sparse matrices with head annotations. These matrices indicated individual locations, which were transformed into continuous density maps using a Gaussian filter. The Gaussian filter smooths images by assigning higher weight to pixels near the center of a Gaussian kernel, creating a blurred effect while preserving important features.

Each head annotation was convolved with a 2D Gaussian kernel to produce smooth density maps. The appropriate standard deviation for the Gaussian kernel was determined by finding the average distance to the nearest neighbors using a KDTree structure. Density maps were generated with the scipy.ndimage.gaussian_filter function and stored in HDF5 files for efficient handling. Visual inspections of the density maps ensured their accuracy and alignment with the images.

Model Training

Model Architecture

This study aims to create a density-based estimation model capable of generating density maps from input images. These maps offer a visual representation of crowd density, allowing for in-depth analysis of crowd behavior. To accomplish this goal, the study utilizes the CrowdNet Architecture as shown in Figure. 5, which is a Convolutional Neural Network (CNN) designed for crowd analysis tasks.

This model comprises multiple convolutional and pooling layers, followed by fully connected layers that capture spatial relationships and extract important features from the input images. Non-linear activation functions like ReLU (Rectified Linear Unit) introduce flexibility and help the model capture complex patterns in the data.

Figure 5. CrowdNet Model Architecture: Illustration a deep neural network architecture for crowd counting

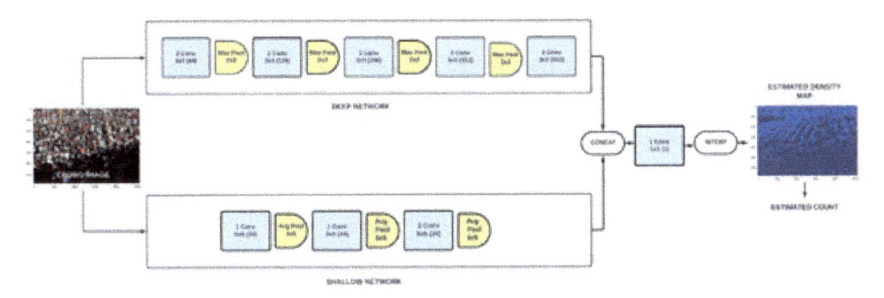

Training Procedure

The model training uses a Stochastic Gradient Descent (SGD) optimizer with momentum to minimize the Mean Squared Error (MSE) Loss Function between predicted and ground truth density maps. Training spans multiple epochs with mini-batch gradient descent for efficiency. Hyperparameters like learning rate, batch size, and regularization are fine-tuned through cross-validation to prevent overfitting.

The goal is to ensure CrowdNet generalizes well across diverse crowd scenes. Training on datasets such as ShanghaiTech and the Mall Dataset helps CrowdNet handle various crowd dynamics and environments, enhancing its reliability in real-world scenarios. Here's a detailed breakdown of the implementation:

Initializing VGG Weights

The init_weights_vgg function initializes a custom model's weights using the trained VGG16 weights. It stacks convolutional layers with 3x3 filters, max-pooling layers, and fully connected layers. The convolutional layers use a stride of 1 and 'same' padding to match the input size, with the number of filters increasing to capture more complex features. Max-pooling reduces spatial dimensions while retaining key information. ReLU activation functions add non-linearity, enabling the model to learn intricate patterns. By leveraging VGG16's pretrained weights, the custom model can quickly adapt to new tasks, even with limited training data.

Constructing the CrowdNet Model

The CrowdNet function defines the crowd-counting model's architecture, selecting layers and configurations to extract complex features from images. Key components include:

- **Convolutional Layers:** Using Keras' Conv2D, these layers apply filters to capture essential patterns and features. Batch normalization is also employed to stabilize and speed up training.
- **Batch Normalization:** This layer normalizes activations to ensure consistent input distributions, promoting faster convergence.
- **Dilated Convolutions:** Implemented via Conv2D, these convolutions expand the receptive field without increasing parameter count, improving contextual accuracy in crowd counting.
- **Optimizer:** The Stochastic Gradient Descent (SGD) optimizer is used to update model parameters and minimize the loss function, enhancing performance through iterative gradient adjustments.

Instantiating the CrowdNet Model

Instantiating the CrowdNet model with the *CrowdNet()* function initiates the model. By calling the *CrowdNet()* function, an instance of the model is created, ready to undergo training and evaluation.

Displaying Model Summary

The model summary provides a comprehensive overview of the CrowdNet architecture, providing insights into the model's structure, layer conFigureuration, and parameter counts. This summary is for understanding the model's complexity, facilitating model inspection, and aiding debugging efforts, as shown in Table. 1

Table 1. Model Summary: Architecture of the CrowdNet Model

Layer (type)	Output Shape	Param #
conv2d 11 (Conv2D)	(None, 224, 224, 64)	1792
conv2d 12 (Conv2D)	(None, 224, 224, 64)	36928
max pooling2d 3 (MaxPoolong2D)	(None, 112, 112, 64)	0
conv2d 13 (Conv2D)	(None, 112, 112, 128)	73856
conv2d 14 (Conv2D)	(None, 112, 112, 128)	147584
max pooling2d 4 (MaxPooling2D)	(None, 56, 56, 128)	0
conv2d 15 (Conv2D)	(None, 56, 56, 256)	295168
conv2d 16 (Conv2D)	(None, 56, 56, 256)	590080
conv2d 17 (Conv2D)	(None, 56, 56, 256)	590080
max pooling2d 5 (MaxPooling2D)	(None, 28, 28, 256)	0
conv2d 18 (Conv2D)	(None, 28, 28, 512)	1180160
conv2d 19 (Conv2D)	(None, 28, 28, 512)	2359808
conv2d 20 (Conv2D)	(None, 28, 28, 512)	2359808
conv2d 21 (Conv2D)	(None, 28, 28, 1)	513

Total params: 7635777 (29.13 MB)
Trainable params: 7635777 (29.13 MB)
Non-trainable params: 0 (0.00 Byte)

Table 1 outlines the architecture of our ensemble model, specifically detailing the output shapes and the number of parameters at each layer. The model consists of multiple convolutional layers, followed by max-pooling layers, with a total of approximately 7.6 million trainable parameters. The final layer reduces the output to a single channel, preparing it for tasks like crowd density estimation where the output is a density map.

Initializing Data Generator

The initialization of the data generator, represented by the *train gen object*, streamlines the flow of training data to the model. Using the *image generator* function, the generator produces batches of training data, each consisting of an input image and its corresponding ground truth density map. This efficient Data Pipeline ensures the model receives a steady stream of diverse training samples, which is essential for robust learning.

Training the Model

Training the CrowdNet model involves exposing it to the training data iteratively, allowing it to learn from experience and refine its predictive capabilities. The training process is organized by setting the number of epochs, steps per epoch, and verbosity level, which helps improve the model. Throughout training, the model adapts its parameters to minimize the specified loss function, which helps to estimate crowd counts accurately.

Saving the Trained Model

The model training pipeline's final step involves preserving the model's state for future use, ready to be deployed in real-world applications or further analyzed.

RESULTS, CONCLUSION AND FUTURE SCOPE

In this section, the results obtained from the implementation and evaluation of the proposed crowd-counting model are presented. Discussions regarding the model's performance across various datasets, robustness, and generalization capabilities are being analyzed in this section. Furthermore, the implications of these findings and potential avenues for future research is discussed. Finally, conclusive remarks are drawn summarizing the study's contributions, limitations, and recommendations for further advancements in the field of crowd analysis and computer vision.

Results & Analysis

Ground Truth Density Maps

In our study, density maps are meticulously generated for the images in both Part A and Part B of the Training and Testing Dataset. These maps use a color spectrum to visually represent varying crowd densities across the observed areas, with different colors indicating different levels of crowd concentration. This approach is critical as it helps in distinguishing between regions with high and low crowd densities, providing a spatially detailed understanding of the crowd distribution within each image.

The generation of these density maps plays a pivotal role in enhancing the performance of our crowd-counting model. By incorporating spatial density information, the model gains a deeper insight into the distribution patterns of crowds, which allows for more accurate differentiation between densely packed and sparsely populated regions. This spatial awareness is particularly beneficial in accurately estimating crowd densities and identifying areas with unusual crowd concentrations.

Additionally, the inclusion of density maps in our Training and Testing Dataset enables a comprehensive evaluation of the model's performance across different crowd scenarios. The enhanced spatial information offered by the density maps ensures that our model can better adapt to real-world situations, leading to more reliable and actionable insights for crowd monitoring and management.

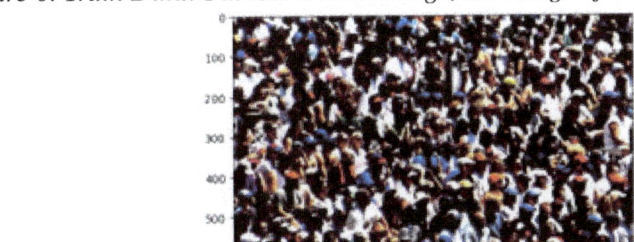

Figure 6. Train Data: Part A: Crowd Image; An image of a dense crowd

Figure 7. Train Data: Part A: Corresponding Density Map; A ground truth density map

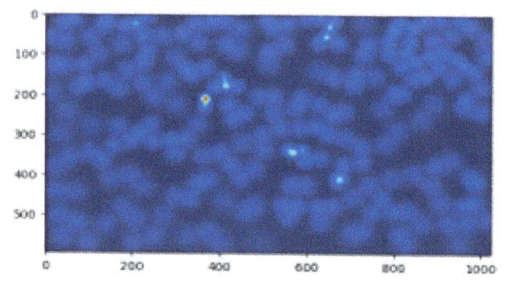

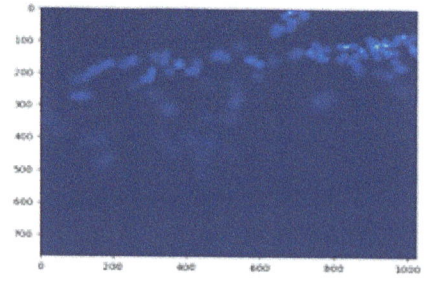

Figure. 6 and Figure. 7, showcases a ground truth density map derived from the Part A Train dataset along with its corresponding Image, specifically. This is similarly performed on all the images in the dataset.

Figure 8. Train Data: Part B: Crowd Image; An image of a sparse crowd

Figure 9. Train Data: Part B: Corresponding Density Map; A ground truth density map

Figure. 8 and Figure. 9, showcases a ground truth density map derived from the Part B Train dataset along with its corresponding Image, specifically. This is similarly performed on all the images in the dataset.

Figure 10. Test Data: Part A: Crowd Image - An image of a dense crowd

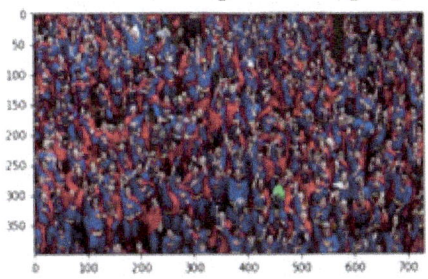

Figure 11. Test Data: Part A: Corresponding Density Map - A ground truth density map

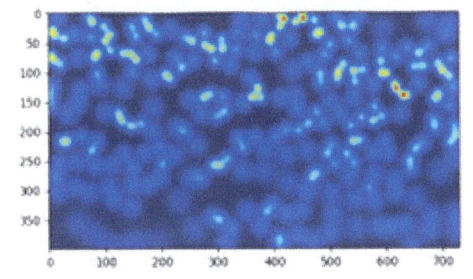

Figure. 10 and Figure. 11, showcases a ground truth density map derived from the Part A Test dataset along with its corresponding Image, specifically. This is similarly performed on all the images in the dataset.

Figure 12. Part B: Crowd Image (IMG 85.jpg) An image of a sparse crowd

Figure 13. Part B: Corresponding Density Map (IMG 85.h5) A ground truth density map

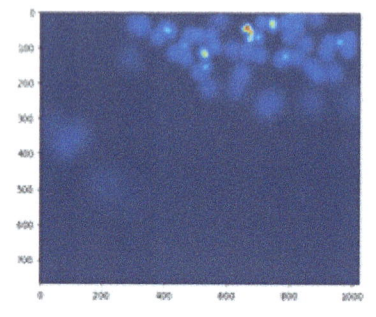

Figure. 12 and Figure. 13, showcases a ground truth density map derived from the Part B Test dataset along with its corresponding Image, specifically. This is similarly performed on all the images in the dataset.

Crowd Density Estimation

In our study, crowd density estimation is systematically performed for the images in both Part A and Part B of the dataset. This process involves precisely quantifying the number of individuals present in each image to provide a comprehensive understanding of crowd dynamics and distribution. By analyzing each image and estimating the crowd density, we aim to capture a detailed representation of how people are distributed across various environments, ranging from high-density urban scenes to more sparsely populated areas.

To achieve accurate crowd density estimation, we employ advanced deep learning techniques that use the detailed annotations and density maps associated with the dataset. These techniques are designed to process the spatial distribution of individuals within each image. The ability to estimate crowd density accurately is crucial for understanding the variations in crowd concentrations, identifying potential hotspots of congestion, and assessing the overall crowd dynamics within different scenes.

The insights gained from this estimation process are invaluable for developing and refining our crowd-counting model. By accurately quantifying crowd densities, our model can better differentiate between densely packed and more open areas, leading to improved performance in both training and testing phases. Overall, the crowd density estimation process is central to our study, providing the foundational data needed to advance our model and ensure its applicability in diverse real-world scenarios.

Figure 14. Train Data: Part A: Predicted Crowd Count (IMG 26.jpg) An image of a dense crowd and the corresponding ground truth density map

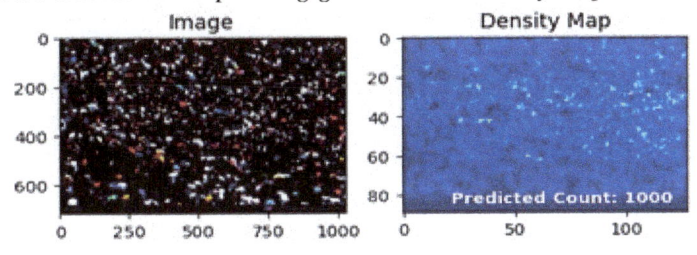

Figure. 14, showcases the Crowd Count predicted from the Part A Train dataset alongside its corresponding image. Specifically, the image has an actual count value of 1076, mentioned for evaluation purposes.

Figure 15. Train Data: Part B: Predicted Crowd Count (IMG 148.jpg) An image of a dense crowd and the corresponding ground truth density map

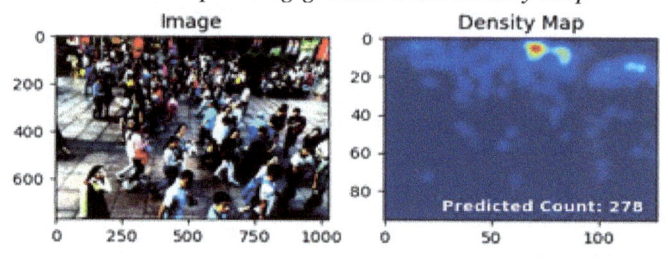

Figure. 15, showcases the Crowd Count predicted from the Part B Train dataset alongside its corresponding image. Specifically, the image has an actual count value of 281, mentioned for evaluation purposes.

Figure 16. Test Data: Part A: Predicted Crowd Count (IMG 7.jpg) An image of a dense crowd and the corresponding ground truth density map

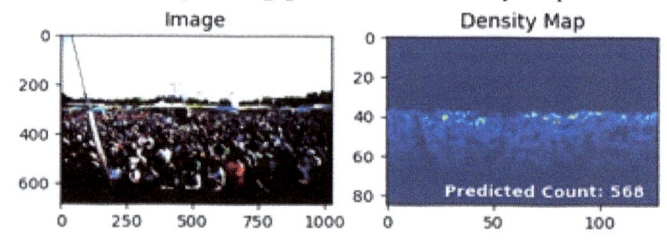

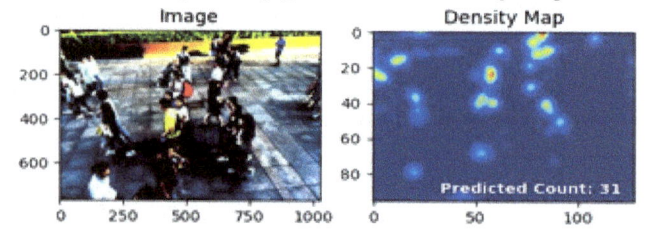

Figure. 16, showcases the Crowd Count predicted from the Part A Test dataset alongside its corresponding image. Specifically, the image has an actual count value of 566, mentioned for evaluation purposes.

Figure 17. Test Data: Part B: Predicted Crowd Count (IMG 14.jpg) An image of a sparse crowd and the corresponding ground truth density map

Figure. 17, showcases the Crowd Count predicted from the Part B Test dataset alongside its corresponding image. Specifically, the image has an actual count value of 28, mentioned for evaluation purposes.

Predicted vs Actual Count

In addition to the images considered in the previous section, other images have been incorporated to provide a comprehensive comparison between predicted and actual counts. The predicted counts is being compared with the actual counts for both Part A and Part B of the study, and a close correspondence is being observed between the two, indicating the effectiveness of the crowd density estimation model.

Figure 18. Train Data: Part A: Predicted Crowd Count (IMG 87.jpg) An image of a dense crowd and the corresponding ground truth density map

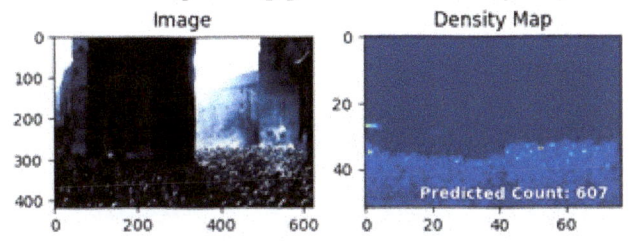

Figure 19. Train Data: Part A: Actual Crowd Count (IMG 87.jpg) An image of a dense crowd and the corresponding ground truth density map

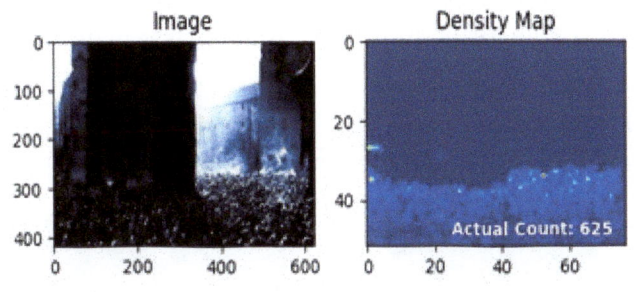

Figure. 18 and Figure. 19, showcases the Predicted Crowd Count an its Actual Crowd Count from the Part A Train dataset along with its corresponding Image, specifically. This is similarly performed on all the images in the dataset.

Figure 20. Train Data: Part B: Predicted Crowd Count (IMG 87.jpg) An image of a dense crowd and the corresponding ground truth density map

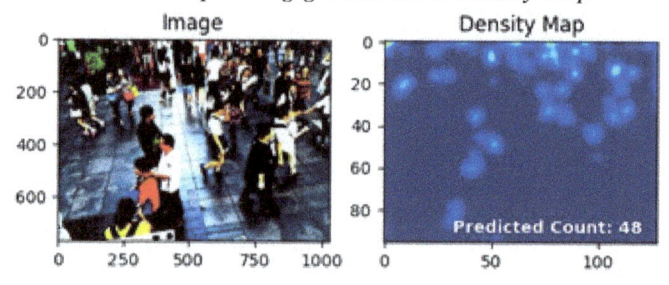

Figure 21. Train Data: Part B: Actual Crowd Count (IMG 87.jpg) An image of a dense crowd and the corresponding ground truth density map

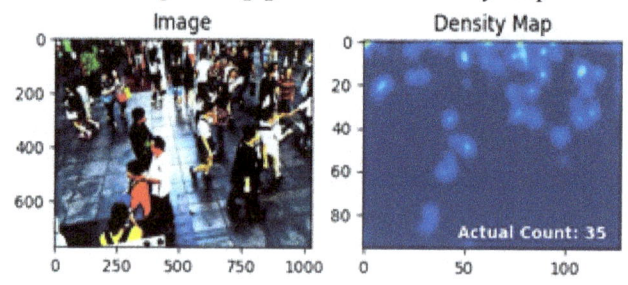

Figure. 20 and Figure. 21, showcases the Predicted Crowd Count an its Actual Crowd Count from the Part B Train dataset along with its corresponding Image, specifically. This is similarly performed on all the images in the dataset.

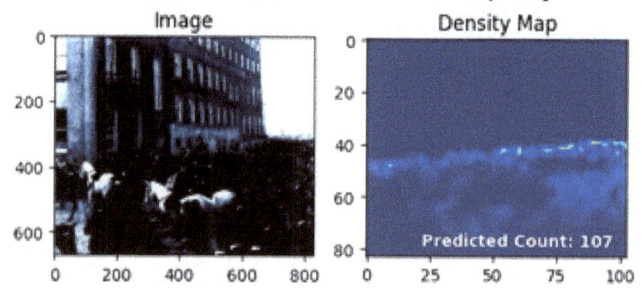

Figure 22. Test Data: Part A: Predicted Crowd Count (IMG 45.jpg) An image of a dense crowd and the corresponding ground truth density map

Figure 23. Test Data: Part A: Actual Crowd Count (IMG 45.jpg) An image of a dense crowd and the corresponding ground truth density map

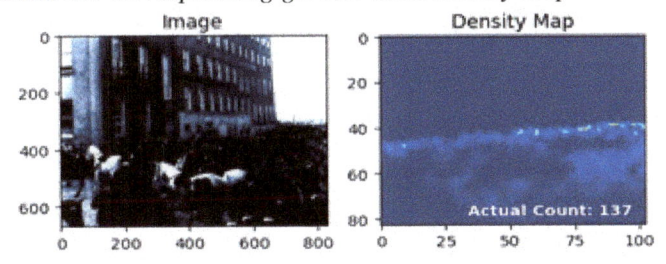

Figure. 22 and Figure. 23, showcases the Predicted Crowd Count an its Actual Crowd Count from the Part A Test dataset along with its corresponding Image, specifically. This is similarly performed on all the images in the dataset.

Figure 24. Test Data: Part B: Predicted Crowd Count (IMG 172.jpg) An image of a sparse crowd and the corresponding ground truth density map

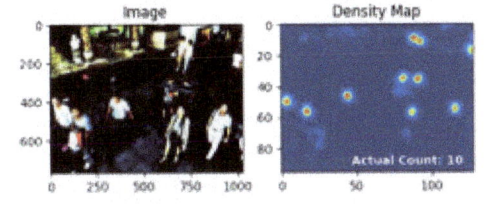

Figure 25. Test Data: Part B: Actual Crowd Count (IMG 172.jpg) An image of a sparse crowd and the corresponding ground truth density map

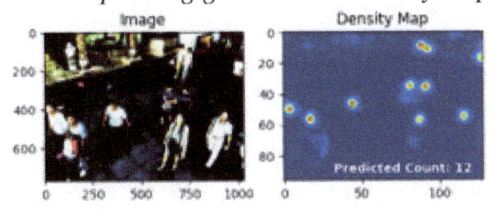

Figure. 24 and Figure. 25, showcases the Predicted Crowd Count an its Actual Crowd Count from the Part B Test dataset along with its corresponding Image, specifically. This is similarly performed on all the images in the dataset.

Comparative Analysis of Crowd Density Scenarios

Figure 26. Part A: Predicted vs Actual Count Scatter Plot

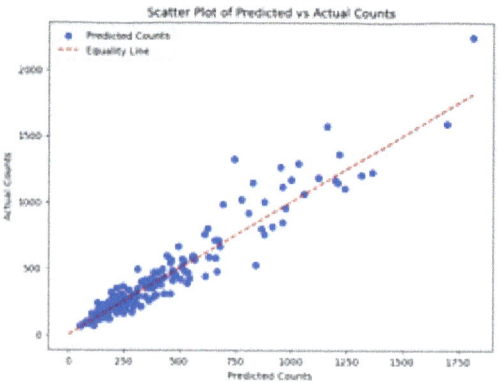

Figure 27. Part B: Predicted vs Actual Count Scatter Plot

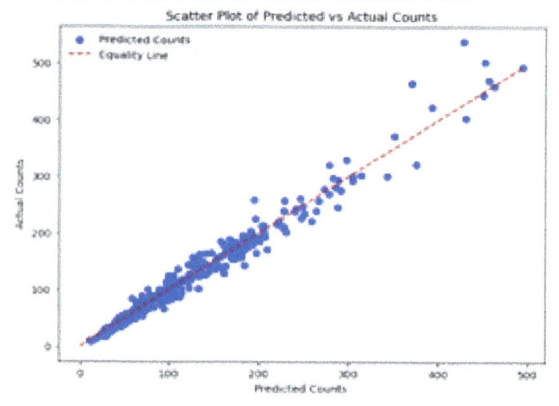

The scatter plot as shown in Figure. 26 illustrates the comparison between predicted and actual crowd counts for Part A of the study. Each data point represents an individual image, with the x-axis indicating the predicted crowd count generated by the crowd density estimation model and the y-axis representing the corresponding actual crowd count obtained from ground truth annotations. The equality line, represented by the red dashed line, is a visual guide that indicates where the predicted count is equal to the actual count. Including this line in the scatter plot provides a reference for evaluating the performance of the model visually.

The scatter plot as shown in Figure. 27 illustrates the comparison between predicted and actual crowd counts for Part B of the study. Each data point represents an individual image, with the x-axis indicating the predicted crowd count generated by the crowd density estimation model and the y-axis representing the corresponding actual crowd count obtained from ground truth annotations. The equality line, represented by the red dashed line, is a visual guide that indicates where the predicted count is equal to the actual count. Including this line in the scatter plot provides a reference for evaluating the performance of the model visually.

Safety-Threshold Based Categorization

Figure 28. Part A: Safe Crowd: An image of a sparse crowd and the corresponding ground truth density map

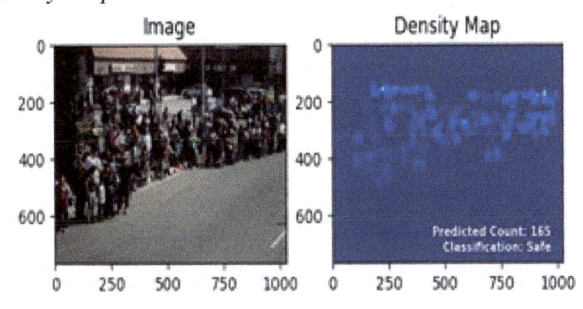

Figure 29. Part A: Unsafe Crowd: An image of a dense crowd and the corresponding ground truth density map

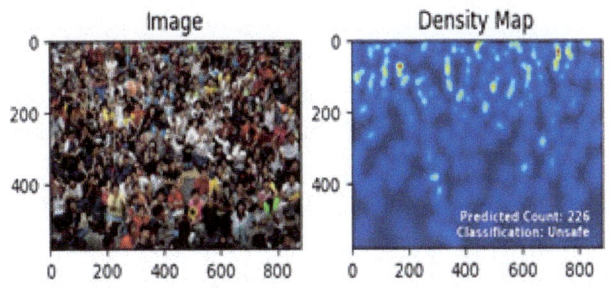

In our study, the visual representations in Figure 28 and Figure 29 provide a clear distinction between "safe crowd" and "unsafe crowd" scenarios based on crowd density estimations. Each image is categorized to illustrate different levels of crowd density and their implications for safety.

Figure 28 depicts a "safe crowd" scenario, where the density of individuals is moderate. In this image, the individuals are spaced adequately, indicating a well-managed crowd with sufficient room for movement and minimal risk of overcrowding. The crowd count in this scenario is below the safety threshold set at 200 individuals for this study. This threshold has been determined to represent a manageable level of crowd density that minimizes potential safety risks. The moderate density and adequate spacing suggest that the crowd is within safe limits, reducing the likelihood of incidents or emergencies related to overcrowding.

In contrast, Figure 29 illustrates an "unsafe crowd" scenario, characterized by a higher density of individuals. In this image, the crowd is more tightly packed, which can lead to potential safety hazards due to the risk of overcrowding. The density in this scenario exceeds the safety threshold of 200 individuals, indicating that the crowd density is high enough to pose significant safety concerns.

The Safety Threshold, set at 200 individuals in our study, serves as a reference point for evaluating crowd density and associated safety risks. However, this threshold is not fixed and can be adjusted based on specific real-world scenarios and requirements. Factors such as the type of event, venue capacity, and crowd behavior may necessitate modifications to the safety threshold to better align with the context of different environments. By adapting the threshold, our study aims to provide a flexible and practical approach to crowd safety management.

Summary of Crowd Density Scenarios

In our study, Table 2 provides a comprehensive comparative analysis of crowd density scenarios, categorized into "safe" and "unsafe" crowds. Each row in the table corresponds to a specific image, allowing for easy reference and evaluation of the crowd scenarios depicted.

The columns in Table 2 include several critical parameters. The Image Name column identifies the specific image being analyzed, making it easy to reference and track individual scenes. The Count column provides the estimated number of individuals present in each image, which is crucial for determining the density level and assessing whether the crowd scenario is categorized as "safe" or "unsafe." Finally, the Category column classifies each image based on its crowd density, indicating whether it falls within safe limits or exceeds the threshold for safety. This categorization helps in understanding the implications of crowd density for safety and guides the interpretation of the data.

Overall, Table 2 serves as a valuable tool for analyzing and comparing different crowd density scenarios in our study. The comparative analysis provided by Table 2 is essential for identifying trends, assessing model performance, and making informed decisions about crowd management strategies.

Table 2. Part A: Tabular results comprehensive comparative analysis of crowd density scenarios

IMG_118.jpg	**238**	**Unsafe**
IMG_119.jpg	246	Unsafe
IMG_12.jpg	320	Unsafe
IMG_120.jpg	131	Safe
IMG_121.jpg	1205	Unsafe
IMG_122.jpg	1146	Unsafe
IMG_123.jpg	112	Safe
IMG_124.jpg	221	Unsafe
IMG_72.jpg	584	Unsafe
IMG_73.jpg	152	Safe
IMG_74.jpg	849	Unsafe
IMG_75.jpg	665	Unsafe
IMG_76.jpg	480	Unsafe
IMG_77.jpg	205	Unsafe
IMG_78.jpg	270	Unsafe
IMG_79.jpg	178	Safe
IMG_8.jpg	1324	Unsafe
IMG_80.jpg	156	Safe
IMG_81.jpg	351	Unsafe
IMG_82.jpg	215	Unsafe
IMG_83.jpg	212	Unsafe
IMG_97.jpg	481	Unsafe

Image Name	**Count**	**Category**
IMG_1.jpg	172	Safe
IMG_10.jpg	498	Unsafe
IMG_100.jpg	384	Unsafe
IMG_101.jpg	211	Unsafe
IMG_102.jpg	218	Unsafe

continued on following page

Table 2. Continued

Image Name	Count	Category
IMG_103.jpg	428	Unsafe
IMG_104.jpg	1164	Unsafe
IMG_105.jpg	258	Unsafe
IMG_106.jpg	1226	Unsafe
IMG_107.jpg	293	Unsafe
IMG_108.jpg	176	Safe
IMG_109.jpg	374	Unsafe
IMG_11.jpg	1064	Unsafe
IMG_110.jpg	1018	Unsafe
IMG_111.jpg	449	Unsafe
IMG_112.jpg	255	Unsafe
IMG_113.jpg	66	Safe
IMG_114.jpg	141	Safe
IMG_115.jpg	1187	Unsafe
IMG_116.jpg	286	Unsafe
IMG_117.jpg	1589	Unsafe

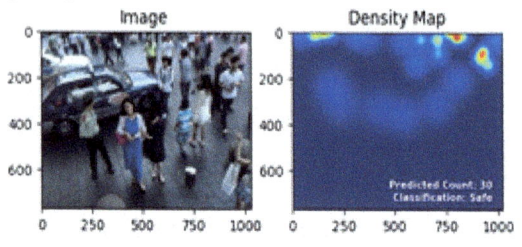

Figure 30. Part B: Safe Crowd: An image of a sparse crowd and the corresponding ground truth density map

Figure 31. Part B: Unsafe Crowd: An image of a dense crowd and the corresponding ground truth density map

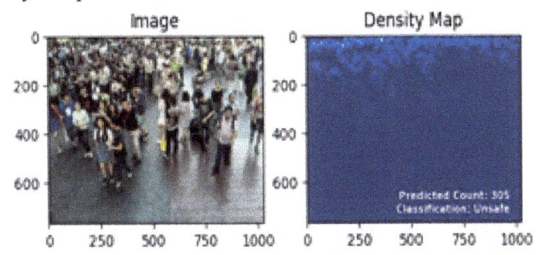

The visual representation as shown in Figure. 30 and Figure. 31 is categorized based on crowd density, with one image depicting a "safe crowd" scenario and the other representing an "unsafe crowd" scenario. In the "safe crowd" image, the density is moderate, with individuals adequately spaced, suggesting minimal safety risks, as the crowd count is below the safety threshold set at 200 individuals for this study. Conversely, the "unsafe crowd" image depicts higher density, indicating potential safety hazards due to overcrowding. The Safety Threshold can be modified based on the real-world scenario

Summary of Crowd Density Scenarios

Table 3 in our study provides a comprehensive comparative analysis of crowd density scenarios, categorizing them into "safe" and "unsafe" crowds. This table is essential for systematically evaluating various images based on their crowd density characteristics and understanding how these densities influence safety. Each row in Table 3 corresponds to a specific image from our dataset, offering a structured view of different crowd scenarios.

The table includes several key columns that are critical for this analysis. The Image Name column identifies each image, allowing for easy reference and retrieval of the visual scenes being analyzed. This helps in tracking and validating the images during the comparative process. The Count column provides the estimated number of individuals present in each image, which is a crucial metric for determining crowd density. By presenting this count, the table enables us to assess whether the crowd falls into the "safe" or "unsafe" category based on predefined safety thresholds.

Additionally, the Category column classifies each image according to its crowd density, indicating whether the scenario is considered "safe" or "unsafe." By organizing and presenting this data, Table 3 helps in interpreting the relationship between crowd density and safety, assessing the performance of our crowd-counting model, and making informed decisions about crowd management strategies.

Table 3. Part B: Tabular Results

IMG_72.jpg	51	Safe
IMG_73.jpg	89	Safe
IMG_74.jpg	259	Unsafe
IMG_75.jpg	539	Unsafe
IMG_76.jpg	118	Safe
IMG_77.jpg	213	Safe
IMG_78.jpg	278	Unsafe
IMG_79.jpg	89	Safe
IMG_8.jpg	59	Safe
IMG_80.jpg	466	Unsafe
IMG_81.jpg	128	Safe
IMG_82.jpg	41	Safe
IMG_83.jpg	55	Safe
Image Name	Count	Category
IMG_1.jpg	22	Safe
IMG_10.jpg	173	Safe
IMG_100.jpg	157	Safe
IMG_101.jpg	34	Safe
IMG_102.jpg	65	Safe
IMG_103.jpg	54	Safe
IMG_104.jpg	42	Safe
IMG_105.jpg	217	Unsafe
IMG_106.jpg	161	Safe
IMG_107.jpg	471	Unsafe
IMG_108.jpg	133	Safe
IMG_84.jpg	53	Safe
IMG_85.jpg	59	Safe

Mall Dataset

Figure 32. Predicted and actual crowd count (seq 001922.jpg) An image of a sparse crowd

Figure 33. Predicted and actual crowd count (seq 000046.jpg) An image of a sparse crowd

Figure. 32 and Figure. 33, showcases the Predicted Crowd Count an its Actual Crowd Count from the Mall dataset along with its corresponding Image, specifically.

Comparative Analysis

Table 4. Comparative analysis

Study	Model	Datasets Used	Evaluation Metrics	Reported Performance	Key Findings
Patwal et al.	Various CNNs	UCF_QNRF, ShanghaiTech, UCF_CC 50	MSE, MAE, Accuracy	High accuracy with challenges in occlusion and varying densities	Emphasizes improvements in CNN architectures but notes limitations in handling diverse densities
Bhuiyan et al.	Various CNNs	Hajj Event	Accuracy, Real-Time Processing	High accuracy in specific events but limited generalizability	Highlights need for robust algorithms in complex scenarios
Saleh et al.	Direct and Indirect Approaches	Various Datasets	Accuracy, Processing Efficiency	Effective with indirect methods, challenges with high-density and occluded crowds	Direct methods struggle with high-density scenarios; indirect methods show improvements
Bhangale et al.	DCNN	CCTV Video Feeds	Accuracy, Real-Time Performance	High accuracy in real-time scenarios, issues with shadows and false positives	Focuses on real-time performance but requires further robustness
Marsden et al.	GMM, SVM	Custom Datasets	Anomaly Detection Accuracy	Effective anomaly detection, emphasis on holistic features	Useful for behavior anomalies rather than density estimation
Cheong et al.	CNNs, BGS, SSD	Public Spaces	Accuracy, Real-Time Performance	High accuracy in real-world environments, emphasizes real-time analytics	Effective in practical applications and public spaces
Wang and Xu	Spatio-Temporal Models	Various Datasets	Anomaly Detection Accuracy	Robust for anomaly detection in low-density crowds	Focused on anomaly detection rather than density estimation
Zitouni et al.	Motion and Appearance Saliency	Large-Scale Crowds	Accuracy, Compact Representation	Effective in big data scenarios, challenges with compact representation	Provides solutions for large-scale crowd data analysis

continued on following page

Table 4. Continued

Study	Model	Datasets Used	Evaluation Metrics	Reported Performance	Key Findings
Xiang et al. and Liu et al.	Deep Learning-Based Method	Exhibition Centers	Accuracy, Real-Time Processing	High accuracy with real-time processing using GPUs	Addresses specific challenges in exhibition centers
Khan et al., Menouar et al., Hamila et al. [10]	LCDNet	Video Surveillance	Real-Time Accuracy, Efficiency	Compact and efficient, suitable for real-time applications	Emphasizes lightweight models for resource-constrained environments
Nayak and Dash [11]	CNNs, LSTMs	Various Datasets	Accuracy, Model Integration	Effective integration of spatial and temporal models	Highlights benefits of combining advanced techniques for crowd dynamics
Our Study	CrowdNet and VGG16 ensemble	ShanghaiTech, Mall datasets	Scatter plots, error rates	Accurate, diverse conditions	High accuracy across diverse conditions

CONCLUSION

This study addresses the intricate challenge of crowd counting in images by leveraging advanced deep learning techniques to achieve both accuracy and efficiency in crowd density estimation. Through a comprehensive process of research, development, and comparative analysis, we have devised a novel approach that shows promising advancements in this domain. At the heart of this approach is an ensemble model of a convolutional neural network (CNN) model called CrowdNet with a pre-trained VGG model, enhanced by integrating Gaussian filter density estimation techniques. This sophisticated model enhances crowd-counting accuracy by combining several crucial components: meticulous image preprocessing, robust feature extraction, and precise density map generation.

Beyond the technical innovations, this study contributes significantly to the field by evaluating and comparing various methodologies for crowd counting. By reviewing existing literature and analyzing different approaches, the study identifies the strengths and limitations of prevailing techniques, which informs the development of our approach. This comparative analysis ensures that our model builds on successful aspects of previous research while addressing existing gaps and challenges.

The study also emphasizes the critical importance of ethical considerations in the development and deployment of computer vision systems. As technological advancements continue, ensuring the protection of privacy, fairness, and transparency is paramount. The study advocates for the implementation of robust data security measures, such as advanced encryption techniques, and the development of anti-discrimination algorithms to prevent bias in AI systems. Transparent AI practices are essential for maintaining public trust and ensuring responsible innovation. Adhering to GDPR principles—such as data minimization, purpose limitation, and ensuring data subject rights—demonstrates a commitment to protecting individual privacy and fostering confidence in AI technologies. By addressing these ethical concerns, the study underscores the importance of creating AI systems that are not only effective but also ethically sound.

In conclusion, this study represents a significant leap forward in both the technical and ethical dimensions of crowd counting and deep learning-based image analysis. The accurate estimation of crowd density provided by our model offers valuable insights and practical benefits for authorities and organizations, enhancing their ability to manage crowds effectively and allocate resources efficiently. This research lays a solid foundation for future work in the field, aiming to refine and adapt the approach to meet the demands of diverse real-world contexts and continue advancing the state of the art in crowd analysis.

Scope for Future Work

The completion of this study marks a significant milestone in advancing the application of deep learning-based computer vision for real-time object detection. It sets the stage for numerous future advancements that can further enhance the capabilities and applications of these technologies. One promising avenue for future work involves integrating computer vision systems with Internet of Things (IoT) technologies. This integration could revolutionize various sectors by creating more responsive and intelligent systems. For instance, combining real-time object detection with IoT sensors can lead to the development of advanced transportation systems that dynamically adapt to changing conditions. By incorporating IoT sensors, these systems could gather and analyze environmental data in real-time, allowing them to respond more effectively to complex scenarios such as fluctuating traffic patterns or sudden obstacles. This synergy between computer vision and IoT could greatly improve system adaptability, efficiency, and overall performance.

Future research should also focus on refining the convolutional neural network (CNN) architecture and expanding the dataset to boost model performance further. Experimenting with various CNN architectures, hyperparameters, and optimization techniques is essential for enhancing both the accuracy and efficiency of the model.

For example, exploring alternative network structures or tuning hyperparameters could yield improvements in detection performance and processing speed. Additionally, data augmentation strategies, including rotation, scaling, and the addition of noise, should be employed to increase dataset diversity and robustness. These techniques can help the model generalize better to new and unseen scenarios. Extending the model to handle video sequences and incorporating temporal information through spatiotemporal feature extraction and recurrent neural networks (RNNs) could also improve its accuracy in dynamic and rapidly changing environments. By integrating these advanced techniques, the system's ability to perform in real-world conditions could be significantly enhanced.

Ethical considerations in the deployment of AI technologies remain a critical focus. As computer vision systems become more embedded in daily life, addressing issues related to privacy, fairness, and transparency is essential for maintaining public trust and ensuring responsible development. Future work should include implementing robust data protection measures to safeguard user information and ensure compliance with privacy regulations. Additionally, efforts should be made to ensure fairness in algorithmic decisions by developing and testing anti-bias mechanisms. Providing clear and accessible explanations of AI processes is also crucial for transparency. Tailoring systems to specific domains and conducting user studies can further refine the model's relevance and usability, ensuring that technological advancements align with ethical standards and meet societal needs. By addressing these ethical considerations, future research can contribute to the development of AI systems that are not only technically advanced but also socially responsible.

REFERENCES

Bhangale, U., Patil, S., Vishwanath, V., Thakker, P., Bansode, A., & Navandhar, D. (2020). Near real-time crowd counting using deep learning approach. *Procedia Computer Science*, 171, 770–779. DOI: 10.1016/j.procs.2020.04.084

Bhuiyan, M. R., Abdullah, J., Hashim, N., & Farid, F. A. (2022). Video analytics using deep learning for crowd analysis: A review. *Multimedia Tools and Applications*, 81(19), 27895–27922. DOI: 10.1007/s11042-022-12833-z

Chen, K., Loy, C. C., Gong, S., & Xiang, T. (2012). *Feature mining for localised crowd counting*. BMVC., DOI: 10.5244/C.26.21

Cheong, K. H., Poeschmann, S., Lai, J. W., Koh, J., Acharya, U. R., Yu, S. C. M., & Tang, K. J. W. (2019). Practical automated video analytics for crowd monitoring and counting. *IEEE Access : Practical Innovations, Open Solutions*, 7, 183252–183261. DOI: 10.1109/ACCESS.2019.2958255

Ha, K. M. (2023). Reviewing the Itaewon Halloween crowd crush, Korea 2022: Qualitative content analysis. *F1000 Research*, 12, 829. DOI: 10.12688/f1000research.135265.1 PMID: 38037564

Khan, M. A., Menouar, H., & Hamila, R. (2023). LCDnet: A lightweight crowd density estimation model for real-time video surveillance. *Journal of Real-Time Image Processing*, 20(2), 29. DOI: 10.1007/s11554-023-01286-8

Li, N. (2023). Ethical considerations in artificial intelligence: A comprehensive discussion from the perspective of computer vision. SHS Web of Conferences, 179, 04024. DOI: 10.1051/shsconf/202317904024

Marsden, M., McGuinness, K., Little, S., & O'Connor, N. E. (2016). Holistic features for real-time crowd behaviour anomaly detection. arXiv. /arxiv.1606.05310DOI: 10.1109/ICIP.2016.7532491

Nayak, M. M., & Dash, S. K. (2020). A review paper on crowd estimation. International Journal of Advanced Research in Engineering and Technology (IJARET), 2020. https://ssrn.com/abstract=3878696

Patwal, A., Diwakar, M., Tripathi, V., & Singh, P. (2023). Crowd counting analysis using deep learning: A critical review. *Procedia Computer Science*, 218, 2448–2458. DOI: 10.1016/j.procs.2023.01.220

Saleh, S. A. M., Suandi, S. A., & Ibrahim, H. (2015). Recent survey on crowd density estimation and counting for visual surveillance. *Engineering Applications of Artificial Intelligence*, 41, 103–114. DOI: 10.1016/j.engappai.2015.01.007

Sharma, N., & Garg, R. D. (2022). Real-time computer vision for transportation safety using deep learning and IoT. *2022 International Conference on Engineering and Emerging Technologies (ICEET)*, Kuala Lumpur, Malaysia, 1-5. DOI: 10.1109/ICEET56468.2022.10007226

Wang, J., & Xu, Z. (2015). *Crowd anomaly detection for automated video surveillance*. IEEE Xplore., DOI: 10.1049/ic.2015.0102

Wickramasinghe, C. S., Marino, D. L., Amarasinghe, K., & Manic, M. (2018). Generalization of deep learning for cyber-physical system security: A survey. IECON 2018 - 44th Annual Conference of the IEEE Industrial Electronics Society, Washington, DC, USA, 745-751. DOI: 10.1109/IECON.2018.8591773

Xiang, J., & Liu, N. (2022). Crowd density estimation method using deep learning for passenger flow detection system in exhibition center. *Scientific Programming*, 2022(1), 1990951.

Zhang, Y., Zhou, D., Chen, S., Gao, S., & Ma, Y. (2016). Single-image crowd counting via multi-column convolutional neural network. *2016 IEEE Conference on Computer Vision and Pattern Recognition (CVPR)*, Las Vegas, NV, USA, 589-597. DOI: 10.1109/CVPR.2016.70

Zitouni, M. S., Dias, J., Al-Mualla, M., & Bhaskar, H. (2015). *Hierarchical crowd detection and representation for big data analytics in visual surveillance*. IEEE Xplore., DOI: 10.1109/SMC.2015.320

Chapter 5
Cyber-Physical Systems and the Future of Urban Living:
Decision-Making Challenges and Opportunities

Rituraj Jain
https://orcid.org/0000-0002-5532-1245
Marwadi University, India

Damodharan Palaniappan
https://orcid.org/0009-0003-0721-3068
Marwadi University, India

Kumar J. Parmar
https://orcid.org/0000-0002-2502-5680
Marwadi University, India

T. Premavathi
https://orcid.org/0009-0003-0172-2021
Marwadi University, India

Kushal Gaddamwar
https://orcid.org/0009-0009-9318-1616
PDPM IIIT Jabalpur, India

Yatharth Srivastava
https://orcid.org/0009-0003-1134-4722
The LNM Institute of Information Technology, India

ABSTRACT

Cyber-physical systems, where software and physical infrastructure are intertwined are promising technologies for re-imagining urban life. As CPS apply tightly coupled networks of sensors and actuators, supported by advanced data analytics, the field has great potential for addressing many aspects of city life and improving the quality of life of city inhabitants. Global urban development and itself increasing complication of the processes involved have produced considerable demand for

DOI: 10.4018/979-8-3693-5728-6.ch005

fresh approaches to various local problems, extending from transport management to energy consumption, disease prevention, and preservation of natural resources. Moreover, CPS can be helpful in conservative consumption of resources: energy, water supply, and others in cities. Addressing these challenges via adaptive policies and frameworks, ethical imperatives, as well as special attention to the principles of smart design, will be vital for the successful translation of the potentials of CPS for improved efficiency, sustainability, and equity in urban settings.

1. INTRODUCTION TO CYBER-PHYSICAL SYSTEMS (CPS) AND URBAN LIVING TRANSFORMATION

Urbanization has advanced over the years and so has the increase in complexity of these emerging today's urban world problems including traffic jams, high energy consumption and pollution. Cyber-physical systems have become revolutionary in managing the City's technical networks and human environment through converged computational applications, communication, and physical support systems, thus improving the life of the City's residents. Cyber-physical systems are intelligent systems and are defined as systems that integrate computational algorithms and communication technologies with physical processes.

1.1 Overview of Cyber-Physical Systems (CPS)

Cyber-physical systems are made of a set of physical entities which interact with each other, for instance, sensors, and actuators, but also comprise computational and communicational qualities. These systems use data analysis and machine learning alongside real-time decision making over actual physiologic processes toward the formulation of improved physical systems alongside the general system robustness (Alshammari et al., 2021). CPS interconnectivity of cyber and physical elements allows coordinating information sharing and employing the control strategies to reconsider the existing urban issues concerning traffic circulation, energy consumption, and environmental impacts. The potential that can be seen from cyber-physical systems in defining and redesigning the way people live in cities is possible is unprecedented. Through integration of CPS, cities can find new and better approaches to management of infrastructure, provision of services and growth of quality of life for citizens Ali et al., 2020.

1.2 Evolution of CPS in Urban Environments

The implementation of cyber-physical systems in the context of cities has expanded in the last few years due to innovation in different technologies such as IoT, cloud computing, and data analytics (Alshammari et al., 2021) (Sharevski & Oteafy, 2018). Previous discussions of CPS in cities were more limited, and centered on niche markets like transportation and energy. While CPS capabilities have evolved over time and become more sophisticated, the use cases have broadened to sectors such as water, waste management, safety, and governance. As the number of sensors has increased, connectivity has improved and more data has become available, CPS solutions have become better in terms of their applicability to urban environments. Such systems utilise such features as data-driven decision making, automation, and real-time response to the multifaceted and interconnected issues characteristic of today's cities.

1.3 The Role of CPS in Future Cities

These complex technology infrastructures are expected to take an essential position in defining the way societies will increasingly live in the urban environment soon. CPS provide the incredible potential to improve the availability, reliability, robustness and performance of urban systems through the integration of computational intelligence, communication networks and physical infrastructure to create a Sustainable Intelligent Environment for citizens of cities.

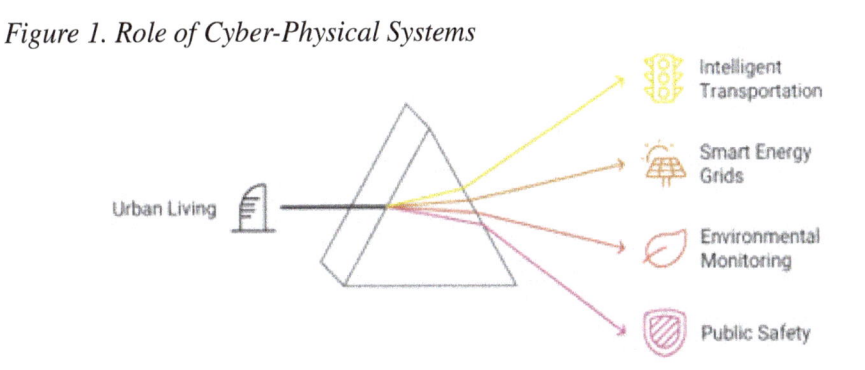

Figure 1. Role of Cyber-Physical Systems

Read the Figure 1 Potential Role of Cyber-Physical System in the future cities and how are they going to do that. Key potential applications of CPS in future cities include - Intelligent transportation systems: CPS can enhance traffic control, minimize traffic density and facilitate the utilization of both autonomous transit and mixed mode transport solutions (Akhuseyinoglu & Joshi, 2020). Smart energy grids: CPS can potentially improve the performance, robustness, and practicability of urban energy systems, promote the penetration of distributed generations and DSM, among others. Environmental monitoring and management: Through CPS, real–time resource usage, quality of air and efficiency of disposal, management can be enhanced contributing to the creation of sustainable and habitable cities.

Public safety and emergency response: CPS, can help in the increase of awareness of these situations, response time to crises, and coordination of people with the responsibility of responding to disasters hence increasing the resilience of cities (Sharevski & Oteafy, 2018). To expand the importance of CPS for future cities requires further studies and development to avoid several within limitations such as data protection and security, ethic issues, and closed system and put into considerations about the equality of access.

1.4 CPS: A Driving Force Behind Urban Transformation

Cyber-physical systems' application in urban contexts is a key enabler of urban change, providing the basis for improving the efficiency and overall quality of life in cities. Considering the issues described, CPS can find the best approaches for utilizing the urban resources, contributing to the development of urban systems, increasing delivery of public services, and increasing the quality of life of individuals living in cities (Jin, 2023). Through the proper utilization of CPS, cities can effectively respond to various acute issues that need to be solved related to traffic problems, energy consumption, pollution, and safety. These elements having been integrated, real-time monitoring, adaptive optimisation and data driven control for the physical infrastructure of the city pave way for smart and resilient systems in the city (Ali et al., 2023). In the following sections a deeper analysis of the opportunities and challenges to be faced when implementing CPS in urban environment about the implications for the future of urban life is to be given based on the information from the given sources (Upreti et al., 2022).

2. OPPORTUNITIES FOR CPS IN URBAN INFRASTRUCTURE OPTIMIZATION

Many opportunities in terms of making the urban infrastructure more efficient and better capable of meeting the needs of its inhabitants is given by the increasing numbers of cyber-physical systems in the cities' everyday life. Among the most appropriate fields of usage of CPS in urban spaces, the intelligent transportation systems deserve to be mentioned.

2.1. Intelligent Transport Systems

The cyber-physical systems can provide strategies for traffic management and optimization of traffic density by using real time traffic information, enhanced transporter self-automation, and concepts for integrating multi modal transport (Sharevski & Oteafy, 2018). Figure. 2, demonstrate that how urban transportation is being improved through Cyber-Physical Systems.

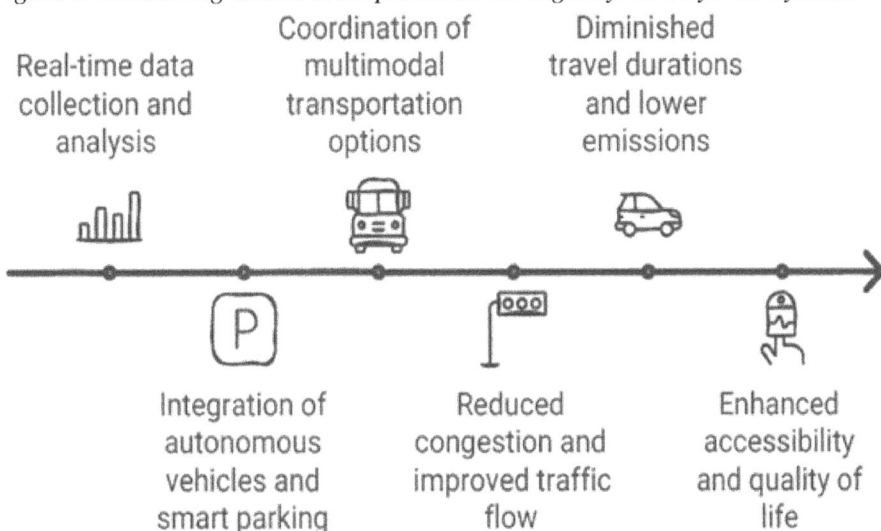

Figure 2. Enhancing Urban Transportation through Cyber-Physical Systems

- Real-time data collection and analysis
- Coordination of multimodal transportation options
- Diminished travel durations and lower emissions
- Integration of autonomous vehicles and smart parking
- Reduced congestion and improved traffic flow
- Enhanced accessibility and quality of life

Cyber-physical systems play a crucial role in improving traffic management and public transit in urban environments - Real-time data collection and predictive analytics: CPS use several sensors and other connected devices to acquire data on traffic conditions in terms of vehicle positions, velocities and density directly in real time. These details can be processed through complex computation and artificial intelligence review to control traffic flows and identify areas of congestion that would require prevention or rectification (Nie et al., 2023). Autonomous vehicles and smart parking: CPS can contribute to the efficient management of environment that allows the usage of autonomous transportation instead of traditional car-driven one. Along with this, through the CPS empowered smart parking system, the drivers can easily locate the parking slots more easily thus reducing the congestion (Haque et al., 2023).

Multimodal transportation coordination: Integrated with a CPS, these modes of transportation can be better managed and synchronized to deliver optimized and efficient multi-modal ridesharing for city citizens. Introducing cyber-physical systems to transport systems for cities provides many benefits such as reduced journey time, reduced pollution levels, increased accessibility and an improved quality of life for citizens. These are capabilities which, when powered by CPS, strengthen the transport system in municipalities and reduce transport congestions hence boosting mobility among the population.

2.2. Smart Grids and Energy Efficiency

It's also essential to note that cyber-physical systems contribute decisively to the advancement of smart urban energy grids to create accountability, dependability, and longevity of energy distribution. Key applications of CPS in urban energy systems include - CPS for Energy Distribution and Demand Response: Cyber-physical systems could also boost performance and reliability of the energy systems in smart cities because it has the capability to offer efficient real-time command and control plus demand response demands. The CPS integrated smart grids can minimize energy losses, incorporate renewable resources and enable consumers to gain better control over energy consumption (Khan et al., 2021). Renewable Energy Integration via Smart Grids: Intelligent networks that advanced CPS technologies can manage the easy incorporation of the green energy types, including solar and wind energy, into the metropolitan energy system. CPS-enabled systems are capable of efficiently controlling and managing the integration of renewable energy into the grid infrastructure, as well as the use of available renewable energy sources (Nazari-Heris et al., 2023).

Reducing Energy Wastage and Carbon Footprint: With help of optimization of the energy distribution, demand management and the integration of renewable energy CPS can contribute significantly to minimize energy waste and bring down the carbon footprint of the cities. Implementing solutions guided by the CPS can assist municipalities as well as the residents of the city to live enviously sustainable energy efficient lives with positive impact on the green environment, thereby promoting the sustainable economy (Zheng, 2022). The enhancement of cyber-physical systems provides cities with a lever to increase the sustainability of its energy structures, and consequently the energy performance of cities.

2.3 E-Governance Platforms and Urban Services

Smart technologies are revolutionizing the functions of cities and restructuring broader paradigms of urban administration and public participation. Due to the concurrent adoption of digital technologies with physical structures and operations, CPS enables municipalities to provide common services efficiently and adapt to the people's demands quickly (Grace et al., 2022; Dabla et al., 2022). CPS allows the development of smart city systems that integrate people, institutions, and objects into an active and complex network. For instance - Smart Street Lighting: Smart lighting is also equipped with sensors to notify streetlights' outages at the same time it necessitates far fewer maintenance efforts and enhances safety on route. Waste Management: Waste bins might feature sensors used to indicate the degree of filling, thus, improving the efficiency of routes that collect waste and decreasing fuel usage and emissions.

Citizen Feedback Systems: The use of smart apps such as mobile applications, as well as online social platforms can actively engage citizens to report various problems, make inputs and contribute to decision making processes. Through effective application of CPS, the municipalities can be able to deliver the public services more effectively, increase operational efficiencies as well as engage citizens more effectively with a resultant cumulative positive effect on the citizens' quality of life. CPS produces a huge volume of data from different urban systems and offer potentials of monitoring and understanding city functioning and citizens' activities. (Xiong et al., 2021) By leveraging data analytics and machine learning, city governments can - Optimize Resource Allocation: Examine traffic flow to optimize locations for public transport or find the location for new parks or gardens. Enhance Public Safety: Police should utilize crime statistics to efficiently distribute resources to address crime and deploy crime prevention measures. Improve Urban Planning: Simulate the effects of major construction projects and policy measures on throughput, pollution levels, and other factors affecting liability.

CPS can help government staff to reduce a high number of administrative tasks that take most of their working time, so they will have time to deal with those important and creative tasks that will benefit the people. Examples include - Online Permitting and Licensing: Simplified compliance with their application procedures, and decreased paperwork and duration of their processing. Automated Invoice Processing: Increasing velocity and decreasing variability when it comes to economic affairs. Smart Parking Systems: Parking payment, control, and space status using parking technology solutions. Owing to the use of CPS technologies, cities should be able to advance the ways in which governance and service delivery accomplish and interact with citizens for the provision of better living and sustainable urban environments (Alshammari et al., 2021).

3. INTEGRAL URBAN SYSTEMS SUSTAINABLE AND RESILIENT

Cps can facilitate adaptive resilient systems which can ease the coping and recovery mechanisms with different challenges such as natural disasters, public health crises and socio-economic shocks (Kapucu et al., 2021).

3.1 Natural Disaster Preparedness and Response

Integrated technology in cyber-physical systems is hence becoming crucial for developing sustainable cities with adequate capacity for planning and response to natural calamities. The integration of sensors, communication networks and data analysis in CPS offers timely alerts, efficient disposition of efforts in managing crises and reduced consequences related to these disasters to human lives and structures. Early warning systems are vital organs in overcoming the effects of natural disasters such as Earthquakes Floods etc as figure 3 indicates how CPS providing early warning during natural disasters.

Figure 3. CPS Early Warning Systems

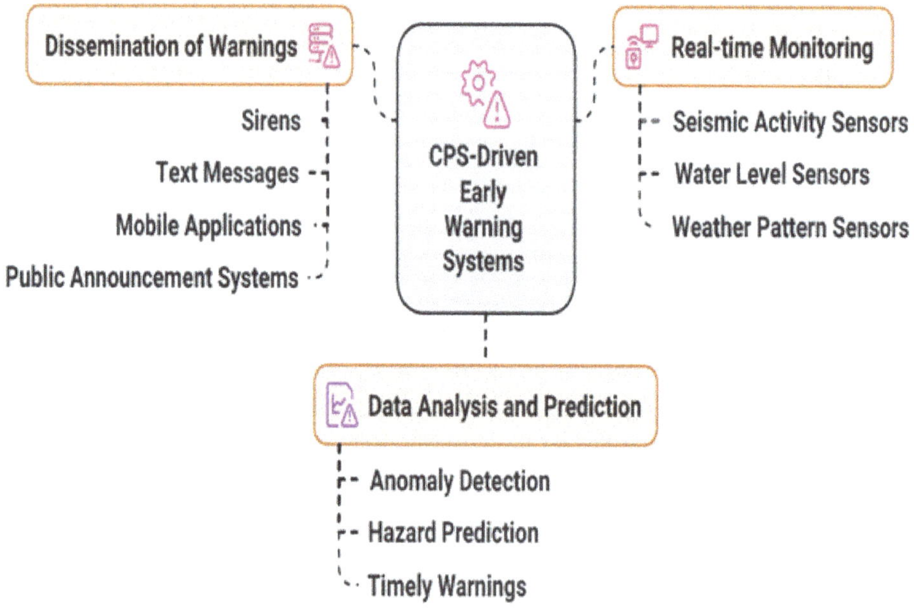

CPS plays a vital role in these systems by - Real-time Monitoring: With help of the networks of sensors, such important parameters as seismic activity, water level, weather conditions can be monitored permanently. According to (Šarak et al., 2020), sensor networks can be deployed for early warning systems, especially for torrential floods (Dabla et al., 2022). Data Analysis and Prediction: Sophisticated software evaluates the sensor information to look for signs of abnormalities, foresee dangers and alert concerned professionals and the populace at large. (Perera et al., 2020) reviews the use of data analysis in flood early warning systems. Dissemination of Warnings: CPS is a system that helps disseminate the warning in a short span of time using tools such as sirens, text, mobile APP, and public address systems. They explain that with the development of communication technology information concerning events can be made available to the public in real time as cited by (Jozinović et al., 2020).

During a natural disaster, CPS can significantly enhance emergency response efforts by - Situational Awareness: Live data feeds from sensors and drones allow first responders to get a clear picture of the situation on the ground to make better decisions. Resource Allocation: Considering emergency motives and their whereabouts, CPS can enhance the distribution of human resources and other needs such as equipment's, and products. Communication and Coordination: They found that through CPS, coordination and communication between different response teams were effectively realized and it enhanced first response's efficiency and quickness. This section would benefit from showing concrete examples of cities employing CPS for emergency readiness. The following examples of cases/case studies where CPS architectures and frameworks are implemented in everyday public emergency response, such as fire alarm/other disaster situations (Khan et al., 2021).

Japan's Earthquake Early Warning System: Has an extensive array of seismometers that record an earthquake and prevent the public from getting hurt or property damaged by giving warning signals few seconds before the shaking occurs (Gunawan et al., 2021). The Netherlands' Flood Control System: A complex of dams and dike and a network of sensors controlled by a central operation center help the country to prevent floods (Maurya et al., 2021). Singapore's Intelligent Flood Monitoring and Warning System: Uses information from sensors, weather forecasts, and records of prior floods to help assay damage potential and alert inhabitants (Imen & Chang, 2016). That way, you will be able to illustrate certain points and examples that would help you build an appealing story about the CPS and its functions in the context of natural disasters.

3.2. Infrastructure Monitoring and Maintenance

CPSs refer to the integration of computation and networking technologies into human physical processes and can greatly improve the surveillance, diagnostics, control, and management of key infrastructures in urban settings like transportation, power and water, and construction. Decision making is now more accurate and efficient, as information for the continuous monitoring and maintenance of critical infrastructures including transportation networks, utility grids and building and structures are provided by cyber-physical systems (Riggs et al., 2023). Through many integrated sensors, CPS can allow to perform condition-based monitoring of degradation and environmental conditions that influence the state of structures and facilities. These sensor data are then integrated into analytical and modelling applications which enable an organization to determine when a machine will need maintenance or repair work and make the necessary actions efficiently and economically. By the combination of sensors, communication networks and data analysis, CPS can support real-time monitoring and smart asset management of city's key infrastructures (Kee et al., 2022). There have been studies demonstrated where info can be used to employ sensor networks to monitor for the performance of structures such as bridges, buildings and other structures. Such sensor networks capture information on other parameters such as vibration, strain, corrosion, and the environment so that early indications of issues arising can easily be detected and maintenance promptly effected (Cronin et al., 2022). Since CPS allows predicting equipment conditions and identifying problems in real time, this technology can increase the lifespan of critical infrastructure, minimize the chances of catastrophic failures, and improve public safety (Kee et al., 2022). When maintained this way and through its CPS capability the lifespan of public assets are seen as being elongated to eliminate costs resulting from premature failures or emergency repairs.

3.3 Crisis Management and Urban Security

There is a potential for cyber-physical systems to make a significant contribution to existing systems that manage crisis situations and protect cities by focusing on public safety, emergency, and critical infrastructure. Real time sensors could relay information to the first responders to help them make an informed decision during disasters. CPS can also assist in better distributing the emergency staff, apparatus, and supplies with real-time requirements and their positions (Kannan et al., 2023). In addition, CPS for effective and efficient multi–team communication to enhance the rate and frequency of response. CPS Systems integrate with surveillance systems in urban settings, thus alarming and preventing security threats, and facilitating public safety outcomes.

4. SUSTAINABLE RESOURCE MANAGEMENT IN URBAN ENVIRONMENT

Recent advances in cyber-physical systems demonstrate potentiality for resource management and learned resource optimization in the urban context such as energy, water and waste. As illustrated in figure 4, the role of Cyber-Physical Systems interfaced to Urban Resource Management.

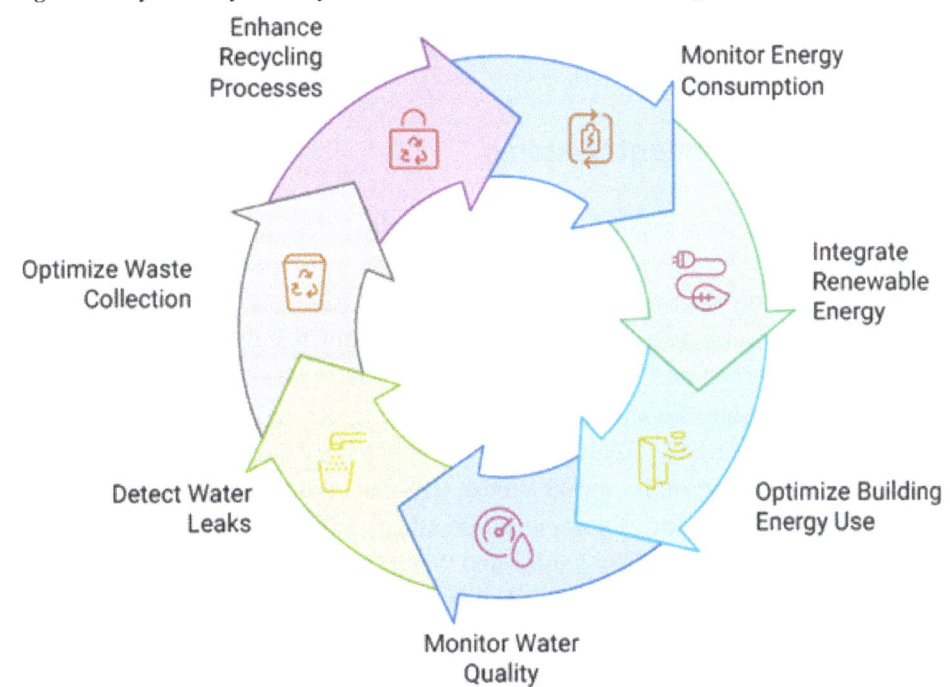

Figure 4. Cyber-Physical Systems in Urban Resource Management

4.1. Energy Consumption Optimization

Cyber-physical systems should be seen as a powerful tool for new strategies of energy consumption in urban areas. CPS facilitate the interconnection of smart grid apparatus, renewable power solutions, and intelligent building control systems to minimize energy inefficiency and encourage efficient energy consumption. The applications of smart grid incorporating CPS include real-time control of energy generation, distribution, and consumption. With the help of sensors and control systems energy flows can be balanced, and load and demand can be reshaped to match their base and peak levels (Hercegová et al., 2021). By integrating CPS, it becomes easy to incorporate solar and wind energy into the urban energy network. These systems can observe meteorological conditions, the level of energy production and the synchronised management of reconnecting renewable energy into the grid. Finally, with the support of CPS, the energy consumption of intelligent building can be automated and optimized through Intelligent Building Management Systems. The available technologies include occupancy sensors and control systems at the building level that regulate the lighting, heating, ventilation, and air conditioning (HVAC), and other mechanisms of the building; these optimize the use of light, temperature and power in accordance with occupancy, weather conditions, and energy demands, thereby saving up to 25 percent of the energy used and at the same time, reducing carbon emissions equally.

4.2 Water Management Systems

Integration of cyber-physical systems potentially offers great opportunities for renewing urban water management in view of water scarcities, water quality, and wastewater. Smart Water Grids for Efficient Distribution and Leakage Detection - Using CPS for water distribution networks also means that the CPS can monitor water flow rates, pressures, and quality always, possibly to identify when a particular pipe is leaking, or how best to distribute water using the network (Palermo et al., 2022). CPS in Wastewater Treatment and Recycling - By employing combination of sensors, data mining, and control systems, CPS can improve the performance and environmental contribution of wastewater treatment and water reuse in urban areas (Zainurin et al., 2022). Flood Prevention and Water Conservation Techniques - With the help of CPS, it is possible to track the weather indicators, namely water levels, rainfall, or earth's moisture content for early flood prevention and foot sustainable water management in urbanized cities and towns (Adedeji & Hamam, 2020). Thus, utilizing CPS as an underlying infrastructure for smart city, the cities can improve the durability, reliability and sustainability of water management systems to meet the increasing water shortage problems and have long term vision for smart water supply in cities.

4.3. The Issues Related to Waste Management and Recycling

Today, it is possible to discuss the contribution of cyber-physical systems to enhance waste and recycling management in urban settings. Through CPS the intelligent waste collection and transportation can be achieved and the levels of wastes in bins and dumpsters around the city can be monitored to facilitate efficient waste collections routes, less Fuel consumption and efficient operation of waste collections.

Moreover, it can also work on improvements of more sophisticated waste sorting and recycling centres using such technologies as automation, computer vision and robotics to increase the effectiveness, profitability and credibility of the waste sorts and materials received. The following are some potential applications of CPS in urban waste management - CPS for Optimized Waste Collection and Disposal Routes - Smart waste bins and effective algorithms to determine optimal Waste Management routes can cut emissions, save fuel costs, contribute to the reduction of the carbon footprint in developed cities (Ghahramani et al., 2022). Smart Recycling Systems to Enhance Sustainability - By fully incorporating CPS in recycling plants, automatic sorting, processing and quality controls can be performed to increase the recycling rates and the volumes of recycled quality material produced (Ahmed et al., 2021). Data-driven Solutions for Waste Reduction - CPS can give substantial information regarding waste generation behaviour which would allow city planner of coming up with specific approaches towards waste minimization with a focus in circular economy. The CPS can help the cities to improve their waste management and recycling features, and provide them with more sustainable environment in the future.

5. SOCIETAL IMPLICATIONS OF CPS ADOPTION IN URBAN SETTINGS

Thus, the use of the principles of cyber-physical systems in the context of urban areas has great opportunities and apparent threats and questions related to societal aspects must also be considered. There are several studies in recent related to understanding the impact of COVID-19 (Khan et al., 2021) (Alshammari et al., 2021).

5.1. Privacy and Data Security Issues

The use of sensors, cameras and connected systems in CPS urban ecosystems is promoted and privacy and data security is threatened. A CPS generates systems collect personal information and location data, making their protection and privacy beyond such systems paramount avoiding unauthorized access, exploitation, or breaches. All crucial data to and from CPS-generated systems must undergo robust access control protocols, data snapshot encryption, and strict data governance polices to mitigate on privacy and security risks. This for adaptation of CPS in urban areas can result in enhanced data acquisition and monitoring raising issues of privacy rights. Authorities at the city level, private firms, and organizations must consider that collected data is sensitive, so it is necessary to create straightforward rules and legislation to proceed with citizen data privacy. It is therefore important to look to a multi-layered solution to addressing the problems at issue to mitigate the risks of data security that CPS presents. Some of the measures that need to be applied include the use of encrypted data transfer, authentication procedures, and the frequent assessment of new security threats and their integration into the concentration system.

CSs in urban contexts also have implications on equity and accessibility in view of their emergence in cyber-physical systems. Although CPS can be beneficial in enhancing the quality of life of every citizen, the disparities in ACIR as well as DL make it possible to develop a digital divide through which some sections of the community fail to enjoy the benefits of such systems. The interconnectivity and complexity of CPS in smart cities make them vulnerable to cyber-attacks, which can have severe consequences for critical urban infrastructure and public safety (Chaudhuri & Kahyaoğlu, 2023). Addressing these cybersecurity challenges requires a comprehensive approach, including - Secure communication between CPS and well implemented cryptographic mechanisms to safeguard CPS information.: A systematic procedure for both detecting threats that may potentially be launched and monitoring them for potential attacks. That city authorities together with the technology suppliers and cybersecurity personnel come up with perfect approaches to the actual security issues. Tackling the privacy and security risks of implementing CPS in urban environment is the key to the proper and ethical advancement of smart city concepts that would benefit all the inhabitants.

With the increased implementation of CPS in urban settings it is high time laws be adopted to control the probable advantages of intelligent systems to citizen's privacy. Cities can only build the right expectations for the use of CPS generated data by developing and implementing good data management policies that define the use, collection and storage policies of the said data. According to Park et al., 2023, citizens should be included in the process of formation of CPS that will help to convince people and reduce their concerns. Moreover, citizens must be informed about CPS technologies, benefits and risks and engage them more actively when constructing smart cities. It will also be crucial to strive to close the gap between developed and developing nations in terms of CPS technological accessibility to build harmonized and sustainable city environments to meet the future's needs (Gupta & Degbelo, 2022).

5.2. Digital Divide and Inclusivity

It also leads to doubts about the possibilities of a growing digital divide given the growing use of cyber-physical systems in urban landscapes. This can especially be attributed to the unfair distribution of technological resources and digital literacy that thus remove some segments of the population from the potential benefits of such systems. Regarding this issue, cities must employ approaches to integrate all people into CPS-enabled urban space and make everyone an active participant and beneficiary of smart city strategies (Mesa, 2023). Thus, the key idea that underlies cyber-physical systems may become a placeholder for changing the nature of other important aspects of life that has high potential for development in low-income and developing areas but can become a challenge at the same time. Such areas are characterized by inadequate infrastructure and resource, low digital literacy that makes the integration of CPS a challenge. The first of them concerns the absence of stable, widely spread access to the Internet, which is vital for CPS applications. Lack of clear, reliable, and strong internet connection or irregular or low internet connectivity in the CPS target areas would dent the usefulness of CPS. The cities need to come up with the best solution to offer CPS so that there will be equitable distribution of the benefits obtained since there is a digital divide. Lastly, the high costs of CPS deployment and particularly its maintenance presents a major challenge, especially to low-income urban areas. Financial resources and funding are among the major challenges that city planners and other decision makers face in developing countries. In addition, the issue of Digital Divide playing out as disparities in the use of technology and skills in the use of technology increases the difficulty of CPS adoptions among Low-income households. Closing this gap and supporting those at the bottom of society is essential to guaranteeing CPS will be an effective tool for change in all the inhabitants of cities around the globe regardless of their status or location.

Technology Enrolment: A Closed Shop of CPS or an Open Sesame of Opportunity? Because the existence of a digital divide is a significant challenge to an open CPS environment, cities cannot simply allow CPS to develop haphazardly but must take deliberate measures to ensure that individuals in need are provided for. It means upgrading and deploying extensive and durable digital assets like fast internet connection in underprivileged and low-income areas. Furthermore, cities should invest in literacy enhancement programs plus training that enhances CPS technology to marginalized populations enhancing way they both benefit from the technologies. Cities should involve community Based organizations, schools plus technology providers in the realization of these strategies to ensure that they fit the local population's needs of the cities. In this way, the operation of the CPS in cities and the provision of the necessary infrastructures are aimed not only to cover all sectors of the population that do not have access to the internet but are also presented as a way of eliminating the digital divide and making the CPS-enabled city available and useful for all its inhabitants, regardless of their income level and/ or their technical literacy. To illustrate the potential for equitable CPS integration, consider the following case studies - San Francisco has been an example of a city where local government officials responded for the need of closing the digital gap and launched a wireless broadband network to connect those left behind. Such measures were combined with raising digital competencies and community participation in CPS-related services like smart mobility and safety, for which CPSs were deployed for CPS-enabled services in the apartment block and its surroundings (Pietrosémoli et al., 2022).

Barcelona is another example of a city that took a wise approach to smart city development with an aim at people-centred smart city solutions. The implemented CPS in the city include CityOS with limited and easily understandable interfaces and APIs for multilingual capability for all the citizens to blend with the change in Smart city interface (Khan et al., 2021). These case studies show that, when planning, implementing and governing cyber-physical systems, cities have the possibility to reduce the digital gap and make urban futures less distant more sustainable.

5.3. The preference, perception and acceptance of the CPS

The effectiveness of using cyber-physical systems in the urban environment also includes the perception that people have about such systems. Some of these concerns include privacy, security of data and job losses arising from CPS automation which must be handled if CPS technology is to enjoy sustainable support from citizens (Alshammari et al., 2021). The Figure.5 refers CPS is improving with Public Perception.

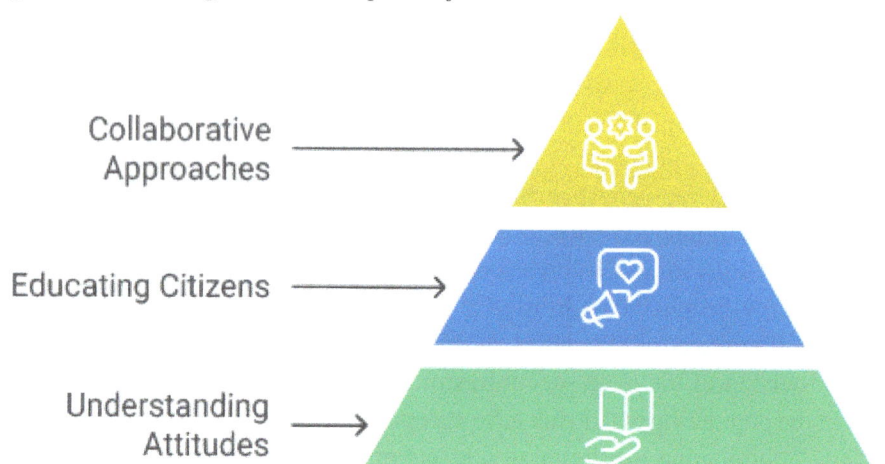

Figure 5. Enhancing Public Perception of CPS

The following paragraphs outline key considerations for enhancing public perception and acceptance of cyber-physical systems in smart cities - Understanding Public Attitudes Towards CPS Technologies - To be effective, cities must invest heavily in basic research and public engagement to ascertain the population's attitudes, risks, and expectations associated with CPSs. Questionnaire may help get information on the fundamentals that affect acceptability of CPS for example; privacy, security and usefulness of CPS (Quincozes et al., 2022).

Educating Citizens on the Benefits and Risks of CPS - For increasing the acceptation, cities need to invest in bettering awareness through explaining the perceived advantages of the CPS such as effectiveness, sustainability, and quality, as well as the nuisance perused by citizens including privacy, security and joblessness (Keshk et al., 2021). Collaborative Approaches for CPS-Driven Urban Development - Meaningful and active citizen participation into the process of designing and deploying CPS-enabled urban solutions could go a long way in fostering trust and guarantee that the resultant technologies are the kinds of technologies the society wants. The essence of co-designing CPS applications with citizens is that the engagement of the public to participate in the transformation process enhances ownership of the cities (Samalna et al., 2022).

6. CHALLENGES IN IMPLEMENTING CYBER-PHYSICAL SYSTEMS IN URBAN AREAS

The shift to a CPS-informed city environment means there are many issues affecting the use of these technologies that cities need to consider to facilitate the implementation of CPS effectively.

6.1 Technological Challenges

The technological challenge that the heterogeneity of devices, protocols and systems in a CPS ecosystem represents for a city is very high. The CPS integration's coherence depends on addressing key challenges such as integration, data protection, and network interoperability in an urban system (Quincozes et al., 2022). Moreover, differences in scale and density of information produced by CPS-based infrastructures and services may saturate the existing mechanisms of data collection and processing, and put extra pressures on cities to obtain necessary computing resources and algorithms. One of the major issues regarding the CSP technologies for cities is the ability to scale and interconnect the CPS technologies. CPS can encompass diverse gadgets, protocols, and systems and the devices must coordinate and communicate harmoniously. Cities need to propose ways of solving scaling concerns to accommodates the increasing amount and sophistication of CPS-originated information as well as CPS integration with other systems and structures of legacy nature. Alleviating these technological challenges is fundamental to realizing CPS's efficiency and applicability to urban settings (Samalna et al., 2022), (Khan et al., 2021).

Cyber-physical systems in cities need to be safe and have backups as they are necessary in infrastructure and services that citizens depend on. This is achieved through having proper backup systems, redundancy, and failover provisions essential for continued functionality of urban systems in the case of system breakdowns. In this context, when CPS is constructed with future disruptive events in mind, all cities will be able to provide citizens with important services without any disruption or interruption, as required. One of the major issues in smart city development is the inclusion of traditional infrastructure combined with new cyber-physical systems. Transportation system, utility and social infrastructure in cities are mostly in their primitive stages and have incompatible features with new CPS-integrated CPS systems. The transition towards a CPS-integrated urban environment requires cities to resolve problems of compatibility, data sharing and integration of systems. The management of infrastructure provides profound insights to integrate and host modern CPS technologies in urban scenarios while sustainable retrofitting of legacy structures is a key challenge towards CPS integration.

6.2. Fiscal Structures

The rapid evolution of complex and smart systems comprising CPS in reducing the rates of their implementation and adoption for addressing the issues that affect cities has been way ahead of the formulation of corresponding policies and laws. Local authorities are required to create qualified policies or rules that will help them regulate the data privacy and security as well as other general ethical dilemmas regarding the CPS technologies. The Urgency of Having Coordinated Structures of Legal Requirements Pertaining to CPS Adoption, Urban CPSs require full legal structures to enhance their implementation and growth in the society. Cities have an obligation for own population protection to provide clear rules and restrictions regarding essential topics, for example, data protection and security, and proper use of CPS technologies. With little supervision, the advancement of CPS could overshadow proper policies suitable for such a technique, and this has dangers and other adverse effects. Strong legal policies will enable the proper regulation of CPS while safeguarding the liberties of the people to reap maximum benefits of the technologies for city living (Park, 2023).

Managing Regulatory Challenges across Various Geographical Locations, Cities face the relatively difficult topic of providing solutions on how to regulate Cyber Physical Systems in cities under legal frameworks. Notably, the legal aspects of interconnected CPS technologies encompass data privacy and security and the ethical use of CPS technologies require guidelines and regulations. The policymakers should therefore engage with the industry players to develop policies that should be able to adapt to the ever-evolving CPS technology. Through, eliminating regulatory issues cities can guarantee citizen rights as well as the use of cyber-physical system to benefit cities.

Policy recommendations for CPS deployment in the future, To make effective usage of CPS in urban regions there is a need for the policy makers to come up with legal frameworks that address the following aspects. These frameworks should provide standards and norms governing privacy and securing the data, and appropriate use of the CPS technologies. This is because most of the CPS advancements are innovative, and creating regulatory standards that would be able to address the future innovations it would require collective efforts of both government and industries. By overcoming these regulatory challenges, cities can realize the advantages of these novel technologies; citizens' rights can be protected and CPSs can be incorporated into urban communities democratically.

6.3 Ethical and Moral Considerations

These applications of cyber-physical systems are in the contemporary urban setting a contentious moral and ethical dilemma that cities must face. With increased CPS enabled automation in cities, the ethical aspect of these technologies should concern cities. Since automated systems are optimized to perform tasks as smoothly as possible, devoid of interactions with actual people, they are likely to make decisions that hurt efficiency and allow adverse consequences or negatively influencing the existence of minority groups. As noted by (Samalna et al., 2022). When it comes to CPS technologies, cities need to have ethical guidelines so that citizens can feel more comfortable with cities deploying these technologies for their benefit. Those who are developing and implementing cyber-physical processes in urban environments should be equally responsible and transparent to retain the public's trust and present benefits to the entire society. There is a growing rely on CPS-driven decision making that leads to numerous problems of fairness, transparency and accountability especially in the urban society. Managers of cities and policy makers also need to ask questions regarding CPSs and bias, or discriminations when used in the decision-making loops. Some features make life easier to live in an automated world where the systems are in-built with societal biases, not only amplifying the same but also enhancing the inequalities that are present in the society. Governments must build strong ethical criteria that regulate the CPS usage so that these technologies become the fair, non-prejudiced tools for cities (Dieterle et al., 2022). It is also important to state that the increased use of cyber-physical systems in urban environment requires transparency and W&P accountability. Cities must guarantee that the decision-making process and algorithms that power these

CPS-enabled technologies are auditable and well understood such that the public can oversee them. Finally, defined paths for citizens to turn to and opportunities to appeal CPS-based decisions not only remain important for people's trust and democratic values but are also missing in the work of the considered governmental body. Cyber-physical systems in cities raise some significant ethical and moral issues in practicing urban planners and policymakers. As detailed plans with more MOS, farsighted rules, establishing rights to social equity, fairness, and accountability needed, the CPS's opportunities will be maximized, and adverse effects and side effects will be minimized.

6.4 Financial and logistical challenges of implementing CPS in developing cities

When it comes to CPS deployment in cities that are still in the process of their formation, there are some major financial and logistic obstacles. The former is costly in upfront investment for technology structure, sensors, and data management frameworks challenges city budgets. Sustaining strategic investments gets a challenge through competing demands within these cities. Moreover, practical challenges that include incorporating CPS into outdated or substandard structures escalate the implementation. Some of the challenges include... Insufficient skill personnel for CPS management and maintenance. Further, delivering sustainable internet connectivity and power in the unserved regions still pose a major challenge. Addressing these issues of financial and logistic constraints is important if CPS is to be made sustainable in developing city contexts.

7. CASE STUDIES: CPS-DRIVEN URBAN TRANSFORMATION

The considered case demonstrates how cyber physical system can give examples and explanation to urban transformation.

7.1 Smart Cities – Starting Point

There is also a growing list of cities across the world that are pioneering concepts of cyber-physical systems within their city systems auxiliary services. Case Study 1: The case examined is Songdo, South Korea – CPS in smart infrastructure. The city under discussion is Songdo, South Korea, where all the described cyber-physical systems are used to update urban structures. CPS technologies have been installed across Songdo as a purpose-built smart city for resolving multiple parameters of building automation and energy control to traffic signals and waste disposal. Cultural, social, and physical CPS solutions in the city have been integrated to support improved resource utilization, the quality of public services and the lives of Songdo residents.

Case Study 2: Barcelona Spain – CPS in Public service as well as the transportation segment. Barcelona, city of Spain is one of the pioneers adopting cyber-physical systems in improvement of its public services and transports system. Integrated application of CPS technologies in different urban sectors has seen Barcelona transform its municipal functions into efficient, effective, and sustainable operations. From managing wastes and distributing energy to developing the city transport system and traffic flow, various CPS based projects in Barcelona have improved the live hood of the residents. For example, the smart parking system will employ sensors and real-time information to direct drivers to the parking slots that are vacant thus minimizing traffic and pollution. As with smart street lights, while implementing building automation systems in Barcelona have also helped the city to consume 30% less energy and lessen CO_2 emissions. Such CPS-powered transformations in Barcelona are an example for other cities willing to leverage technologies of the fourth industrial revolution to improve future urban living, efficiency and sustainability. CPS technologies in the two discussed cities – Songdo and Barcelona – indicate what CPS might turn cities into in the future, the extant literature also reveals what CPS's widespread application entails for cities, including benefits and drawbacks.

7.2. CPS in Emerging Economies

Exploring the role of CPS in the urban environment in developing countries, three peculiarities about integrating cyber-physical systems in such environment can be identified.

Case Study 3: Smart City Development in India (Some cases: Pune, Jaipur), This paper also seeks to explore the prospects and issues associated with smart cities aimed at the Indian context practical to the cities of Pune and Jaipur. India has stated a plan to develop 100 smart cities for implementation of CSPs for integrated infrastructure and services into every city. Two of the cities that have been clearly leading this smart city transformation are the cities of Pune and Jaipur, and these cities have employed CPS technologies to solve some of the major challenges facing cities such as traffic flow, waste, and resource management. Even though social welfare program-driven initiatives in smart cities have helped to scale up new models that are effective to some extent, they expose some of the problems that developing economy nations experience when implementing effective large scale smart cities; problems such as funding, underlying infrastructures, and citizens' participation. This paper seeks to understand the main lessons and recommendations that can be of help in the successful future implementation of smart cities in India and other developing nations.

Case Study 4: Introducing: Brazil's Smart Urban Initiatives, This present paper therefore focuses on the intelligent city development that Brazil has embarked on to face various challenges in its cities with the use of cyber-physical systems. Brazil has gone further to integrate CPS technologies in the likes of transportation, public service and infrastructure. Some of the prominent initiatives include deploying of smart traffic management, smart waste management, energy efficient building automation system in cities like Rio de Janeiro and Sao Paulo. However, the case study also analyses distinctly Brazilian challenges to the expansion of these CPS-driven initiatives, such as funding, digital environment, and multi-level governance. Such analysis might inform other developing economies that intend to leverage cyber-physical systems for the enhancement of their urban space, based on the lessons from the Brazilian case (Samalna et al., 2022).

7.3. Lessons Learned and Best Practices

In the present chapter several case studies have been provided and discussed, and these have brought out the positive as well as the negative. The case studies have revealed several key lessons and best practices that can inform the successful implementation of cyber-physical systems in urban environments - Prioritize Citizen Engagement and Digital Inclusion: CPS-driven urban transformation involvement of citizens is most important if CPS-driven urban transformation is to succeed. Ensure Robust Digital Infrastructure: To adopt CPS technologies, reliable and pervasive communications, assured data handling, and systems integrate ability are key. Adopt a Holistic, Collaborative Approach: CPS implementation cannot be done in isolation but requires meaningful collaboration and the development of plans across sectors as well as the understanding of the urban environment. Emphasize Sustainability and Resilience: The city solutions enabled by CPS should capture requirements based on long-term perspectives from the perspective of the city development to account for new requirements and challenges. Foster Innovation and Agility: Contrary to centralized organizational hierarchy, cities must adopt innovation and experimentative culture to employ CPS technologies optimally.

While the case studies have highlighted the transformative potential of cyber-physical systems in urban environments, the successful scaling and broader deployment of CPS-driven initiatives remains a significant challenge (Samalna et al., 2022). To address this, cities must - Develop Scalable and Replicable CPS Models: Promote the identification and sharing of best practices, frameworks as well as tools to be used that can be used universally in numerous cities. Foster Cross-City Collaboration and Knowledge Sharing: Organization of effective networks and forums to pool together and disseminate information arising from experiences as well as innovative CPS applications from cities across the world. Promote Public-Private Partnerships: Engage private actors in CPS technological development and deployment using their skill set, capital, and innovation assets as a driver for more CPS integration in urban environments. Invest in Education and Workforce Development: CPS should be supported by a capable workforce capable of creating, deploying and sustaining CPS based urban solutions.

Establish Supportive Policy and Regulatory Frameworks: Formulate measures concerning establishment and interaction of cyber physical systems while considering the limits concerning data privacy, Security and Ethical framework. Through solving the mentioned critical challenges cities will create environment for the spread and scaling the CPS Technologies, which will enable the real meaningful change in the future of urban environment.

8. STRATEGIES FOR NAVIGATING CPS CHALLENGES IN URBAN DEVELOPMENT

There is a transformative capability in the use of cyber-physical systems for the urban ecosystem but it comes with distinct complexities (Samalna et al., 2022). Mitigating these is essential for fostering positive process, positive impact, and positive experience.

8.1 Put in place sound legal necessities

Sound CPS RLs that are flexible and well-defined are fundamental in informing acceptable CPS development and deployment. This involves - Government and Industry Collaboration for CPS Standards: The idea of a set goal consistent data safety expectations and integration and idea of an ethical use of such technology must be a collective effort amongst policymakers, technology suppliers, and urban anthropologists (Khan et al., 2021). Best Practices for Legal Frameworks Addressing CPS Challenges: Current legal systems do not necessarily encompass new CPS issues like confidentiality of information, unbiased algorithms, and responsibility in the AI-dependent systems. It is necessary to create new legal instruments or modify the existing ones.

8.2 The CPS Intervention and Ethical Issues

Ethical concerns should always form the core of every phase in CPS development and deployment. Key areas of focus include - Ensuring Ethical Use of Data and Automation: To ensure that the huge amount of data produced by CPS is not abused, strong data management policies are required. This includes; promoting transparency and accountability in the usage of personal data and allowing the citizen control over it. Strategies for Inclusive and Equitable Urban CPS Development: CPS should offer its positive effects to span across various divides; subgroups in the society. This entails the strategic neutralization of biases that follow data as well as algorithms, making CPS benefits accessible to everybody and particularly involving everyone from the disadvantaged communities in the CPS design and implementation processes.

8.3 Design Principles for Inclusive and Resilient CPS Systems

Designing CPS systems for long-term resilience and inclusivity requires a forward-thinking approach - Designing CPS Systems for Marginalized Communities: Make sure that the needs of the minority during the process of developing the framework has been addressed. This refers to physical access for people with disabilities, accommodating for low digital literacy levels, and for language barriers as well. Adaptive CPS Design to Handle Future Urban Challenges: Because of urbanization, cities change their appearance and character all the time. CPS should be made modular so that new technologies can be added to it and accommodate future issues such as climate change and population growth. Promoting Public Participation in CPS Urban Planning: Public involvement is essential. When the public is engaged in the planning and implementation of CPS, citizens will trust that the systems in place meet their needs and that the benefit to all stakeholders is optimized. To this end, addressing these challenges actively will help cities take advantage of CPS to address issues of efficiency, equity, and sustainability within the compact urban space.

CONCLUSION

In conclusion, cyber-physical systems' application in urban ecosystems create innovation possibilities and involve relevant risks. This way, the key opportunities can be identified as the increased usage of technologies that will enhance the possibility of obtaining more efficient, sustainable, and resilient cities. However, CPS deployment entails a broader and system-agnostic, organizational systems thinking and CPS-client cooperation that focuses on more than short-term rewards and isolated deliverables while exploring and securing sustainability, innovation, and agile working. Prominent issues in CPS in the context of urban development include developing sound safeguard regulations, evaluating ethical concern in data and automation, and how to make CPS systems progressive and inclusive. When these issues are addressed ahead of time, the power of CPS can be harnessed for the improvement of equity, efficiency and sustainability of cities. Future research and development for CPS technologies will involve issues of data handling, system adaptability, and new strategies for involving the public in the design process for making our future cities smarter. Technology, policy and city building experts must come together for the CPS revolution to be properly harnessed in a way that brings positive impacts in society about diversity. By adopting this cooperation approach can be built the sustainable efficient and equitably cities in the future.

REFERENCES

Adedeji, K. B., & Hamam, Y. (2020, November 17). Cyber-Physical Systems for Water Supply Network Management: Basics, Challenges, and Roadmap. *Sustainability (Basel)*, 12(22), 9555–9555. DOI: 10.3390/su12229555

Ahmed, A A., Nazzal, M A., & Darras, B M. (2021, October 29). Cyber-Physical Systems as an Enabler of Circular Economy to Achieve Sustainable Development Goals: A Comprehensive Review. Springer Science+Business Media, 9(3), 955-975. DOI: 10.1007/s40684-021-00398-5

Akhuseyinoglu, N B., & Joshi, J. (2020, September 1). A constraint and risk-aware approach to attribute-based access control for cyber-physical systems. Elsevier BV, 96, 101802-101802. DOI: 10.1016/j.cose.2020.101802

Ali, M., Naeem, F., Adam, N., Kaddoum, G., Huda, N. U., Adnan, M., & Tariq, M. U. (2023, January 1). Integration of Data Driven Technologies in Smart Grids for Resilient and Sustainable Smart Cities: A Comprehensive Review. Cornell University. https://doi.org/DOI: 10.48550/arXiv.2301

Ali, N., Hussain, M., Kim, Y., & Hong, J. (2020, January 1). A Generic Framework For Capturing Reliability in Cyber Physical Systems. Cornell University. https://doi.org//arxiv.2010.05490DOI: 10.48550

Alshammari, K., Beach, T., & Rezgui, Y. (2021, August 11). Cybersecurity for Digital Twins in the Built Environment: Research Landscape, Industry Attitudes and Future Direction. Hilaris, 15(8), 382-387. https://publications.waset.org/10012172/pdf

Chaudhuri, A., & Kahyaoğlu, S B. (2023, March 8). CYBERSECURITY ASSURANCE IN SMART CITIES: A RISK MANAGEMENT PERSPECTIVE. Taylor & Francis, 67(4), 1-22. DOI: 10.1080/07366981.2023.2165293

Chohlas-Wood, A., Coots, M., Goel, S., & Nyarko, J. (2023, July 24). Designing equitable algorithms. *Nature Computational Science*, 3(7), 601–610. DOI: 10.1038/s43588-023-00485-4 PMID: 38177749

Cronin, L., Eshkevari, S. S., Matarazzo, T. J., Milardo, S., Dabbaghchian, I., Santi, P., Pakzad, S. N., & Ratti, C. (2022, January 1). Identifying Damage-Sensitive Spatial Vibration Characteristics of Bridges from Widespread Smartphone Data. Cornell University. https://doi.org//arxiv.2211.01363DOI: 10.48550

Dabla, P. K., Sharma, S., Dabas, A., Tyagi, V., Agrawal, S., Jhamb, U., Begos, D., Upreti, K., & Mir, R. (2022). Ionized Blood Magnesium in Sick Children: An Overlooked Electrolyte. *Journal of Tropical Pediatrics*, 68(2), fmac022. Advance online publication. DOI: 10.1093/tropej/fmac022 PMID: 35265997

Dabla, P. K., Upreti, K., Singh, D., Singh, A., Sharma, J., Dabas, A., Gruson, D., Gouget, B., Bernardini, S., Homsak, E., & Stankovic, S. (2022). Target association rule mining to explore novel paediatric illness patterns in emergency settings. *Scandinavian Journal of Clinical and Laboratory Investigation*, 82(7–8), 595–600. DOI: 10.1080/00365513.2022.2148121 PMID: 36399102

Damaševičius, R., Bačanin, N., & Misra, S. (2023, May 16). From Sensors to Safety: Internet of Emergency Services (IoES) for Emergency Response and Disaster Management. *Multidisciplinary Digital Publishing Institute*, 12(3), 41–41. DOI: 10.3390/jsan12030041

Dieterle, E., Dede, C., & Walker, M E. (2022, September 27). The cyclical ethical effects of using artificial intelligence in education. Springer Nature. DOI: 10.1007/s00146-022-01497-w

Ghahramani, M., Zhou, M., Mölter, A., & Pilla, F. (2022, July 15). IoT-Based Route Recommendation for an Intelligent Waste Management System. *IEEE Internet of Things Journal*, 9(14), 11883–11892. DOI: 10.1109/JIOT.2021.3132126

Gilad, M., Fishbein, D., Nave, G., & Packin, N. G. (2023, March 31). Science for policy to protect children in cyberspace. *Science*, 379(6639), 1294–1297. DOI: 10.1126/science.ade9447 PMID: 36996216

Grace, V., Mofokeng, J T., Olutola, A A., & Morero, M. (2022, March 22). An evaluation of the challenges encountered by the South African police service with regard to the fourth industrial revolution. Ümit Hacıoğlu, 11(2), 447-453. DOI: 10.20525/ijrbs.v11i2.1606

Gunawan, G., Nasution, B B., Zarlis, M., Sari, M., Lubis, A R., & Solikhun. (2021, April 1). Design of Earthquake Early Warning System Based on Internet of Thing. IOP Publishing, 1830(1), 012010-012010. DOI: 10.1088/1742-6596/1830/1/012010

Haque, M., Kumar, V. V., Singh, P., Goyal, A. A., Upreti, K., & Verma, A. (2023). A systematic meta-analysis of blockchain technology for educational sector and its advancements towards education 4.0. *Education and Information Technologies*, 28(10), 13841–13867. DOI: 10.1007/s10639-023-11744-2

Hercegová, K., Baranovskaya, T. P., & Efanova, N. (2021, January 1). Smart technologies for energy consumption management. *EDP Sciences*, 128, 02005–02005. DOI: 10.1051/shsconf/202112802005

Imen, S., & Chang, N. (2016, April 1). Developing a cyber-physical system for smart and sustainable drinking water infrastructure management. DOI: 10.1109/ICNSC.2016.7478983

Jin, X. (2023, September 1). Art Interactive Design of Public Leisure Space Environment Based on PSO Algorithm. Taylor & Francis, 43-58. DOI: 10.14733/cadaps.2024.S7.43-58

Jo, S., Han, H., Leem, Y., & Lee, S. (2021, September 3). Sustainable Smart Cities and Industrial Ecosystem: Structural and Relational Changes of the Smart City Industries in Korea. *Sustainability (Basel)*, 13(17), 9917–9917. DOI: 10.3390/su13179917

Jozinović, D., Lomax, A., Štajduhar, I., & Michelini, A. (2020, May 31). Rapid prediction of earthquake ground shaking intensity using raw waveform data and a convolutional neural network. Oxford University Press, 222(2), 1379-1389. DOI: 10.1093/gji/ggaa233

Kannan, N., Upreti, K., Pradhan, R., Dhingra, M., Kalimuthukumar, S., Mahaveerakannan, R., & Gayathri, R. (2023). Future perspectives on new innovative technologies comparison against hybrid renewable energy systems. *Computers & Electrical Engineering*, 111, 108910. DOI: 10.1016/j.compeleceng.2023.108910

Kapucu, N., Ge, Y., Martín, Y., & Williamson, Z. (2021, October 12). Urban resilience for building a sustainable and safe environment. Elsevier BV, 1(1), 10-16. DOI: 10.1016/j.ugj.2021.09.001

Kee, K., Lau, S., Lim, Y. S., Ting, Y. P., & Rashidi, R. B. (2022, February 1). Universal cyber physical system, a prototype for predictive maintenance. [IAES]. *Institute of Advanced Engineering and Science*, 11(1), 42–49. DOI: 10.11591/eei.v11i1.3216

Keshk, M., Turnbull, B., Sitnikova, E., Vatsalan, D., & Moustafa, N. (2021, January 1). Privacy-Preserving Schemes for Safeguarding Heterogeneous Data Sources in Cyber-Physical Systems. *IEEE Access : Practical Innovations, Open Solutions*, 9, 55077–55097. DOI: 10.1109/ACCESS.2021.3069737

Khan, F., Ramasamy, L. K., Kadry, S., Nam, Y., & Meqdad, M. N. (2021, August 1). Cyber physical systems: A smart city perspective. [IAES]. *Institute of Advanced Engineering and Science*, 11(4), 3609–3609. DOI: 10.11591/ijece.v11i4.pp3609-3616

Komane, B. L., & Mathonsi, T. E. (2023, January 1). Design of a Smart Waste Management System for the City of Johannesburg. Cornell University. https://doi.org//arxiv.2303.14436DOI: 10.48550

Liu, S., Trivedi, A., Yin, X., & Zamani, M. (2022, January 1). Secure-by-Construction Synthesis of Cyber-Physical Systems. Cornell University. https://doi.org/DOI: 10.48550/arXiv.2202

Lu, J., Riley, C., Gurumurthy, K. M., & Hentenryck, P. V. (2023, January 1). Revitalizing Public Transit in Low Ridership Areas: An Exploration of On-Demand Multimodal Transit Systems. Cornell University. https://doi.org//arxiv.2308.01298DOI: 10.48550

Maurya, J., Pant, H., Dwivedi, S., & Jaiswal, M. (2021, May 1). FLOOD AVOIDANCE USING IOT. *IJEAST*, 6(1). Advance online publication. DOI: 10.33564/IJEAST.2021.v06i01.021

Mesa, D. (2023, February 21). Digital divide, e-government and trust in public service: The key role of education. *Frontiers of Medicine*, 8, 1140416. Advance online publication. DOI: 10.3389/fsoc.2023.1140416 PMID: 36895333

Moik, B., Bobek, V., & Horvat, T. (2021, June 30). India's National Smart City Mission. *Analysis of Project Dimensions Including Sources of Funding.*, 13(1), 50–59. DOI: 10.32015/JIBM/2021.13.1.50-59

Mushtaq, A., Haq, I. U., Sarwar, M. A., Khan, A., & Shafiq, O. (2022, January 1). Traffic Management of Autonomous Vehicles using Policy Based Deep Reinforcement Learning and Intelligent Routing. Cornell University. https://doi.org/DOI: 10.48550/arXiv.2206

Nazari-Heris, M., Esfehankalateh, A. T., & Ifaei, P. (2023, June 15). Hybrid Energy Systems for Buildings: A Techno-Economic-Enviro Systematic Review. *Energies*, 16(12), 4725–4725. DOI: 10.3390/en16124725

Nie, T., Qin, G., Wang, Y., & Sun, J. (2023, January 1). Towards better traffic volume estimation: Jointly addressing the underdetermination and nonequilibrium problems with correlation-adaptive GNNs. Cornell University. https://doi.org//arxiv.2303.05660DOI: 10.48550

O'Shaughnessy, M. (2023, January 1). Five policy uses of algorithmic transparency and explainability. Cornell University. https://doi.org/DOI: 10.48550/arXiv.2302

Ostad-Ali-Askari, K., Gholami, H., Dehghan, S., & Ghane, M. (2021, February 1). The Role of Public Participation in Promoting Urban Planning. *American Journal of Engineering and Applied Sciences*, 14(2), 177–184. DOI: 10.3844/ajeassp.2021.177.184

Palermo, S. A., Maiolo, M., Brusco, A. C., Turco, M., Pirouz, B., Greco, E., Spezzano, G., & Piro, P. (2022, August 19). Smart Technologies for Water Resource Management: An Overview. *Sensors (Basel)*, 22(16), 6225–6225. DOI: 10.3390/s22166225 PMID: 36015982

Park, H., Oh, H., & Choi, J. K. (2023, January 1). A Consent-Based Privacy-Compliant Personal Data-Sharing System. *IEEE Access : Practical Innovations, Open Solutions*, 11, 95912–95927. DOI: 10.1109/ACCESS.2023.3311823

Park, S. (2023, January 1). Smart Cities: Striking a Balance Between Urban Resilience and Civil Liberties. Cornell University. https://doi.org//arxiv.2303.14597DOI: 10.48550

Patel, S B., & Lam, K. (2023, March 1). ChatGPT: the future of discharge summaries?. Elsevier BV, 5(3), e107-e108. DOI: 10.1016/S2589-7500(23)00021-3

Perera, D., Seidou, O., Agnihotri, J., Mehmood, H., & Rasmy, M. (2020, December 16). Challenges and Technical Advances in Flood Early Warning Systems (FEWSs). IntechOpen. DOI: 10.5772/intechopen.93069

Pietrosémoli, E., Rainone, M., Zennaro, M., & Mikeka, C. (2022, January 1). Massive RF Simulation Applied to School Connectivity in Malawi. Cornell University. https://doi.org/DOI: 10.48550/arXiv.2207

Quincozes, S. E., Mossé, D., Passos, D., Albuquerque, C., Ochi, L. S., & Santos, V F D. (2022, March 1). On the Performance of GRASP-Based Feature Selection for CPS Intrusion Detection. *IEEE Transactions on Network and Service Management*, 19(1), 614–626. DOI: 10.1109/TNSM.2021.3088763

Riggs, H., Tufail, S., Parvez, I., Tariq, M., Khan, M. A., Amir, A., Vuda, K. V., & Sarwat, A. I. (2023, April 17). Impact, Vulnerabilities, and Mitigation Strategies for Cyber-Secure Critical Infrastructure. *Sensors (Basel)*, 23(8), 4060–4060. DOI: 10.3390/s23084060 PMID: 37112400

Samalna, D A., Ngossaha, J M., Ari, A A A., & Kolyang. (2022, December 22). Cyber-Physical Urban Mobility Systems. IGI Global, 11(1), 1-21. DOI: 10.4018/IJSI.315662

Šarak, E., Dobrojević, M., & Sedmak, S. (2020, January 1). IoT based early warning system for torrential floods. *Faculty of Agronomy in Čačak*, 48(3), 511–515. DOI: 10.5937/fme2003511S

Sharevski, F., & Oteafy, S. (2018, January 1). Security for Cyber-Physical Systems: Leveraging Cellular Networks and Fog Computing. Cornell University. https://doi.org//arxiv.1806.11053DOI: 10.48550

Statti, A., & Torres, K M. (2020, January 1). Digital Literacy: The Need for Technology Integration and Its Impact on Learning and Engagement in Community School Environments. Taylor & Francis, 95(1), 90-100. DOI: 10.1080/0161956X.2019.1702426

Upreti, K., Syed, M. H., Khan, M. A., Fatima, H., Alam, M. S., & Sharma, A. (2022). Enhanced algorithmic modelling and architecture in deep reinforcement learning based on wireless communication Fintech technology. *Optik (Stuttgart)*, 272, 170309. DOI: 10.1016/j.ijleo.2022.170309

Wang, K., Belt, M C D., Heath, G., Walzberg, J., Curtis, T L., Berrie, J., Schroeder, P., Lazer, L., & Altamirano, J. (2022, November 1). Circular economy as a climate strategy: current knowledge and calls-to-action. DOI: 10.2172/1897625

Xiong, G., Li, Z., Wu, H., Chen, S., Dong, X., Zhu, F., & Lv, Y. (2021, January 26). Building Urban Public Traffic Dynamic Network Based on CPSS: An Integrated Approach of Big Data and AI. *Applied Sciences (Basel, Switzerland)*, 11(3), 1109–1109. DOI: 10.3390/app11031109

Yuan, C., Ding, S., Wang, Y., Feng, J., & Ma, N. (2023, January 1). Emergency Resource Layout with Multiple Objectives under Complex Disaster Scenarios. Cornell University. https://doi.org/DOI: 10.48550/arXiv.2304

Zaidan, E., Ghofrani, A., Abulibdeh, A., & Jafari, M. A. (2022, March 2). Accelerating the Change to Smart Societies- a Strategic Knowledge-Based Framework for Smart Energy Transition of Urban Communities. *Frontiers in Energy Research*, 10, 852092. Advance online publication. DOI: 10.3389/fenrg.2022.852092

Zainurin, S. N., Ismail, W. Z. W., Mahamud, S. N. I., Ismail, I., Jamaludin, J., Ariffin, K. N. Z., & Kamil, W. M. W. A. (2022, October 28). Advancements in Monitoring Water Quality Based on Various Sensing Methods: A Systematic Review. *International Journal of Environmental Research and Public Health*, 19(21), 14080–14080. DOI: 10.3390/ijerph192114080 PMID: 36360992

Zheng, X. (2022, July 27). Application of CPS Under the Background of Intelligent Construction. *Application of CPS Under the Background of Intelligent Construction.*, 6(4), 22–33. DOI: 10.26689/jwa.v6i4.4190

Chapter 6
Artificial Intelligence and Cybersecurity Prospects and Confronts

Anya Behera
https://orcid.org/0009-0005-2809-5931
Alliance University, India

A. Vedashree
Alliance University, India

M. Rupesh Kumar
https://orcid.org/0000-0002-6229-6885
Alliance University, India

Kamal Upreti
https://orcid.org/0000-0003-0665-530X
Christ University, India

ABSTRACT

The advancement of artificial intelligence has made robust cybersecurity essential. As governments implement various policies to protect citizens, the utilization of AI by governmental agencies has significantly increased. Despite this, there is a significant gap between theoretical knowledge of AI and cybersecurity as separate fields and their practical integration. The challenge lies in the rapid evolution of both AI and cybersecurity, which often outpaces the legislative process, rendering many regulatory efforts outdated before they are fully realized. Due to the lack of dedicated laws for these dynamic areas, achieving optimal results is difficult. Therefore, it is crucial to explore how AI can be leveraged to improve cybersecurity. By using appropriate safeguards, AI can autonomously protect against threats like

DOI: 10.4018/979-8-3693-5728-6.ch006

Copyright © 2025, IGI Global. Scientific Publishing. Copying or distributing in print or electronic forms without written permission of IGI Global is prohibited.

viruses, misuse, and hacking attacks. This paper examines the role of technology in AI and cybersecurity and investigates how AI can optimize cybersecurity, the opportunities & challenges of AI in cybersecurity, regulatory bodies, and strategies to merge these fields.

1. INTRODUCTION

Artificial intelligence is broadly defined as computer systems that exhibit capabilities traditionally associated with human intelligence, including perception, understanding, learning, reasoning, and problem-solving. AI encompasses the development of these systems to perform tasks that typically require human intelligence, such as decision-making and object detection. (Tripathi and Ghatak, 2018). AI encompasses domains such as machine learning, deep learning, neural networks, natural language processing, object detection, knowledge-based expert systems, solving complex problems, increasing accuracy, and performing high-level computations.

Cybersecurity is an evolving field of study that focuses on protecting networks, systems, and data from malicious cyber threats. Cyber threats like theft, disruption, damage, and unauthorised access. It involves using a variety of technologies, processes, and strategies such as to assure integrity, obtainability of digital facts and figures as well as the confidentiality to protect sensitive information from unauthorized access or misuse. Cybersecurity encompasses a range of technical and non-technical activities and measures designed to safeguard the infrastructure of cyberspace. This protection extends to devices, software, and the information they contain and communicate, guarding them against all possible threats (Cavelty, 2018). Through the help of encryption algorithm, the data's can be well protected through converting it into a format which is not easily readable. To protect and secure the ongoing data safeguarding networks are needed. To protect, identify, secure and response immediately to any attempt of illegal access or any sceptical activities, the customers should be given knowledge and education of best practices so that they can deal, identify, and respond to any suspicious and potential risk arising in the network. As technology is evolving day by day and more people are getting connected through online platforms, cyberattacks have become more frequent and sophisticated. It is essential that organizations should start protecting their data, networks, systems, and users from cyber threats. This requires a comprehensive approach to cybersecurity that would include legal protection, preventive measures, and responsive plans.

2. ROLE OF TECHNOLOGY IN CYBERSECURITY AND AI

Technology is growing day by day thus giving scope for cybersecurity to grow in different areas of businesses as well as for common man. Due to the sudden growth in cybersecurity, malicious threat has become a part of it and finds way to exploit vulnerable systems thus, robust security systems are needed to protect networks and data from unauthorised use and infringement. Cybersecurity is not only related to finance but also relates to the confidential information about a particular person. Customers confidential documents, personal information and financial documents are guarded by cybersecurity and if there is any infringement then it might lead to damage of reputation, facing of legal consequences, financial loss, and identity theft. To secure the networks from probable cyber-attack proactive measures should be taken up by the organisation. This includes implementing measures such as encryption, authentication, access control and monitoring to detect any suspicious activity on their networks.

AI system and Technologies uses computer vision and Natural Language Processing to interpret and analyse visual data and human language. AI applications such as translators, Chatbots, virtual assistants and many more applications are already in use. AI systems can analyse user preference, behaviour, and historical data. AI-powered recommendation engines promote suitable content or goods based on individual user profiles which is visible in platforms such as, e-commerce websites streaming service, and social media platforms. Cybersecurity can be defended by implementing daily patching of software vulnerabilities, implication of authentication method and encryption of confidential information. As technology advances and more data is shared over the internet, it becomes increasingly important for organisation, companies, and industries to invest in cybersecurity solutions that can protect their assets.

Artificial Intelligence progressed well in healthcare sector such as treatment planning, medical photo, medical diagnostics, patient data, medication development, and genomic information. They use machine learning algorithm to get the accurate result and help the medical sector. It gives assistance to people with disabilities and impairment by developing computer vision applications, text-to-speech technology, and voice recognition systems. By developing technologies like identification of potential threats, network traffic patterns and divergence, and real time response to security issues it can control cybersecurity thus playing important role in this sector.

The advent of AI-powered models it can analyse, flaws and manage huge amount of data and can also predict future attacks. Facilitates self-driving automobiles, robots, and drones. Helps in judiciary for analysing, navigating, storing, and deciding judgements. By stimulating innovation, increasing productivity, and opening new job opportunities, AI could stimulate economic growth. It makes it possible to create new goods, business and services model that have a positive effect on many different industries and economics. Along with this it is very useful in astronomy, drug development, genetics, and climate modelling. These are the roles played by Artificial Intelligence in advancing the said sectors.

3. PROGRESS OF CYBERSECURITY AND AI IN MODERN ERA

Cybersecurity is expanding in today's digital age. It is a set of techniques used to protect programs, systems, and networks from any malicious digital attacks. Notable progress can be seen in this field in relation to cybersecurity and Artificial Intelligence, Cloud security, Privacy enhancing Technology, Internet of Things (IoT), Automation & Orchestration and Quantum-resistant Cryptography. Currently, we can see a lot of growth of Artificial intelligence in Cybersecurity. When we see the usage of Artificial Intelligence in Cybersecurity it shows us that using AI powered tools it is automatically able to detect any kind of threat, recognise pattern and give immediate response to the threat. Along with this it is also helping in analysing large sets of data. Machine learning algorithm can recognise various new threats by improvising the threat detection accuracy.

Similarly, when we look upon Cloud Security has gained more popularity as usage of Cloud Computing and storage has become the need of the hour. Sturdy and rigid safeguard measures have been adopted by the cloud service providers to protect the data. Security measures like controlling of all the access, monitoring tools and encryption. Along with this all the organisations are now trying to focus on protecting of cloud-based data. Due to the sudden rise of Internet of Things Security devices it has led to formation of new security challenges. Protection of interconnected network of systems, sensors, and devices are the major focus areas of IoT. For IoT devices, efforts have been initiated so that security standards can be implemented, establish encryption mechanism and safe firmware updates (Morovat & Panda, 2020).

Partnership among different organisation, institutions, Public Sector, Private Sector, and Industries along with threat intelligence information sharing has made early detection of cyber risk possible, responding to cyber threat and reduction of cyber threats more achievable. All Exchange of threat indicators, best practices and attack techniques are facilitated by the Information sharing platforms and initiatives. Severe usage of technology along with cybersecurity has raised the question of privacy which has resulted in the development of privacy-enhancing technologies. To protect sensitive and confidential data few techniques are used such as Differential privacy, Homomorphic encryption, and securing multi-party computation. But this technique would not affect the analysing or processing part. User awareness and training programs are now mostly focused by the organisation after identifying the role of humans in cybersecurity. To reduce the chances of cyber-attack the users should be educated so that they can deal with different attacks like phishing, social engineering and thus adopt safe online practices (Zaidi, 2021).

Cryptographic algorithm is under severe potential threat because of the rise of quantum computing. To this currently efforts are put in research and development for creating quantum resistant cryptographic algorithm which will act as a resistant to attack from quantum computers. Automation and Orchestration technologies are continuously in use to streamline and accelerate security operations. To enable speedy and more effective response, security automation tools can be used which would help in automatic threat detection, carry out routine security task, and interact with different security systems. All over the world, government of each country have implemented laws and regulations to protect confidential data, critical infrastructure and ensure compliance.

Artificial Intelligence has shown advancement and has come up with new technologies like Machine Learning, Deep Learning, Robotics, Healthcare, Computer vision, Autonomous Vehicle, Natural Language processing, Smart Assistants, Finance, Generative Models, and Reinforcement learning. Machine learning technique and algorithm has the capabilities to enhance the system to learn directly from those data which is fed to the system and thus improving the performance over the time. To focus on areas like natural language processing and computer vision machine learning came up with a subset of it called Deep learning. AI has played crucial role in healthcare sector by giving accurate and better results than the doctors. In case of dermatologist, it can analyse and predict skin allergies with time but Artificial Intelligence by using machine learning it can analyse, assist early detection, predict disease outcomes and other skin problems in seconds by using medical images. AI model is used in virtual assistant to understand and answer to the questions of the human such as Google Assistant, Siri, and Alexa.

This application is designed by using Machine Learning and Natural Language Processing. New music, text and images are created by using generative models like Variational autoencoders and Generative adversarial networks. AI is vastly used in robotics as AI technologies make the robots efficient enough to perform complicated task like interacting with humans, healthcare, agriculture, navigate environments and manufacturing. During covid robots were used to do contactless deliveries of medicine to the covid suffering patient and were also capable to check temperature. Autonomous vehicles are developing because of AI enabled technologies implemented by them. With the use of AI algorithm, autonomous vehicle without the intervention of human can understand the surrounding and, makes decision. AI has developed in almost all the sectors and thus using AI enable systems it can procure better outcomes.

4. RECENT DEVELOPMENTS IN AI DRIVEN CYBERSECURITY TOOLS

While AI-driven cybersecurity tools are increasingly prevalent in the commercial sector, cutting-edge research is pushing the boundaries of what's possible, revealing transformative potential beyond incremental advancements. Investigating these frontier concepts can unlock new levels of awareness, insight, and control, potentially revolutionizing security system design and threat management.

Adversarial Machine Learning

A key focus in recent research is enhancing threat detection models against adversarial attacks—sophisticated manipulations designed to mislead algorithms. Researchers are developing advanced techniques to expose these vulnerabilities, retraining models with adversarial examples to build greater resilience. This approach not only strengthens defenses by highlighting weaknesses but also generates insights from creative attack simulations that traditional methods might miss. By emulating persistent, simulated threats, adversarial machine learning aims to fortify AI systems against the evolving landscape of algorithmic warfare.

Meta-Learning

Adapting cybersecurity models to new threats often involves time-consuming data preparation, which delays response to emerging attacks. Meta-learning offers a solution by enabling models to learn how to learn from diverse security tasks. This approach allows for rapid development of specialized detectors from small datasets, facilitating swift adaptation to zero-day threats. Meta-learning also enables automated customization, scaling personalized security measures to protect against evolving threats efficiently.

Multi-Agent Systems

With the exponential growth of data and complexity across varied technology and business environments, centralized analytics can struggle to keep pace. Multi-agent systems address this challenge by distributing tasks among specialized AI agents. These agents operate autonomously, handling data collection, modeling, and response in parallel, which enhances scalability and contextual understanding. By coordinating outputs from multiple agents, organizations gain comprehensive situational awareness and can deploy tailored responses more effectively. This approach creates a dynamic, widespread security framework that surpasses the capabilities of traditional monolithic systems.

5. OPPORTUNITIES OF AI IN CYBERSECURITY

The AI mechanisms employed in cybersecurity include machine learning, deep learning, deep neural networks, natural language processing, supervised learning, unsupervised learning, and reinforcement learning. Machine learning, a subset of artificial intelligence, enables computers to learn from experience and direct data through computational methods. Three types of Machine learning are Supervised learning, Unsupervised Learning and Reinforcement learning (Arulkumaran, Deisenroth & Brundage, 2017). Next, we will understand about how AI is being used in cybersecurity to stop, prevent, detect, determine, and respond to the cyberattacks.

5.1 Approaches of AI in Cybersecurity:

- A user's identification can be easily stolen by doing cyber-attack such as phishing. Sensitive data or information and disruption or destroying of the data can take place when the hackers send a malicious link to the user and upon clicking the link malware gets activated immediately thus in resulting the access of all the confidential data. Examples of phishing attacks such as Dictionary attack and Bruce-force attacks. Now, it can be easily detected whether the email or website is a phishing trap with the help of AI-based systems.
- A new phishing detection technique has been proposed by Researchers from the University of North Dakota by using machine learning, reinforcement learning and modified neural network in return which will help in differentiating between the phishing or legitimate email by understanding the structure of the email. By using risk minimization approach and Monte Carlo algorithm Fengetal tried to identify the phishing websites (Feng et al., 2018). The Mimecast's Cyber Graph gave three major capabilities related to the prevention of phishing attack or impersonation which were blocking trackers, identifying pattern, and alerting users.
- When the IT companies are having large set of data inventory it is difficult to analyse each component and give security thus making the breach risk more complicated. To detect any sort of breach, risk or attacks AI tools can be used. These AI-models can be used for early detection of hacking and send alerts to create security this results in protection of the organisations from suffering huge losses, prepare themselves from future attacks and create cyber resilience. This entire concept is known as Breach risk Prediction.

- Spam emails are unnecessary mail which keeps on popping up in the email box. These emails may contain obscene, vulgar, or unsecured content which may disrupt the working of the computer system leading to security issues known as Spam detection. AI -based algorithm can be used to separate legitimate email from the spam mails (Feng et al., 2016).
- Malware Detection is a type of cyberattack which uses worms, trojan horses and viruses to create disruptions. Malware attacks can be prevented and identified by using AI -algorithms. AI models by understanding the nature of the known malware can detect the new variants and block them from further disruptions. AI by using data mining and Machine learning can easily identify, classify, and detect the malware attacks. One scholar to detect unknown malwares had used support vector machine as ML classifiers and k-nearest neighbours. Bio-inspired computation was also used by the researcher to prevent malware and to increase the efficiency of the malware detection genetics algorithm was implied (Morel 2011).
- Automated Response system can be developed by using AI models so that it can give quick responses to any cyberattack in real time. These models are designed in such a way that it can identify, mitigate, and prevent attacks and damages by reducing the response time (Eperjesi 2022).
- Privacy and data protection can be made efficient by using AI. AI will help in automatic data protection process and analyse the future threats of data breach. They use different types of privacy measures which will protect all the sensitive and confidential data but giving access to all the required data.
- Vulnerability assessments also use AI which helps in analyses and scan of software potential weaknesses. Thus, before the cybercriminals exploit or take undue advantage of the vulnerabilities the organisation can detect and fix the vulnerabilities. AI can automate vulnerability scanning, identification, and prioritisation process. Since the potential impact of AI powered models can interpret and analyse potential vulnerabilities and system configuration. This model will help the security team in efficient and early detection of vulnerabilities and fix the same.
- For fighting cybercriminals and detecting of complicated scams and hacks the AI has developed an advanced and new malware analysis tool. Even AIOps Platforms tries to fight cyber threat even before they take place. It takes less time to detect and block any such threats.
- Software exploitation can take place when the vulnerabilities are week and easily detectable and exploited by the attacker. Some software developments are costly and difficult to release in the market such software vulnerabilities are integer overflow, cross site during design and development process, SQL Injection, buffer overflow, and cross site scripting. Thus, detecting, and fixing problem goes simultaneously. It is difficult for the computer to read each code, if they are introduced to the vulnerabilities then the software bugs can be easily fixed.

- Normal computers do not have the capabilities to do so thus AI-based computer system can easily fix the software bugs by detecting them. Bruce Schneier says that the internet can be regarded as the most complex machine mankind ever built. We barely understand how it works, let alone how to secure it. (Moral 2011).
- Improvising AI application security using AI techniques. They mainly focused on the use of probabilistic reasoning, Bayesian algorithm and knowledge-based system which helps in detecting of software exploitation (Zhu, 2018).
- Authentication mechanisms are being improvised with the help of AI tools and techniques. AI technique can be seen in voice recognition, facial recognition and biometric authentication which makes them look sophisticated, secure, and difficult to scam. These AI-powered authentication methods provide more sturdy and secure access control resulting in reduction of risk of identity theft and unauthorised access.
- Human mostly get tired doing complex and strategic works; AI can release humans from doing complex and repetitive work by performing routine and repetitive task in cybersecurity operations by doing this it will free the cybersecurity expertise and allow them to focus on higher value task that requires human knowledge and expertise. It will result in reduction of human error and improvising operational efficacy. Rather than using traditional human approaches AI model having the AI technique can be used for handling huge amount of volume of data in large scale. This scalability enables organisations to provide immediate response to security mishaps, manage and analyse huge volume of data set and intense monitoring capability.
- The usage of Artificial Intelligence in cybersecurity it shows us that using AI powered tools it is automatically able to detect any kind of threat, recognise pattern and give immediate response to the threat. Along with this it is also helping in analysing large sets of data. Machine learning algorithm can recognise various new threats by improvising the threat detection accuracy.

When we are looking at the interconnection of cybersecurity and Artificial Intelligence it is visible to us that both are mutually benefited of each other. While the focus of cybersecurity is to provide protection and assure safe and responsible development of AI technology, on the other side AI plays a pivotal role in making the cybersecurity defences much stronger and efficient. There are few interconnections such as AI for threat detection and prevention, Behavioural Analytics, Adversarial machine learning, Privacy and data protection and Vulnerability management. Machine learning and deep learning are AI-powered technologies which are repeatedly used in cybersecurity for the detection and prevention of cyber threat. With the help of AI algorithm patterns can be easily recognised and it will also detect inconsistency that may lead to any malicious, suspicious, or potential cyber-attacks. This helps in fast reaction and responds to any potential cyber threat thus, by improving the overall efficacy and working of cybersecurity defences. AI strengthen cybersecurity by providing improved risk management, advanced threat detection and faster incident response. In return integrity, security, and privacy of all the sensitive data and AI system is taken care by cybersecurity (Rehman 2022).

6. CHALLENGES OF AI IN CYBERSECURITY

There are lot many opportunities which can be seen when AI and Cybersecurity works mutually. But still there are few gaps which are:

- **Adversarial attack:** Specially crafted data or malicious data are added to the AI model which would manipulate and deceive the model this attack is known as an Adversarial attack. They have this ability to exploit vulnerabilities in AI algorithm and circumvent security measures. To cause false positives/negatives or to detour AI-based security measures the attackers purposefully create design inputs which would lead to an attack. Research is still going on for developing a sturdy defence against adversarial attack and this is also an ongoing challenged faced by AI and Cybersecurity (Anthi, 2021).
- **Data Quality and Bias:** AI powered cybersecurity system is developed using AI algorithms. This algorithm is trained with large amount of data which would include structured data, unstructured data, different qualities of data and bias data. This data can create huge impact on the security system. When the algorithm is trained with biased inputs then it would give results which are inaccurate, incomplete, discriminatory, and uneven in nature. For ensuring unbiased, complete, and accurate cybersecurity applications, addressing the bias and diverse representative datasets are needed (Jalil, Siegel and Madnick, 2019).
- **Explainability and Transparency:** Deep learning algorithm has been incorporated in AI models which is mostly complex in nature and very difficult to understand and interpret. Thus, this becomes a biggest challenge as it is not clear enough and it is unable to explain as in how AI system is able to make decisions or detect potential biases and vulnerabilities. To achieve trust and accountability it is required to build an AI model which is explainable and transparent in the relation to cybersecurity. When we are dealing with Cybersecurity and AI it is critical to justify and explain any reasoning behind the working of an AI based security decision as it always includes confidential and sensitive data. To understand how AI models give decision or detects malicious activities, explainability becomes an important criterion. But for now, it is one of the challenges which is faced by Cybersecurity and AI.
- **Scalability and resource requirements:** Processing power and Computational resources plays a major role in implementation of AI in

Cybersecurity. The analysis of vast amount of data based on real time is only possible because of Computational resources. Huge datasets and high-speed network traffic is needed along with the Computational capabilities and infrastructure must fulfil the high demand of AI powered Cybersecurity. This is a challenge which is faced by AI and Cybersecurity.

- **Ethical Consideration:** Privacy invasion, possibility of discrimination and automated decision making takes place when AI is incorporated in Cybersecurity. It is difficult to assure that ethical use is taking place or not. Weather they are looking upon the ideals of fairness, transparency, openness, and accountability.
- **Unforeseen Vulnerabilities:** Incorporation of AI into cybersecurity may led to potential new attack or vulnerabilities. When there is an unforeseen vulnerability, it may end up in targeting AI models and algorithm which may lead to unauthorised access to AI models, data poisoning and model evasion. To safeguard and maintain AI-powered cybersecurity system's integrity and security it is important to understand and address the situation.
- **Data privacy and Security:** AI models and system should be incorporated in the Cybersecurity measures for better functioning of the measures like data encryption, secure data handling practices and access control to prevent unauthorised access and data breaches. For the proper functioning of AI, it is dependent on huge amount of data including confidential and sensitive data. The privacy issue arises when the collection and processing of these data takes place. Suitable data protection measures should be taken by the organisations to protect the confidential data.
- **Laws, Standards, and regulation:** There is a rapid development of AI in Cybersecurity, which is unable to pace with the existing consolidated regulation and standards. To ensure responsible, ethical, and secured implementation of AI technology appropriate guidelines and safeguards must be established. As the technology is growing day by day, the consolidate laws and technology are unable to match each other. Regulations, relevant laws, and company standards should be complied when AI is incorporated into Cybersecurity.

7. REGULATORY BODIES RELATED TO CYBERSECURITY

- Cyber Regulations Appellate Tribunal (CRAT) was established by Central Government of India under section 62 of the Information Technology Act,2000. The major intention behind setting up of this tribunal was to give this authority, power to act as a chief governing body which will be involved

in work such as finding of facts, collection of cyber evidence and witness examination. Its functioning as well as jurisdiction is very much restricted. CRAT gets all its power from Code of Civil Procedure and Civil Court such as the power to review the final judgement of court, by producing affidavits they can receive evidence, make sure all the evidence should be submitted in the court, can issue summon for witness examination and can also declare, approve or dismiss defaulter's application as ex-parte (The Intact one, 2023).
- Insurance Regulatory and Development Authority (IRDAI) regulates and controls the insurance sector of India. Recently, in 2022 they mainly focused on the security areas related to the insurers. Thus, bringing up an improvised guidelines related to cybersecurity. Data leaks, fraud related to transactions, intellectual property rights infringement and ransomware attack are the major risk areas which are focused by IRDAI. Strict risk assessment plan is essential for insurance firms to improve mitigation methods of internal and external threats, different types of fraud as well as to prevent ransomware attacks and thus to implement a sturdy business plan. Fine beyond 1 Lakh rupees can be imposed on both the insurers and businesses if they are involved in any kind of violation.

Using the new Information and Cyber Security for Insurers Guidelines, the IRDAI:

- Mandates that insurance companies appoint a Chief Information Security Officer (CISO)
- Establishes an information security committee
- Develops plans for managing cyber crises
- Creates and implements cybersecurity assurance programs
- Enforces proper data protection methods
- Maintains processes for risk identification and risk mitigation

A national based nodal agency was set up in 2004 to collect, analyse, forecast, and disseminate non-critical cybersecurity particularly on phone-based incidents. The agency is responsible for cybersecurity incident reporting and notifying to other government agency and issuing them safeguards and guidelines for controlling, managing, and preventing any sort of incident related to cybersecurity. This nodal agency is known as Computer Emergency Response Team (CERT-In, 2023).

The missing mobile phone is a saleable commodity in case it is in working stage and the IMEI (International Mobile Equipment Identity) number is essential for reselling the equipment. CERT has launched a web-based portal where the missing mobile phone is informed and blocked. Thereby the phone is made invalid by blocking the IMEI number. This is a case of physical security for the mobile phones. Information Technology (The Indian Computer Emergency Response Team and Manner of Performing Functions and Duties) Rules, 2013 provides all the rules and regulations related to CERT-In.

"CRET-In is operational since January 2004. The constituency of CRET-In is the Indian Cyber community. CRET-In is the national nodal agency for responding to computer security incidents as and when they occur. India does not have a consolidated act for cybersecurity, it follows Information Technology Act and different rules and regulations to promote cybersecurity. Indian Computer Emergency Response Team regulates the IT Act of 2000 which got passed by the parliament of India, it safeguards law of Indian cybersecurity, e-commerce, private sector, cybercrimes, e-banking, data protection and e-governance. Under Section 72A of the IT Act fine of Rs. 5,00,000 or up to three years of imprisonment or both can be penalised on any person or intermediaries who without the consent of the owner, with an intention to cause damage discloses confidential data. For example, under section 43A of the IT Act, Indian Businesses and organisations must have reasonable security practices and procedures to protect sensitive information from being compromised, damaged, exposed, or misused.

The IT Act was amended in 2008 and Information Technology Amendment Act, was also passed in the same year. This amendment mostly provided the improvised framework and guidelines for cybersecurity in India. This improvision was more updated and clearer and included validation and authentication of electronic Signature, talks about cybercrime in-depth, Identifies and controls the intermediaries, came up with new set of rules and regulations regarding digital signature, protects e-payment, electronic transaction, decrypting and controlling of electronic records, protects computer from unauthorised use, protects all the private data, sensitive data and information from identity theft, DDoS attacks, malware and phishing, compels the companies to focus and implement on practicing data security and making them liable for any breach of data. Up to 10 years of imprisonment can be imposed on the offenders committing any offence related to cybercrimes. This act has a flaw which is the section 69 of the IT Act, 2008 which says that the Indian government holds the power to surveillance, intercept, block, decrypt or remove any content. This may lead to arising of serious issues related to privacy.

- Information Technology (Reasonable Security Practices and Procedures and Sensitive Personal Data or Information) Rules 2011 (Privacy Rules) plays a major role in the regulation of cybersecurity laws in India. Few things like cheating, update penalties and fees in relation to the violation of cybercrimes, slander, reregulation of intermediaries and publishing of private images with-

out consent were included in the amendment of IT Rules 2011 (Mehndiratta, 2022).
- Reserve Bank of India Act 2018 was introduced to provide laws and regulations for Urban Co-operative banks, proper guidelines related to Cybersecurity and payment operators. It mostly aims to focus on banks security purpose such as making compulsory for all banks to formulate plans related to cyber crisis management, banks to conduct threat assessment audits on a regular basis, to provide protection against phishing and malware by implementing own mail domains with anti-protection technology, to formulate equal and similar guidelines for all banks and payment operators along with recovery measures and it is compulsory for all the banks to follow the rules and regulation for better and efficient functioning of the security system along with smooth working of the bank. Penalty up to 10 lakh rupees can be imposed for non-compliance of the laws by the bank and financial sector.
- Under the Ministry of Finance, Security and Exchange Board of India (SEBI) was introduced in the year 1988 to provide securities to commodity markets. All the needs such as market investor, market intermediaries, issuers of securities along with the customer data, protecting data and transaction are ensured by SEBI. The Ministry of Electronics and information Technology, CERT-In, Department of Telecommunication and National Cyber Coordination Centre are all agencies which are well connected with SEBI. To look after the cybersecurity rules and regulation six committee panel is created which advises SEBI regarding the requirements of cybersecurity. Few organisation-like mutual funds, stockbrokers, Asset management companies and stock exchange comes within the scope of SEBI and is applied on them. Penalties of about 20,000 per day is implied on the companies for non-compliance and this penalty continues till the time compliance is reached.
- To protect the public organisation and the private organisation from cyber-attacks the Department of Electronics and Information Technology brought security safeguards and guidelines in the year 2013 and thus, came up with National Cyber Security Policy 2013. To make policies wherein the cyber ecosystem of India can be protected along with that it aims to provide training and skill development to IT expertise professional who are over 500,000 in number, over the span of five years. The main objectives of National Cyber Security Policy 2013 are "to protect information and information infrastructure in cyberspace, build capabilities to prevent and respond to cyber threats, reduce vulnerabilities and minimize damages from cyber incidents through a combination of institutional structure, people, processes, technology and cooperation". To further improve the efforts of cybersecurity National Cyber Security Strategy of 2020 was formulated. This plan is still pending for re-

view in the National Security Council Secretariat. The aim of the plan is to improvise the audit process and quality along with this to protect and prevent the policymakers, stakeholders and corporate leaders from cyber terrorism and cyber mishaps in the cyberspace.
- In 1860 the Indian Penal Code was established, and it includes provisions related to cybercrime which amounts to all offences committed through electronic sources, offence like defamation, fraud, cheating, and extortion. These cybercrimes do not have any recognition in the IT Act.
- On 16th January 2014 under section 70A of the IT Act, 2000 the National Critical Infrastructure Protection Centre was created by Indian Government in New Delhi. To look after and protect the Critical Information infrastructure this national nodal agency was appointed by the Government of India. It comes under Prime Minister's Office (PMO) and is also known as National Technical Research Organisation. The cybersecurity has been divided into two parts by the Indian Parliament one is the "Non-Critical Infrastructure (NCI) which is dealt by CRET-In, and the other is dealt by NCIIPIC that is "Critical Information Infrastructure". Banking, Insurance sector, Government, Power and Energy, Transportation, Financial services, Telecommunication and information, Strategic and public enterprise are the critical sectors which requires monitoring and reporting by the NCIIPC to critical information infrastructure.
- The Indian Parliament defines Critical Information Infrastructure (CII) as "facilities, systems, or functions whose disruption or destruction would severely impact national security, governance, the economy, and the social well-being. In August 2021, the Revamped Distribution Sector Scheme was sanctioned by the Indian Government to improvise the DISCOMs (Distribution Companies) operation by including AI-based solution to the Cyber infrastructure to increase its efficacy.

8. INDIAN LAWS ON AI AND CYBERSECURITY

Due to the growth of technology and AI, fundings are being initiated by the union government for speeding up the research related to AI, providing skills, knowledge, and training in the field of AI which would result in experts. This funding towards the development of AI is done under the shade of Digital India Programme. The Union Government's Digital India Initiative is aimed at transforming India into a 'digitally empowered society and knowledge economy.' To fit in into the Make in India programme, the Government of India is currently focusing on made in India AI technologies to promote AI in a global scale. Currently, India does not have any specific consolidated rules and regulations related to AI and data protection, but it is relying on the Information Technology Act where in under section 72A and 43A personal information can be protected along with this it gives similar compensation as to GDPR for disclose of personal information in an improper way.

NITI Aayog (Policy Commission) initiated in 2018 few programs which were based on AI applications. This Commission in February 2021 released "Responsible AI" and in August 2021 it released "Operationalising Principles for Responsible AI" both are AI strategy document for India. To analyse and put emphasis on multiple issues of AI related to ethics, the Ministry of Electronics and Information Technology has constituted four committees which are along with this are also focusing on Personal Data Protection Bill,2019. Though this bill is not directly related but can be helpful and implemented for sharing of data, collecting of data or processing of data. Nowadays, the industries are mainly trying to provide better knowledge about AI to their labours and thus AI is expanding (Yashi, 2023).

The National Strategy for Artificial Intelligence will help in covering all other sectors which are not covered or unidentified by this commission. This strategy focuses on privacy, ethics, and security along with that it promotes research, development and adaptation of technologies related to AI. Few regulations can affect the use of AI, when you investigate specific sectors having their own regulations like healthcare, telecommunication or banking they use AI. So now, if we see Reserve Bank of India, it has already sanctioned regulation to make use of AI for protecting the customers and focusing on risk management (Marda, 2018). The IT Minister, Mr. Ashwini Vaishnaw spoke about the growth and development of AI and how the government has taken steps to implement AI in daily life work along with it he has also showed his concern on formulating no new laws related to AI in the country. According to him the government is neither interested in planning to regulate the growth nor create any new laws related to AI. The written submission to the parliament says that "The Government is not considering bringing a law or regulating the growth of artificial intelligence in the country".

The IT Minister also gave a statement in which he spoke about NITI Aayog which says NITI Aayog has released a series of papers on Responsible AI for All. Despite this, the government is not currently planning to introduce legislation or regulations to govern the growth of artificial intelligence in the country. The IT Minister, Mr. Ashwini Vaishnaw answered "plannings are being made as how to harness the potential of AI to offer personalised and interactive citizen-centric services through digital public platforms" when asked with the questions on the regulation of AI. The IT Minister also spoke about the other issues associated with AI were in he said that "AI poses ethical concerns and risks due to issues like bias and discrimination in decision-making, privacy violations, lack of transparency in

AI systems, and questions about responsibility for harm caused by it. These concerns were highlighted in the National Strategy for AI (NSAI) released in June 2018".

The Ministry of Electronics and IT in collaboration with CDAC is coming up with a new project which is focusing on providing of "common computing platform for AI research and Knowledge assimilation" called AIRAWAT that is AI Research, Analytics, and Knowledge Dissemination Platform. IT Minister also stated that "The Proof of Concept (PoC) for AIRAWAT is developed with a 200 petaflops Mixed Precision AI Machine, designed to be scalable to a peak performance of one AI Exaflop." National Knowledge Network in collaboration with AI computing infrastructure will be implemented in scientific communities, technology innovation hubs, industry and startup institution and research labs. These were the few points emphasised by Mr. Ashwini Vaishnaw, the information and Broadcasting, Electronics & Information Technology, Government of India.

The rules and regulations are given in National Cybersecurity Policy, 2013 to safeguard cyber arena in India. This policy is mainly aiming for the strategies, and objectives to ensure the cybersecurity infrastructure and secure information's. This act specifically is not focusing on the protection of AI but indirectly it can be related. To rise against the challenges of cybersecurity the collaboration and partnership of industry, international organisation, government, and academia are needed. These few things were specifically emphasized by this policy.

The Personal Data Protection Bill, 2019 is still an act and has still not been enacted in India. It would specifically not help AI, but it would help in data handling and access. This bill would incorporate those data which would be relevant to both AI and cybersecurity which are data localisation, individual rights, provisions related to consent and cross border data transfers. Few regulations can affect the use of AI, when you investigate specific sectors having their own framework and regulations like healthcare, telecommunication or banking which would indirectly talk about the interconnection between AI and cybersecurity.

9. GLOBAL LAWS ON AI IN CYBERSECURITY

USA

The AI in Government Act of 2020, introduced during the 116th Congress, was approved by the House of Representatives on September 14, 2020, but did not advance to become law. Despite its failure to pass, the Act sparked important discussions about the responsible development and use of AI in the public sector, with elements of the proposed legislation influencing other executive orders and policies. The Advancing American AI Act of 2022 aims to promote the growth and application of AI while aligning with core U.S. principles such as privacy protection, civil rights, and civil liberties. Initially introduced in the Senate in April 2021, the Act was incorporated into the National Défense Authorization Act for Fiscal Year 2023 and took effect on December 23, 2022, with a grace period. Despite its significance, the Act has some limitations, including vague definitions of U.S. principles and regulations, a narrow focus on public procurement, limited scope, and potential bureaucratic complexity.

In addition to federal efforts, several U.S. states have established their own AI-related regulations. California, for instance, enacted the California Consumer Privacy Act in October 2019, which was later expanded into the California Privacy Rights Act, effective January 1, 2023. On the federal level, several organizations oversee AI research, development, and implementation. The National Institute of Standards and Technology (NIST) provides recommendations for managing AI risk, emphasizing high-quality data, transparency, and fairness. The Federal Trade Commission (FTC) has a specialized department dedicated to monitoring AI and emerging technologies to prevent fraudulent and unfair practices. Additionally, the Artificial Intelligence Capabilities and Transparency Act, enacted in December 2021, aims to enhance transparency in government AI systems, reflecting the National Security Commission on AI's recommendations (Hasan, 2024). Regulation such as California Consumer Privacy Act (CCPA) and the General Data Protection Regulation (GDPR) of the European Union (Pupillo et al., 2021).

CHINA

In January 2022, China introduced the Provisions on the Management of Deep Synthesis in Internet Information Services, which took effect on January 10, 2023. This law addresses a range of virtual content—including text, speech, biometric, and non-biometric data—and aims to combat deep fakes and regulate activities related to deep synthesis technologies. It mandates that providers implement safeguards that do not impede user access while ensuring compliance with regulatory requirements (Hasan, 2024).

On July 13, 2023, the Cyberspace Administration of China (CAC) and six other Chinese authorities announced the Interim Measures for the Management of Generative Artificial Intelligence Services. These measures, which came into effect on August 15, 2023, focus on fostering platform development, independent innovation, and international exchange in generative AI. They also stress the importance of acceptable oversight for AI technologies. Article 20 grants Chinese authorities the power to regulate both domestic and foreign generative AI platforms operating in China, while Article 23 outlines the framework for foreign investment in China's generative AI sector.

AFRICA

The African Union (AU) has been actively working with its 55 member states to enhance governance across the continent, with a particular focus on artificial intelligence (AI). As part of these efforts, the AU has established a dedicated working group on AI, created an Africa-specific AI regulatory blueprint, ratified Resolution 473, and adopted the Malabo Convention. This convention, which came into force in June 2023 after a nine-year approval process, is the first binding regional treaty outside Europe focused on privacy and personal data protection. It provides a comprehensive framework addressing personal data protection, electronic commerce, cybersecurity, and cybercrimes within Africa. The Malabo Convention introduces several fundamental rights for individuals, such as the right to be informed, access their data, object to data processing, and request data erasure, as detailed in Articles 9-23.

However, despite its pioneering status, the Malabo Convention has been criticized for lacking clarity on its applicability to data processors or controllers based outside the continent. This contrasts with the EU's General Data Protection Regulation (GDPR), which clearly covers such cases, particularly when processing involves providing goods or services to individuals in the EU or monitoring their behavior. Additionally, the Smart Africa Blueprint on Artificial Intelligence, a key component of the Smart Africa Initiative, offers a framework for African nations to develop and implement AI strategies. The Blueprint outlines the opportunities and challenges of AI in Africa and provides specific policy recommendations to optimize AI benefits while mitigating risks.

10. THREATS IN CYBERSECURITY

In today's interconnected digital environment, organizations and individuals face a wide range of evolving cybersecurity threats. The landscape is continuously shifting, with cyber-attacks growing more sophisticated and diverse. Weekly reports of viruses, hacking attempts, and phishing schemes highlight the ongoing challenges. Despite deploying security measures like firewalls, antivirus software, and filtering tools, many users remain vulnerable to emerging risks, which can be broadly categorized into malicious software, network attacks, and network abuses (Abawajy & Kim, 2019).

The advent of artificial intelligence (AI) has transformed cybersecurity by offering advanced capabilities for threat detection, prevention, and response. Early AI systems used rule-based approaches with predefined signatures to identify known threats. However, as cyber threats evolved, these methods became insufficient for detecting zero-day attacks and polymorphic malware. AI-powered solutions now employ machine learning algorithms to analyze large datasets, detect anomalous behavior, and predict potential threats. This evolution enables organizations to enhance their security posture and stay ahead of emerging cyber threats in real-time. As the digital landscape grows more complex, the field of cybersecurity must continuously adapt to address the shifting nature of threats. AI and machine learning are at the forefront of this battle, providing new tools and strategies to combat increasingly sophisticated and dynamic cyber-attacks (Sengupta & Ayan, 2019). The organizations can better prepare to defend against them, protect their digital assets, and maintain the trust of their stakeholders.

11. INDIAN CASE LAWS

In 1995 case called LIC vs. Consumer Education and Research Centre it majorly did not focus on Cybersecurity or AI, but the highlight of this case was "the obligation of organisations to protect customer data. The Supreme court ruled that insurance companies have a duty to maintain the confidentiality of policyholder's information and ensure its security."

Subramanian Swamy vs. Union of India in 2016, "The Supreme Court of India examined the issues of cybersecurity and the protection of personal data in the context of social media platforms. The court emphasized the need for safeguards against cyber threats and recommended measures to strengthen cybersecurity and privacy protection."

Ram Sewak Sharma vs. State of Jharkhand in 2017, "The Supreme Court of India emphasized the importance of cybersecurity and directed the government to ensure the safety and security of Aadhaar data, a unique identification system. The court highlighted the need for robust cybersecurity measures to protect sensitive personal information."

Union of India vs. K.A Najeeb in 2017, "The Kerala High Court addressed concerns related to the collection and use of personal data through mobile applications. The Court stressed the importance of data protection and privacy and emphasized the responsibility of app developers and service providers to ensure cybersecurity measures are in place."

12. GLOBAL CASE STUDIES - ILLUSTRATING RECENT CYBER ATTACKS

- **Solar Winds Supply Chain Attack**

In December 2020, a sophisticated supply chain attack exposed the vulnerabilities of network management systems through SolarWinds, a leading software provider. Hackers inserted malicious code into updates for SolarWinds' Orion software, which were then distributed to thousands of customers, including government agencies and major corporations. The attack demonstrated how compromising a trusted software vendor can lead to unauthorized access to highly secure environments, highlighting the urgent need for improved supply chain security measures.

- **Twitter Social Engineering Attack**

In July 2020, a significant social engineering attack targeted Twitter, compromising high-profile accounts, including those of Elon Musk, Barack Obama, and Joe Biden. The attackers manipulated Twitter employees into revealing access credentials and internal tools, which they then used to hijack accounts and execute a cryptocurrency scam. This breach exposed the vulnerabilities of social media platforms to human manipulation and emphasized the critical importance of robust security awareness training and effective protocols to counter insider threats and social engineering tactics.

- **Colonial Pipeline Ransomware Attack**

In May 2021, the Colonial Pipeline, a vital infrastructure supplying nearly half of the fuel to the East Coast of the United States, was targeted by a major ransomware attack executed by the cybercriminal group Darkside. The attack infiltrated the pipeline's network, leading to a system-wide shut down that triggered significant disruptions in fuel supplies and caused widespread panic buying. This incident underscored the vulnerabilities in critical infrastructure and the potential cascading effects of cyber-attacks on essential services and national security.

13. ETHICAL CONSIDERATIONS OF AI

- **Bias and Discrimination**

One of the most pressing ethical concerns with AI is the risk of bias and discrimination. AI systems mirror the biases present in their training data. Consequently, if the data used to train these systems is skewed, the AI's decisions will also be biased. This has led to issues such as facial recognition technology being less accurate for women and people of colour and biased algorithms in criminal justice that exacerbate wrongful convictions. To combat these issues, it's crucial to ensure diverse and representative training data and to regularly audit AI systems for bias, making necessary adjustments to mitigate these biases.

- **Transparency**

Transparency is another critical ethical concern. AI systems can be complex and opaque, making it difficult to understand how they make decisions. This lack of transparency is particularly troubling in high-stakes areas like criminal justice and healthcare, where decisions can have significant consequences. Efforts are underway to enhance AI explainability, including developing methods to visualize decision-making processes and implementing regulations that mandate disclosure of how AI systems operate. These measures aim to ensure that AI decisions are understandable and accountable.

- **Privacy**

AI systems often require extensive personal data, raising significant privacy concerns, especially in contexts such as surveillance and targeted advertising. Protecting individual privacy involves giving people control over their data and ensuring companies are transparent about data collection and usage practices. Regulations like the General Data Protection Regulation (GDPR) in the European Union play a crucial role in safeguarding privacy by requiring explicit consent for data collection and use.

- **Autonomous Weapons**

The use of autonomous weapons, such as drones capable of identifying and attacking targets without human intervention, presents significant ethical challenges. The primary concerns are accountability and the risk of unintended consequences. Ensuring the responsible use of autonomous weapons is crucial, and many advocate for international bans and the development of regulations and protocols to govern their deployment and ensure ethical use. By prioritizing these ethical issues, we can harness AI's benefits while mitigating potential risks and ensuring that AI technology contributes positively to society (Ozden, 2023).

14. RECOMMENDATIONS

The organizations should develop comprehensive policies and procedures for responding to cyber threats to minimize the damage caused by these attacks. To safeguard data, systems, and network from all the malicious attack, cybersecurity comprises of technologies, procedures, and responses. Cybersecurity can help protect organisations from malicious actors trying to access confidential data and destructive operations. To protect the confidentiality of the data the companies take cybersecurity measures like firewalls, anti-virus software, security protocols, authentication method, encryption techniques, and intrusion detection systems by making sure only authorised organisation having access to confidential information. For best practices the staffs should be well educated and given full knowledge about the technology of cybersecurity.

- **Adversarial defence mechanism:** Different mechanism should be developed and adopted so that adversarial attacks can be detected and countered. Mechanism like input sanitization, adversarial training, and detection of peculiar activities. Investments should be done in the research and development sector which can encourage the researchers to do continuous research which will result in easy understanding of the vulnerabilities. Due to their ongoing search the researchers can also predict the upcoming attacks in the future.
- **Unbiased and Diverse high-quality data:** It should be ensured that the AI-model should always be trained with unbiased data, high quality data, diverse datasets, and authenticated data. By using strategies like implementation of strict data validation, pre-processing techniques and authenticated diverse data collection the biased and low-quality data can be easily pointed out and corrected.
- **Privacy preserving AI:** To safeguard the confidential personal data accessed by the AI models enhanced privacy techniques should be implemented. Techniques such as federated learning and differential privacy. To assure that the privacy has been well incorporated into the development of AI powered cybersecurity solutions principles like privacy-by-design should be implicated.
- **Ethical rules and regulations**: AI in Cybersecurity should be used ethically thus bringing up an ethical framework and governance mechanism can solve this issue. Implementing guidelines that can solve the issues of transparency, accountability, and fairness in an AI- based security systems. To make it more transparent and ethical stake holders should be a part of the decision-making process.

- **Regulatory compliance:** Should be up to date with the standards of industry, latest legislations and rules related to Cybersecurity and AI. For the secured and assured use of AI in Cybersecurity it is a need to develop compiled framework and standards for this both companies, industry organisation and regulating bodies should contribute resulting it better security regime.

15. CONCLUSION

Besides the advancement of Technology and AI it is becoming difficult to collect, store, process, and handle huge amount of data as AI requires huge amount of data to function. It is getting tough for the cyber analysts to monitor and control the volume, variety, and velocity across the firewalls. Due to the increasing use and access of personal data and sensitive data in the cyberspace there is a need of cybersecurity as it is resulting in a greater number of cyberattacks. Nowadays the firms, organisation and different industries are understanding the need of AI application as it would help them to overcome against cyberattacks. AI technique is getting adopted in cybersecurity, vastly by different organisations, especially who are working in data security, endpoint security and network security. The trend of leveraging machine and deep learning in cybersecurity is rising, with 61% of organizations acknowledging that they cannot identify critical threats without AI.

AI has the potential to strengthen the defences related to cybersecurity by automating mitigation, threat detection, responses, and analysing the patterns of vast data. For reducing the gap which is present between the AI and cybersecurity there is a need in adopting a robust and comprehensive approach. Fostering interdisciplinary collaboration, governance mechanism, implementation of privacy-preserving techniques, assuring of explainability and transparency of AI model, and sturdy AI system for testing and development are the few ways in which the gaps can be fixed. To keep up with AI and cybersecurity simultaneously it is important to update the rules and regulation frequently to stay relevant. The overall cyber threat can be resolved and evolved if AI is integrated well with Cybersecurity as it would mitigate and reduce the risk and protect the cyberspace along with this there is a need of proper rules and regulation for the safeguard of AI in cybersecurity.

When we see this entire paper, it can be concluded that both Cybersecurity and AI are complicated to understand and interpret because it is too much interconnected. In one side cybersecurity gets a lot of advantages from AI as AI provides advanced tools and techniques which increases the efficiency, detects any attack or threat before its occurrence and responds in real-time. Using AI, the potential attack can be predicted, and vulnerabilities can be identified easily. AI provides a lot of opportunities, but it also has challenges which must be corrected such as data privacy concern, biased algorithm, shortage of skill and knowledge, adversarial attack and many more but with the help of AI and cybersecurity in connection it can overcome these problems.

16. FUTURE SCOPE OF ARTIFICIAL INTELLIGENCE

- **Developing Robust AI Models:** AI models should be designed to continuously learn from new data, improving their accuracy and effectiveness over time. Incorporating adversarial examples in training datasets can help AI systems better recognize and defend against sophisticated attacks.
- **Enhancing Data Security:** Ensuring that data used to train AI models is encrypted can help protect it from unauthorized access. Developing protocols for secure data sharing between organizations can improve the quality and breadth of training data available.
- **Ethical AI Development:** AI systems should be transparent in their decision-making processes, and developers should be accountable for their systems' actions. Regular audits and updates to AI systems can help identify and mitigate biases, ensuring fairer outcomes.

REFERENCES

Abawajy, J. H., & Kim, T. H. (2019). Artificial intelligence and cybersecurity: Trends, challenges, and future directions. *Journal of Cybersecurity*, 5(1), 1–14.

Allahrakha, N. (2023). Balancing Cyber-security and Privacy: Legal and Ethical Considerations in the Digital Age. *Legal Issues in the Digital Age*, 4(2), 78–121. DOI: 10.17323/10.17323/2713-2749.2023.2.78.121

Anthi, E., Williams, L., Rhode, M., Burnap, P., & Wedgbury, A. (2021). Adversarial attacks on machine learning cybersecurity defences in industrial control systems. *Journal of Information Security and Applications*, 58, 102717. DOI: 10.1016/j.jisa.2020.102717

Arulkumaran, K., Deisenroth, M. P., & Brundage, M. (2017). *Deep reinforcement learning: a brief survey.* IEEE., DOI: 10.1109/MSP.2017.2743240

California Attorney General. (n.d.). *California Consumer Privacy Act (CCPA).* https://oag.ca.gov/privacy/ccpa

Cavelty, D. M. (2018), *The Routledge Handbook of New Security Studies,* 2.

Conversation on Intellectual Property and Artificial Intelligence. (2019), *WIPO,* 2, https://www.wipo.int/edocs/mdocs/mdocs/en/wipo_ip_ai_ge_19/wipo_ip_ai_ge_19_inf_4.pdf

Cyberspace Administration of China. (2021, August 25). *Regulation on the management of network data security.*https://www.cac.gov.cn/2021-08/25/c_1631480920680924.htm

Eperjesi, A. (2022), Automated Incident Response: Everything You Need to Know, https://securityboulevard.com/2022/10/automated-incident-response-everything-you-need-to-know

European Union General Data Protection Regulation (GDPR). (n.d.). *What is GDPR, the EU's new data protection law?*https://gdpr.eu/what-is-gdpr/

Federal Trade Commission (FTC). (n.d.). *Federal Trade Commission (FTC).*https://www.ftc.gov/

Feng, F., Sun, J., Zhang, L., Cao, C., & Yang, Q. (2016), A support vector machine based naive Bayes algorithm for spam filtering, *IEEE,* https://www.computer.org/csdl/proceedings-article/ipccc/2016/07820655/12OmNzWx0bb

Feng, F., Zhou, Q., Shen, Z., Yang, X., Han, L., & Wang, J. (2018). *The application of a novel neural network in the detection of phishing websites.* Springer.

Feng, F., Zhou, Q., Shen, Z., Yang, X., Han, L., & Wang, J. (2024). The application of a novel neural network in the detection of phishing websites. *Journal of Ambient Intelligence and Humanized Computing,* 15(3), 1–15. DOI: 10.1007/s12652-018-0786-3

Feng, W., Sun, J., Zhang, L., Cao, C., & Yang, Q. A., (2016). support vector machine based naive Bayes algorithm for spam filtering, *IEEE,* https://www.computer.org/csdl/proceedings-article/ipccc/2016/07820655/12OmNzWx0bb

Hasan, M. (2024). Regulating Artificial Intelligence: A Study in the Comparison between South Asia and Other Countries. *Legal Issues in the digital. The Age (Melbourne, Vic.),* (1), 122–149.

IBM. (n.d.). *What is cybersecurity?* https://www.ibm.com/topics/cybersecurity

Indian Computer Emergency Response Team (CERT-In). (n.d.). *CERT-In – National nodal agency for cyber security.* https://www.cert-in.org.in/

Indian Penal Code. 1860, No. 45, The Acts of Parliament, 1862 (India).

Information Technology Act, 2000 S. 62, No. 21, Acts of Parliament, 2000 (India).

Information Technology Act, 2000, S 72A 43A, No. 21, Acts of Parliament, 2000 (India).

Insurance Regulatory and Development Authority of India. https://irdai.gov.in/

IT Governance UK. (n.d.). *What is cybersecurity?* https://www.itgovernance.co.uk/what-is-cybersecurity

Jalali, M. S., Siegel, M., & Madnick, S. (2019). Decision-making and biases in cybersecurity capability development: Evidence from a simulation game experiment. *The Journal of Strategic Information Systems,* 28(1), 66–82. DOI: 10.1016/j.jsis.2018.09.003

Kaspersky. (n.d.). *What is cybersecurity?* https://www.kaspersky.co.in/resource-center/definitions/what-is-cyber-security

. LIC vs. Consumer Education and research Centre, (1995). AIR 1811, SCC (5), 482.

Marda, V. (2018). Artificial intelligence policy in India: A framework for engaging the limits of data-driven decision-making. *Philosophical Transactions. Series A, Mathematical, Physical, and Engineering Sciences,* 376(2133), 20180087. DOI: 10.1098/rsta.2018.0087 PMID: 30323001

Mehndiratta, M. (2022) Information Technology Act 2000, Ipleaders, https://blog.ipleaders.in/information-technology-act-2000

Ministry of Electronics and Information Technology, Government of India. (2013). *National Cyber Security Policy - 2013*.https://www.meity.gov.in/writereaddata/files/downloads/National_cyber_security_policy-2013(1).pdf

Mohammad, S. (2019). *Jalali & Michael Siegel, Stuart Madnick, Decision- making and biases in cybersecurity capability development: Evidence from a simulation game experiment*. Elsevier.

Morel, B. (2011), Anomaly Based Intrusion Detection and Artificial Intelligence, Intechopen, https://www.intechopen.com/chapters/14355

Morovat, K., & Panda, B. (2020). A survey of artificial intelligence in cybersecurity. In 2020 *International conference on computational science and computational intelligence* (CSCI) (pp. 109-115). IEEE. DOI: 10.1109/CSCI51800.2020.00026

National Critical Infrastructure Protection Centre. https://nciipc.gov.in/

National Cyber Security Policy. (2013), 1, 2-4, https://www.meity.gov.in/writereaddata/files/downloads/National_cyber_security_policy-2013%281%29.pdf

National Institute of Standards and Technology (NIST). (n.d.). *NIST – Cybersecurity*. https://www.nist.gov/

. Ozden, C. (2023). AI ethical consideration and cybersecurity. *International Studies in Social, Human and Administrative Sciences-I, 85*.

Pupillo, L., Fantin, S., Ferreira, A., & Polito, C. (2021), Artificial Intelligence and Cybersecurity, CEPS 2, https://www.ceps.eu/wp-content/uploads/2021/05/CEPS-TFR-Artificial-Intelligence-and-Cybersecurity.pdf

Rehman, S. F. (2022). Practical Implementation of Artificial Intelligence in Cybersecurity–A Study. *International Journal of Advanced Research in Computer and Communication Engineering*, 11(11). Advance online publication. DOI: 10.17148/IJARCCE.2022.111103

Securities and Exchange Board of India (SEBI). (n.d.). *SEBI – Protecting the interests of investors in securities*.https://www.sebi.gov.in

Sengupta, S., & Ayan, R. (2019). Ethical Considerations in Artificial Intelligence: A Perspective from India. *Computer Science and Information Technology*, 9(1), 21–29.

Shivanshu, (2024), What is Cybersecurity? Definition and Types Explained, Intellippat, https://intellipaat.com/blog/what-is-cyber-security/#no1

Sontan, A. D., & Samuel, S. V. Adewale Daniel Sontan Segun Victor Samuel. (2024). The intersection of Artificial Intelligence and cybersecurity: Challenges and opportunities. *World Journal of Advanced Research and Reviews*, 21(2), 1720–1736. DOI: 10.30574/wjarr.2024.21.2.0607

. Swamy S. (2016), Union of India, 7, SCC 221.

TechTarget. (n.d.). *Cybersecurity skills gap: Why it exists and how to address it.* https://www.techtarget.com/searchsecurity/tip/Cybersecurity-skills-gap-Why-it-exists-and-how-to-address-

The Intact One. (2023, April 8). *Cyber Regulation Appellate Tribunal.* https://theintactone.com/2023/04/08/cyber-regulation-appellate-tribunal/

The Personal Data Protection Bill. (2019). https://prsindia.org/billtrack/the-personal-data-protection-bill-2019

Tripathi, S., & Ghatak, C. (2018). Artificial Intelligence and Intellectual Property Law, 8, *Christ University Law Journal*, http://journals.christuniversity.in/index.php

. Union of India v. K.A Najeeb, (2021), 3 SCC 713.

U.S. Congress. (2021). *S.1353 – International Cybercrime Prevention Act.* https://www.congress.gov/bill/117th-congress/senate-bill/1353/text

U.S. Senate Committee on Armed Services. (2024). *FY25 National Defense Authorization Act executive summary.* https://www.armed-services.senate.gov/imo/media/doc/fy25_ndaa_executive_summary.pdf

Yashi., (2023). Artificial Intelligence and Laws in India, *Legal Service India*, https://legalserviceindia.com/legal/article-8171-artificial-intelligence-and-laws-in-india.html

Zaidi, K. (2021). Artificial Intelligence and Cyber Law, *Ipleaders*, https://blog.ipleaders.in/artificial-intelligence-cyber-law

Zhu, H. J., You, Z. H., Zhu, Z. X., Shi, W. L., Chen, X., & Cheng, L. (2018). *Effective and robust detection of android malware using static analysis along with rotation forest model.* Semantic Scholar. DOI: 10.1016/j.neucom.2017.07.030

Chapter 7
Ensuring Security in Vehicular Cyber Physical Using Flexray Protocol

Neha Bagga
Guru Nanak Dev University, India & Lovely Professional University, India

Sheetal Kalra
https://orcid.org/0000-0003-0694-7468
Guru Nanak Dev University, India

Parminder Kaur
https://orcid.org/0000-0003-1954-3390
Guru Nanak Dev University, India

ABSTRACT

With evolving automotive technology V2V communication will follow an evolutionary path as well alerts are provided to driver to maintain safety on road and take timely decision for received warnings. V2I helps vehicle to share information with roadside components of Intelligent Transportation Systems. All the communication above is susceptible to be intercepted and wrong messages can be communicated by exploiting the integrity, or act of cyber terrorism can be performed. Erroneous communication done by attackers can lead to opening of airbags while driving, giving wrong indication of turning of vehicle, which in turn can cause loss of human life, damage to vehicles. ECU's are the easiest target for the attackers to gain access into the vehicle as communication protocols like CAN, LIN, FlexRay, MOST and Ethernet are connected to the ECU's. In this chapter authors would be discussing the Flexray protocol specifications which can be exploited to perform attacks and the corresponding potential security and safety effects of these attacks and propose

DOI: 10.4018/979-8-3693-5728-6.ch007

some futuristic security protections.

1. INTRODUCTION

1. Vehicular cyber-physical systems (VCPS) are arrangements of vehicles equipped with integrated computational, sorting, and real capabilities, allowing them to interact with their internal and external environments. Contemporary vehicles contain more than 100 electronic control units (ECUs) and numerous sensors and actuators, which are connected through various communication networks such as CAN, LIN and FlexRay. These integrated components consistently provide continuous control, status monitoring, and safety systems in vehicles. As vehicles continue to advance towards greater levels of automation and connectivity, they are becoming increasingly complex digital real-world structures (FlexRay Consortium, 2005) (Mateus & Königseder, 2014).

Figure 1. Bridging gap between Physical and Cyber World using Vehicular Cyber Physical System

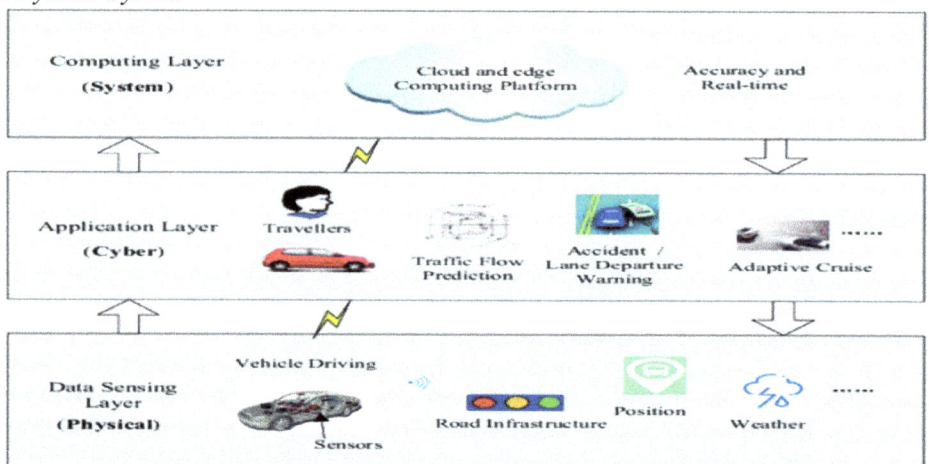

VCPS are safety-critical systems, as any security breaches can directly endanger human lives. At the same time, increased connectivity also exposes them to cyber attacks aiming to take control or manipulate the system. Research shows that currently prevalent bus protocols like CAN are vulnerable, allowing attackers to inject malicious messages or take over ECUs once they gain internal access. Hence security in VCPS is an immensely critical issue that needs to be handled at multiple layers across hardware, networking and software.

FlexRay is one of the newer generation automotive networking protocols designed to be faster, more reliable and secure than CAN. It was developed by the FlexRay consortium initially formed by BMW, Bosch, Freescale, GM, Philips in 2000. Key features that make FlexRay stand out include:

- **Determinism** - Messages are transmitted based on static, pre-determined schedules, enabling predictable communication patterns and timing guarantees, essential for control systems.
- **High speed data rates** - Each channel supports up to 10 Mbps bandwidth, nearly 10-100x faster than CAN bus. This enables data-intensive applications with high sampling rates.
- **Redundancy and fault tolerance** - Dual independent communication channels provide redundant paths between ECUs. Clock synchronization mechanisms make the nodes fault tolerant.
- **Stronger signalling, error detection** - Differential signalling makes it resilient to EMI noise. Extended CRC, coding mechanisms enable robust error detection. (FlexRay Consortium, 2005)

These attributes suit FlexRay for 'drive-by-wire' systems, advanced driver assistance, power train control and other compute-intensive VCPS applications requiring strong reliability, determinism and redundancy. It is already adopted in several production cars for features like electronic stability control.

However, FlexRay alone does not include extensive or dedicated security mechanisms. As connectivity and consequent cyber attack risks grow enormously, security has become an absolute imperative, demanding innovations spanning across communication protocols, hardware, algorithms and software components.

2. FLEXRAY PROTOCOL OVERVIEW

As vehicles become more dependent on electronics and software for essential functions like power control, braking, steering, and driver assistance, the need for specialized in-vehicle communications with reliability, determinism, and redundancy has become paramount. FlexRay stands out as one of the most advanced automotive communication protocols and network architectures available today, specifically designed for safety-critical control systems.

Figure 2. Structure of FlexRay Protocol

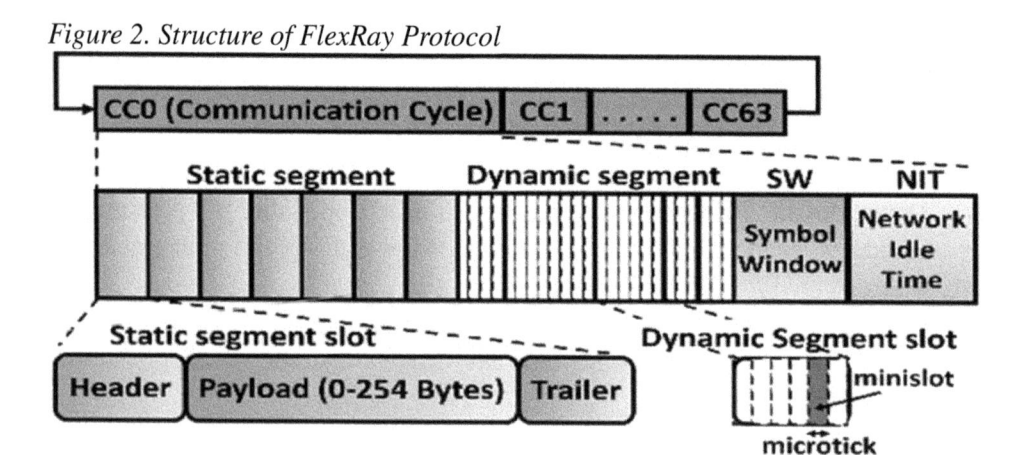

Key Features and Functionality

The Key Capabilities of FlexRay Protocol include:

High-Speed Data Transmission: FlexRay enables data transmission speeds of up to 10 Mbit/s - faster than CAN, LIN, or other legacy bus systems. This high-bandwidth allows transport of larger data sets needed for complex messages, configuration data, embedded logic, firmware updates, diagnostics, enriched multimedia, and autonomous driving functions.

Fault Tolerance & Redundancy: FlexRay networks utilize redundant communication channels in a dual or triple configuration that provides continued operation even in the event of partial network failure or temporary errors. Channels can have independent power supplies and wiring paths to limit single points of failure.

Deterministic Message Timing: The FlexRay protocol ensures deterministic, reliable message timing by using a TDMA-based scheme with static, pre-configured time slots for each node. This ensures components can access the bus predictably with minimal latency jitter, critical for functions like motion control, collisions avoidance, and stability applications.

Precision Clock Synchronization: All nodes in a FlexRay system synchronize precisely to a uniform, distributed clock source to ensure consistent understanding of timing between components distributed across the vehicle. Synchronization occurs down to the microsecond level.

Scalability: FlexRay scales reliably up to 64 nodes per network segment, suitable for complex vehicle topologies across zones and functions. Geographic separation of nodes is also allowed using wiring harnesses, connectors, and hybrid physical/wireless channels.

Composition Flexibility: FlexRay integrates appropriately with different network transports - it can utilize electrical signals over copper wires, fiber optic communication, wireless communication channels, or other transport mechanisms necessary for the vehicle environment and nodes physical locations.

This unique blend of speed, reliability, redundancy, determinism, precision timing, scalability, and communication flexibility is what positions FlexRay as an ideal backbone for vehicle cyber physical systems especially safety-critical applications using complex sensor fusion, automated decision making, and autonomous control.

Frame Structure

All digital communications sent over a FlexRay network segment are formatted as discrete frames containing different fields with specialized functions. Two main types of frames exist:

1. Data Frames
2. Null Frames

Data frames carry information like vehicle sensor measurements, ON/OFF signals for components, computation results, commands for actuators, multimedia streams from driver assistance cameras and microphones, cryptographic signatures, firmware updates, and more. Data frames contain the following key sections:

o **Header Segment:**
o Indicates the start of valid frame
o 11-bit identifier
o 1 bit payload preamble indicator
o Cycle count
o 1 bit header CRC
o **Payload Segment:**
o Up to 254 bytes of data
o Support for variable payload lengths
o **Trailer Segment**
o 24-bit CRC checksum
o 7-bit DTS timestamp

This organization, with starting frame indicators, payload sizes, timestamps, and checksums, provides structure to communication and enable functionality like clock synchronization and corrupted frame detection.

The other frame type - Null Frames - consist of only the header and trailer segments. They do not carry any payload data. Null frames provide an efficient way to maintain timing and synchronization even during periods where components have no substantive information to share. Their inclusion makes the protocol communication pattern more consistent.

In total, a normal FlexRay data frame contains up to 286 bits, transmitted in fixed time slots at even intervals. The strict frame structure and error checking mechanisms enable reliability and system integrity vital for safety systems.

Clock Synchronization & Timing

As discussed briefly already, one of FlexRay's most critical capabilities is precision clock synchronization between all components on the network through a process called Distributed Clock. This enables event ordering, time-triggered coordination, and guaranteed delivery delays across up to 64 devices distributed throughout the vehicle.

At the hardware level, each node contains a bus guardian component with an oscillator crystal serving as a local time base. Bus guardians integrate with two fault-tolerant timing units - the Timer, Input Capture, Output Compare (TICO) ASIL D compliant timer, and the Microtick Timer (MTT) providing microsecond and nanosecond precision time increments. These system timers are synchronized through a two-phase process:

1. **Offset Correction:** In this phase, clock offset differences between nodes are measured and corrected by adjusting local time bases. Special frames called Synchronization Frames transmitted by the host perform these offset measurements. This happens on start-up and then continuously throughout operation.
2. **Rate Correction:** Here clock skew/drift is addressed by controlling oscillator tick rates directly. Synchronization frames from the host regulate rates and frequencies at nodes.

Consistency between distributed timer units is crucial for tasks like high-precision velocity adjustment, triggering simultaneous actuations, or collision avoidance manoeuvring. FlexRay manages this through the interplay between hardware timers and the synchronization algorithm running on top of the communication protocol logic (Shane et al., 2015) (Puhm et al., 2008).

Enforcing this synchronized distributed clock scheme does introduce overhead complexity to provide the strong precision and consistency guarantees. For supporting lower criticality functions that can tolerate more timing variance, FlexRay defines zones called Dynamic Segments where this full synchronization process is optional, allowing dynamic bandwidth allocation rather than pre-scheduled slots. This demonstrates how FlexRay builds in versatility - its dual static and dynamic communication modes allow it to meet both maximum reliability needs as well as more flexible data requirements.

Topology

The physical structure and connectivity of nodes over the communication channels that move frames comprises the FlexRay network topology. Several methods exist for linking the sensors, electronic control units (ECUs), actuators, and microcontrollers together over the dual redundant FlexRay channels. Common approaches include bus, star, cascaded stars, hybrid topologies and rings with loops allowed.

One straightforward way is simply a flat bus architecture - where every component connects to an open communication bus that serves logically like a party telephone line multiple people listen and speak on. However, a weakness with buses is lack of redundancy and fault containment. If the bus is compromised, infected, severed, jammed, or otherwise degraded, global impact spanning the entire system ranging from denial of service to safety failures could result.

An improved approach is using star, tree, or cascaded star topologies. Here different vehicle zones have local hub nodes that aggregate information from area components and relay it higher up a tree hierarchy until reaching a top-level central gateway node situated at the root. Each star's central hub node manages connectivity for its domain through the redundant channels. So for example, cockpit systems may form one sub-tree network, drive-train, stability control and infotainment yet another. Traffic from different sub-trees gets merged and transported to a global vehicle integration hub. Because stars provide switch ability, isolation, and local concentration of connectivity, damage doesn't necessarily propagate globally. This enables better security, safety containment, and fault tolerance (Sander et al., 2008).

Interlinked rings are another topology model sometimes leveraged where lines of systems loop in cyclic fashion rather being chain linked down a tree. Rings allow revolving directional message passing and continued operation when segments are disrupted. Self-healing meshes represent related flexible architectures as well.

Overall FlexRay supports all major topology styles, leaving it to network designers to select optimal physical communication structure and protocols based on number of nodes, criticality, physical locations, redundancy needs, and security considerations. Hard real-time control loops, drive-by-wire systems, or battery management subsystems will likely utilize more compartmentalized, well-guarded communication patterns than multimedia streaming, weather sensors, and cabin microphones for example. These nuanced architectural decisions impact functioning during both normal conditions and adverse circumstances involving component outages, environmental noise, or deliberate attack. FlexRay provides underlying configurability to tailor connectivity resilience for the automotive environment's broad use cases and threat models.

Communication Cycle

With distributed connectivity established through the network topology, frames containing vehicle data get transmitted over the FlexRay channels in a repeating communication cycle. This cycle executes periodically, it governs information flow for the VCPS. Cycles facilitate regular data exchange between vehicle subsystems needed for digital sensing, control, actuation, diagnostics, and processing.

Segment sizes get defined during FlexRay network configuration. Typical splits are around 5% dynamic, 10% symbol window, and 80% static allocation out of the total 10 ms cycle time. But durations can adjust based on bandwidth needs. Busy cycles with lots of sensor readings and vehicle telemetry may have shorter idle periods, for example. Segment splitting allows efficient utilization of bus bandwidth for different types of communication traffic patterns (Herber et al., 2015) (Kim et al., 2015).

In the static segment with its TDMA collisions-free scheme, the host controller calculates schedules with transmission slots of uniform size allocated to each component. This guarantees nodes access to the bus within the timeframe they need - achieving reliable, real-time delivery critical for functions like acceleration control or collision avoidance. In dynamic portions, bandwidth gets divided into mini-slots that components can arbitrate and reserve based on priority level, in support of more intermittently-triggered functions.

Overall this repeating cycle architecture transmits vehicle information reliably while avoiding underutilization of available network bandwidth. FlexRay's dual-channel redundancy adds fault tolerance as well - cycles occur independently on both the main and backup buses. This robust delivery helps fulfil the needs of complex, integrated VCPS systems spanning many critical and non-critical uses.

3. SECURITY REQUIREMENTS AND THREAT MODEL

As connectivity and complexity continue rising inside modern vehicles, ensuring cyber protections evolve in parallel becomes imperative. Security cannot be an afterthought - it must align closely to system architectures during foundational phases. Attempting to retrofit defences onto legacy designs with no security consideration often proves highly challenging or even infeasible. By thoroughly assessing critical requirements, examining likely threat vectors, and enumerating insider versus outsider hazards early on, FlexRay networks can integrate tailored protections upfront for maximum safety and resilience as vehicles advance.

Fundamental Security Properties

Three core security principles establish a useful framework for analyzing protections around vehicular protocols like FlexRay: confidentiality, integrity, and availability. Assessing how well systems uphold these properties in the face of different disruptions helps guide cyber hardening efforts.

Confidentiality entails preventing unauthorized access to sensitive vehicular data through use of encryption, stringent access controls, and securing overall traffic flows. User privacy represents a major confidentiality consideration as vehicles contain immense amounts of personal trip data. Everything from GPS coordinates, camera feeds of passengers, sensor scans capturing biometrics and conversations, entertainment preferences, and interfaces with mobile devices harbours private information that could prove extremely damaging if accessed by criminals or abusive partners for example. Beyond privacy, other confidentiality priority areas include digital keys for access control and authentication, financial information and payment tokens, as well as proprietary automotive intellectual property and trade secrets tied to competitiveness. Insurance companies would love to access comprehensive vehicle telemetry and diagnostics for premium setting. Governments seek this data as well for taxation. Even metadata like connection timestamps, destinations, charger characteristics and energy usage profiles bear sensitivity around user movements and habits. Stringent confidentiality protections through encoding, access restrictions, hardened interfaces and secure storage constitute imperative starting points to counter risks of unauthorized data thefts, surveillance, and inference attacks (Mateus & Königseder, 2014)(Soares et al., 2015)(Sakiz & Sen, 2017).

Integrity centres on safeguarding accuracy and completeness of vehicular information and operations. Mechanisms include error detection checksums, cryptographic signature verifications, behaviour monitoring, activity logging and auditing. Several integrity areas stand out as especially important. CAN bus communication channels must maintain integrity as their compromised messages get consumed across vehicle components - malicious signals injection could fool systems into harming human safety. Sensor readings and vision outputs feed increasingly automated analysis and actuation so must avoid manipulation. Processors rely on stored firmware containing code logic and collected telemetry so require integrity defences against rewrite attacks that could insert bugs or backdoors. Clock synchronization between components enables deterministic coordination so its distortion could undermine vehicle coordination. Signal translation between legacy interfaces as bridging proceeds requires correctness else risk propagating malformed outputs. Remote commands from infrastructure or personal devices need authentication to confirm validity and avoid impersonation attempts. Geo-positioning data and navigation processors constitute prime targets as well given attackers could redirect vehicles dangerously or achieve kidnapping by feeding false destinations. Even Over-the-air software updates capability represents high integrity priority since compromised code redistribution channels could automatically infect large swaths of vehicles before detection. Myriad other integrity risks around ECU computations, in-vehicle database reads/writes, and parametric configuration exchanges necessitate strong verification and sanity checking defenses in the face of rising attacks against road vehicles (Soares et al., 2015) (Irshad et al., 2011).

Lastly availability focuses on ensuring reliable, timely access and functioning of vehicular systems and services, despite disruptions. As with traditional IT systems, redundancy, resilience mechanisms, and recovery protocols help maximize robust accessibility. Responsiveness of always-on sensing and data gathering functionality grows in importance as continuous telemetry feeds increasingly autonomous decision making and vehicle environmental modelling. Overall as vehicular systems take on more critical driving functionality, availability protections become foundational to avoid disruption threats leaving passengers stranded.

Threat Vectors & Attack Surfaces

Unfortunately vehicle systems face multifaceted risks spanning across equipment faults, human errors, environmental disruptions and intentional harm - defenders must address across these threat vectors. Design flaws, faulty components, wiring problems, signal noise, software defects, unexpected electromagnetic pulses, and more can trigger abnormal operations undermining security protections. Assuring robustness requires design analysis and testing practices that model real world conditions, while also building in adequate health self-checks diagnostics, and fail-safe override mechanisms. Environmental resilience deserves focus as well to counter risks like flood damage, connector debris accumulation, vibration stresses or extreme thermal swings. The application of security-informed architecture principles requires the modelling of appropriate protections that correspond to the level of criticality of the application, regardless of whether the application is for mainstream consumer vehicles or off-road mining trucks that are exposed to harsh contaminants.

Among the most concerning risks though stem from intentional attackers trying to extract value or cause harm through FlexRay networks and associated automotive systems. Financial incentives drive most cybercrime today - stolen credit cards, user identities and account credentials, or encryption keys and intellectual property represent common attacker motivations. But harms manifest physically in automotive contexts as well - safety and human lives at stake make motivation vectors more multidimensional. Vandals may seek amusement triggering vehicle malfunctions through hacking. Personal vendettas could turn violent through automobile vulnerabilities. Terrorists utilize vehicle attacks as insidious tools for economic disruption, infrastructure damage or mass casualty events. Conflict zones and military domains incentivize cyber warfare programs to incapacitate opponent transport. Corporate espionage incentives arise around theft of proprietary technology. Myriad motivations drive cyber targeting of automotive attack surfaces.

Corresponding tactics run the gamut as well based on adversary sophistication levels:

- ✓ Packet injection to feed false sensor data, trigger errors, or overwhelm components
- ✓ Memory tampering and flash rewriting for backdoors
- ✓ Cryptographic key theft enabling access escalation or deception
- ✓ Eavesdropping CAN bus or OBD-II interfaces for reconnaissance
- ✓ Reverse engineering firmware images for vulnerabilities
- ✓ Jamming legitimate signals to cause interference
- ✓ Spoofing messages from valid vehicle addresses
- ✓ Manipulating buffers, queues, schedules to induce resource exhaustion, timing disruptions, deadlocks
- ✓ Intercepting GPS inputs or sensor readings for inaccurate location, fake obstacles, etc.
- ✓ Redirecting sensor image feeds for illusion of nonexistent road signs, obstacles, blank wall projections
- ✓ Physical manipulation of electrical interfaces, data buses, storage media connections
- ✓ Wireless infiltration of infotainment systems, dongles, tablets, or Bluetooth apps to traverse air gaps
- ✓ Ransomware or wipers that immobilize vehicle controllers until payment

The automotive attack surface and exposure points continue expanding with increasing interconnections across drive train, stability, braking systems along with sensors, transmitters, wireless modem interfaces, navigation systems and entertainment options. Security-aware design must assess risks across these myriad surfaces to inform where hardening of FlexRay protocols can strengthen protection.

Insider and Outsider Considerations

Threat models provide further dimensionality by scoping insider vs. outsider distinctions which necessitate tailored security controls. Insiders already have some degree of authorized privilege or physical access through approved channels - employees, suppliers, technicians, fleet operators. Outsiders in contrast lack native access and must find external vulnerabilities to penetrate.

Insiders' greater access means they can often circumvent controls by abusing debug modes, proprietary test interfaces, administrative tools, master keys/passwords, specialized diagnostic requests and more. More emphasis must focus on internal compartmentalization, least privilege, behaviour monitoring and process supervision. Locking down unneeded legacy physical ports impedes exploitation. Principle of least privilege inhibits privilege creep spread. Cryptographically robust identity and access management solutions can contain insider lateral traversal through networks.

Outsiders lacking any approved access require external perimeter and boundary defences. Firewalls, gateways, guards, and strict physical access protections take priority to block adversary reconnaissance and initial intrusions. Assuming outsiders successfully penetrate anyway, additional internal mechanisms as mentioned help contain threats and prevent transit to truly sensitive vehicle controllers.

Overall FlexRay's solid access controls, redundancy, and fault containment capabilities establish strong foundations against both insider and outsider threats. But dedicated security analysis throughout the design process must still scrutinize risks across the numerous emerging wireless interfaces, entertainment services, navigation and automation components that expand attack surfaces. This informs where hardening of FlexRay backbone protections requires bolstering to match modern vehicular system complexities (PATAK Engineering, n.d.).

4. FLEXRAY PROTOCOL SECURITY MECHANISMS

With vehicular connectivity and rich sensor data exchange enabling smarter, safer transportation through automation, cyber protections attaining equal intelligence represents imperative cross-functional focus touching hardware, software, cryptography, network architecture, and beyond. Supporting protocols like FlexRay through their decade's long lifecycles mandates adapting fluidly to match risk climate evolution. Various integrated mechanisms across authentication, integrity checking, synchronized timing, threat monitoring, and resilient management help harden FlexRay's defences amid the software defined vehicle revolution.

Multi-Factor Authentication

As connectivity opportunities with external systems and remote infrastructure continue expanding, properly authenticating identities and authorizing appropriate access grows in importance to contain externally-driven threats. Without rigorously confirming message sources and establishing access levels, attackers could manipulate vehicle controllers by simply spoofing identities or utilizing default credentials still left enabled.

FlexRay's host controller provides built-in authentication services for vehicles ECUs and components. Through storing identifiers and challenge response pairs, it cryptographically confirms identities by issuing random nonces (numbers used once) that authorized nodes must mathematically transform using secret parameters to prove themselves. This prevents impersonation unless cryptography gets compromised which represents extremely high difficulty against modern algorithms. Defence-in-depth builds on this through secondary multi-factor authentication prior to enabling critical functionality – biometrics like fingerprints, external tokens, or user passwords further corroborate user or component legitimacy. Rate limiting mechanisms additionally slows down brute force attacks. Combined strong authentication and least privilege authorization provide potent frontline protections for FlexRay and remainder of vehicular systems (Liu et al., 2017) (Fadlullah et al., 2013).

Message Protection & Integrity Checking

Guarding against corrupted or maliciously modified in-transit data represents crucial security priority as well given vehicle controllers fundamentally operate on sensor readings, computational analysis, and timing signals. Myriad points vulnerable to tampering exist across physical communications media like cables and antennas or software interfaces receiving external inputs. Defence requires information integrity protections applied strategically to FlexRay's frame communication structures. Each frame checksums generally utilize cyclic redundancy checks on headers and payloads to reliably detect bit flips indicating errors or manipulation. The 24-bit cyclic redundancy check within trailers provides high probability detection for common transmission mistakes or purposeful tampering. However, traditional math-powered checksums struggle catching crafty adversarial bit alterations that still yield plausible sensor readings yet dangerously incorrect - feeding trajectory calculations slightly off or skewing computer vision results to hide obstacles for instance.

Augmenting reliability and integrity further entails cryptographic protections that mathematically bind sensor outputs irrevocably to their sources through digital signatures and hash chains. Signing frames attaches asset identifiers along with metadata like timestamp, geo-coordinates, serial numbers, or firmware revision useful when investigating incidents later. Tamper evident signing also deters adversaries lacking cryptographic keys from modifying frames undetectably or injecting wholly fabricated outputs instead. Origin authentication confirms frames come from expected transmitter sources rather than imposters. sessFlexRay's existing cyclic redundancy checks with emerging signing techniques provides defense-in-depth against risks like GPS coordinate alterations, sensor spoofing, or timing manipulation attacks that could trick vehicle systems with dangerous effects.

Secure Clock Synchronization

As outlined earlier, precision time synchronization represents foundational FlexRay capability for interconnecting vehicle components across locations. However this strength also introduces critical vulnerability – attackers manipulating no de-clock rates could directly crash deterministic coordination by inducing confusion around event ordering, interrupts firing, priority arbitration, or scheduled resources expecting availability. Subtle localized timing shifts can cascade into complete vehicular functions failure given extensive time sensitive interdependencies. Securing synchronization and rate control against adversarial interference constitutes high priority.

Overall FlexRay's multi-channel redundancy, extensive fault tolerance and deterministic message handling provide helpful primitives securing time synchronization integrity upon which adding other sophisticated protections can build.

Intrusion Detection Systems

Security experts consider intrusion detection systems crucial through providing ongoing visibility into potential misuse, anomalies, exploited vulnerabilities that bypass preventative access controls or malware defences. FlexRay networks represent ripe attack targets given their central vehicular role. Detection requires situational awareness capabilities that add into the various FlexRay interfaces and messaging activity.

Figure 3. FlexRay Intrusion Detection System (IDS) Architecture

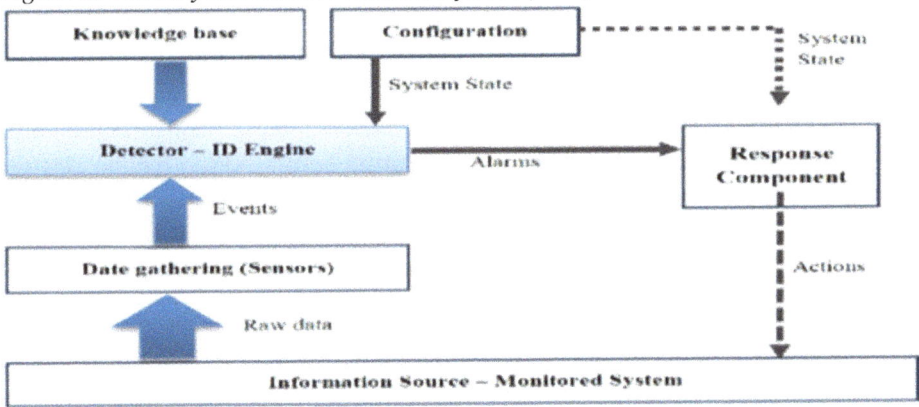

Approaches range from signature-driven monitors checking against known compromise patterns, to machine learning systems automatically flagging statistical outliers deviating from normal patterns. Anomalies indicating potential hacking include things like unfamiliar message sender addresses, abnormal payloads sizes, improper encapsulation formats, unsupported bus protocols, non-standard debug modes toggling; incorrect cyclic redundancy checks values, and unusual sensor data combinations off expected ranges. Analysts set threshold sensitivity levels based on acceptable false positives tradeoffs. Distributed IDS nodes also allow correlating detections across vehicle zones to piece together multi-stage attacks spanning initial reconnaissance through to payload delivery or effects triggering. Response capabilities additionally integrate, enacting tactical measures ranging from alerting drivers, throttling suspect signals, halting affected processes, to gracefully degrading to minimal safe operations while blocking adversarial progress (Liu et al., 2017).

More capable frameworks even enact threat hunting proactively rather than purely reacting to alerts. This involves IDS nodes directly querying component history logs, firmware, or memory for remnants of compromise like injected shell code, modified binaries, or backdoor user accounts. Overall FlexRay networks benefit tremendously from tightly integrated intrusion systems performing ongoing behavioural analytics, memory forensics, model-based detection, signature pattern matching and bootstrapping response flows when anomalies surface.

Secure Network Management

Configuring, monitoring and maintaining dozens of FlexRay controller nodes plus interconnections with sensors/actuators requires networked management interfaces. Typically orchestration happens out-of-band on physically separate channels to avoid interfering with core control or telemetry operations. However these ecosystems themselves require stringent protections given their centralized privilege over hardware and software provisioning inherently risky. The UK public transportation ransomware attacks that encrypted system databases while demanding Bitcoin payment exemplified recent mass transit network management infrastructure exposures. Vehicle contexts face similar risks from compromised diagnostics or firmware update channels for instance. Defence requires applying the same layered protections to the FlexRay management plane itself including granular role based access, authentication, cryptography and channel isolation. Separate reporting capabilities also prove useful for notifying administrators around suspicious management requests or policy violations to investigate before harm manifests from potential intrusions. Further lockdown comes from strictly minimizing unneeded physical bus connectivity reducing attack surface exposure points, with oversight by internal audit teams assessing configuration risks independent from engineers. Overall FlexRay networks warrant complementary secure and resilient management capabilities to parallel the advanced reliable communications natively provided.

Supporting FlexRay's continued decade long viability requires evolving protections in stride with modern connectivity and threats. Cryptography, fault tolerance, access controls comprise integral aspects within FlexRay beneficial for security. Equally critical is applying discipline during architecture design, implementation, testing and operational maintenance to constrain risks that connectivity introduces. Well-integrated authentication, integrity protections, secure timing, vigilant monitoring and resilient management help harden FlexRay robustness amid the software defined vehicle revolution underway. With proactive cyber security prioritization, FlexRay's advanced, reliable in-vehicle communications will continue safely realizing emerging autonomous transportation efficiencies for years ahead.

5. ENHANCING FLEXRAY DEFENSES

Previous sections outlined core security capabilities integrated within FlexRay like access controls, resilience and redundancy that provide baseline protections for vehicular messaging. However modern threat sophistication requires additional defensive layers atop these basics – robust cryptography, firewalls, behavioural monitoring, secure remote access channels and anti-tampering defences help strengthen protection.

Cryptography Pervasively Applied

Cryptography represents fundamental concept enabling modern secure communications from financial transactions to messaging apps to websites. FlexRay networking equally requires pervasive application of encryption, signatures, hash functions and key management fundamentals. To maintain confidentiality and integrity, this approach safeguards against threats posed by unauthorized access using stolen credentials or vulnerabilities in unpatched software, among other scenarios.

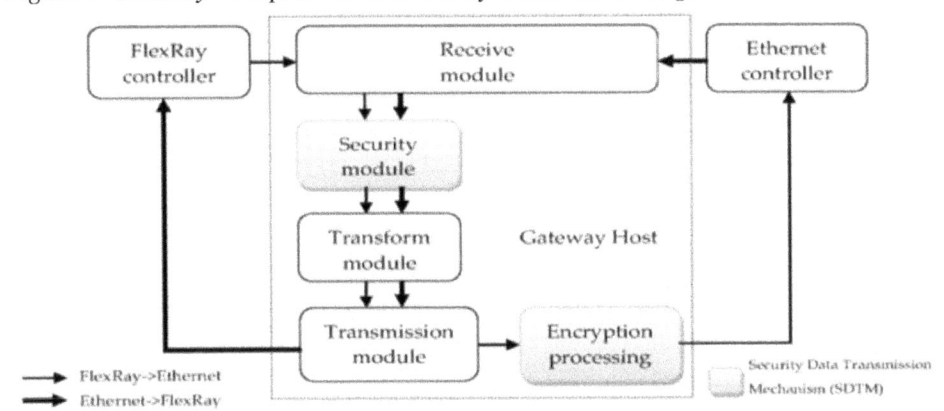

Figure 4. Security Components in FlexRay Hardware Design

Specific cryptographic protections applicable to FlexRay environments include symmetric encryption of vehicle sensor readings and control messages before transmission. This jumbles sensitive telemetry, actuator commands and configuration settings into indecipherable cipher texts, only authorized endpoints can decrypt using shared secret keys. Sensitive user data like biometrics, conversations and entertainment usage requires similar shielding when traversing FlexRay networks between cockpit systems and passenger infotainment units. Alongside encryption to protect confidentiality, digital signatures attached to messages confirm origin authenticity while also detecting tampering if signature verification checks fail. Signed firmware updates prevent malware or backdoor code injection on controller nodes. Signed timing messages avoid rate manipulation attacks. Carefully implemented public key infrastructure enables securely provisioning keys to valid vehicle components at manufacturing time for later authentication and encrypted communications needs.

Around key management itself, secure crypto-processors combined with hierarchical key derivation and sharing mechanisms help limit single point compromise escalation. If one node gets infiltrated, the damage stays localized rather isolated from affecting other zones or requiring full certificate authority reboot. Best practice dictates binding keys directly to device identities through certain parameters that must exactly match during decryption or signing operations - adding supplemental protections preventing copied keys from functioning in unfamiliar contexts (Jadoon et al., 2018).

Overall systematic application of modern cryptography can significantly shrink FlexRay's attack surface by mathematically shielding critical communications, storage and updates that otherwise remain exposed through the increasing wireless interfaces, external connectivity and over-the-air management channels introduced for convenience reasons. Handled meticulously as part of the secure design process rather than ad-hoc retrofitting, cryptography solutions will provide considerable benefit securing vehicle electronics against intrusions for years ahead.

Firewall Isolation and Guard Nodes

Alongside cryptography strengthening internal subsystems directly, network architectural protections like firewalls, gateways and guard systems provide crucial enforcement points blocking, filtering or throttling suspicious traffic attempting to traverse between vehicle domains. Most modern vehicles contain abstracted zones - separate drive train, passenger and infotainment networks for example. As connectivity opportunities continue expanding, adversaries utilize any bridging or routing mechanisms between these traditionally isolated zones to gain lateral access.

Hardened guard systems purpose-built to monitor inter-zone messages and protocols allows blocking suspicious communication patterns like unfamiliar sensor readings or timing requests seemingly trying to manipulate downstream controllers. Threshold rate limiting additionally slows volumetric attacks attempting denial of service by overloading components with excess traffic often a precursor to control signal injection. Gateway nodes able to gracefully degrade or actuate minimum safe operating modes during detected attacks provide another architectural protection mechanism similar to IT domain firewalls cutting connectivity during incidents but adapted for vehicular physical safety requirements. Layered gateways and guards, when combined with tailored rule sets for typical vehicle communication requirements and profiles, show great potential in detecting and preventing anomalies while also fulfilling the real-time messaging necessities that are crucial in automotive applications (National Institute of Standards and Technology, 2020) (Hung & Hsu, 2018).

Anomaly Detection Beyond Guard Nodes

Firewalls typically concentrate on filtering inbound external traffic, while other detection capabilities, such as internal bus traffic analysis, memory forensics, and ECU activity, are often more effective at identifying insider threats, such as compromised devices that spread laterally. Anomaly detection through ECU interface statistics, communication pattern heuristics, and sensor data trend monitoring is crucial for detecting malfunctions or misuse. Behavioural monitors can function independently on individual critical components or collaborate across the entire fleet using centralized threat intelligence to detect early signs of potential harm, such as system crashes, data destruction, or vehicle impairment issues.

Detection models codify expert knowledge around baseline expectations for parameters like voltage input ranges, ambient temperature distributions, computed velocity algorithms accuracy bands, equipment versions and supported interfaces. Deviations then indicate potential malfunctions or hacker experimentation that prompts further investigation. High false positive thresholds avoid distraction but still catch meaningful cases. Anomaly detection integrated natively into FlexRay controllers provides inherent benefit given the central network vantage point for gathering rich telemetry, interfacing numerous zones and analyzing the high volume traffic exchanges crucial to vehicle functioning (Fadlullah et al., 2013). Over the air updates allow improving detection rule sets continuously to address novel attack methods observed against vehicles on the roads today.

Secure Remote Access Channels

Like all complex computing equipment, automotive systems require remote access channels for purposes like diagnostics, software updates and configuration management due to economies unachievable relying solely on physical dealership visits. However security teams rightfully view remote channels as inherently risky attack vectors. Compromise of communications or endpoints could enable large scale vehicle malware distribution, paralytic denial of service through ill-intentioned commands, or access for pivoting laterally deeper behind firewalls. Therefore FlexRay networks warrant extremely stringent protections specifically for external management access flows both in transit and regarding endpoints. Frequent patching, multi-factor authentication, certificate-based infrastructure, compartmentalization, activity monitoring, rate limiting and operator access reviews help manage risks to acceptable levels that remote supportability necessitate by economic realities. The extensive needed controls justify fully separating external management networks from core FlexRay communications physically and cryptographically. Operational technology environments cannot eliminate remote access but can constrain attacks via layered defences.

Anti-Tampering Protections

Lastly, safeguarding vehicle electronics necessitates not only securing external wireless interfaces, but also physically fortifying against attempts to tamper while the vehicles are stationary. Examples of such attempts include bridging physical pins and ports on ECU boards to access diagnostic modes that enable deeper infiltration, as well as using portable device emulators to impersonate authorized devices and inject malicious payloads into unprotected interfaces. Similarly, storage media such as firmware flash chips are often left exposed, allowing for rewrite attacks that can embed persistent malware that remains effective even after reset attempts. Infiltrating backdoor access at the hardware or firmware levels can provide lasting footholds that survive most integrity checks or forensic monitoring at the operating system software layer.

To counteract such low-level compromises, anti-tampering defenses tailored for automobile electronics can be employed. These might include epoxy encapsulation of sensitive controllers, mesh shields over exposed connectors and buses, port access monitoring using motion sensors, secure boot protections against corrupted firmware, cryptographically signed code to detect unauthorized modifications, and the use of Trusted Platform Modules with built-in protections for storing encryption keys and platform measurements to reliably detect tampering attempts. By combining physical hardening approaches with secure boot and integrity checking, robust assurance can be provided against threats that seek to gain persistence through low-level vectors that bypass FlexRay's native communications security controls (A Structured Approach to Anomaly Detection, 2010).

While encryption, access controls and redundancy constitute integral protection aspects within FlexRay critical for security, threat complexity mandates additional defensive layers. Firewalls and inspection guard nodes, anomaly behavioural monitoring, stringent remote access protections and anti-tampering measures combine to offer considerable benefit hardening FlexRay robustness in depth. With proactive adoption during early design stages, multilayered FlexRay cyber protections will continue upholding stability and safety for autonomous transportation systems many years down the road against extremely motivated adversaries.

6. FUTURE DIRECTIONS AND OPEN CHALLENGES

Navigating FlexRay's Future Security Landscape

FlexRay powers intricate in-vehicle connectivity today, but technology continually evolves and security must keep pace against rising digital threats targeting vehicles. Several impactful trends, updates, and open research problems deserve ongoing attention to ensure FlexRay's continued safe utilization over forthcoming decades supporting self-driving transportation.

Emerging Vehicular Connectivity Standards

While currently enjoying widespread vehicle integration, Forward-looking analysis recognizes FlexRay will co-exist alongside newer automotive connectivity standards like CAN FD, Ethernet TSN and Auto-Sar hitting mainstream adoption. Each standard brings unique capabilities related to timing precision, bandwidth, fault tolerance and topologies. However transition periods mean multi-standard interoperability demands rise temporarily. Gateways and bridges translate messaging between FlexRay's strict cyclic redundancy checking approach and potentially more laissez-faire or efficiency oriented protocols. This interlinking if not robustly secured could enable adversaries penetrating one standard's defences to then laterally access more sensitive systems relying on a separate standard. While segmentation and isolation best practices generally discourage connectivity between safety critical power train or autonomous navigation systems and entertainment/internet domains, economics incentives push consolidation and developers seek convenience short-cuts that undo security compartmentalization. Similar gateway risks persisted early on bridging CAN bus to indirectly reach deeper electronic control unit networks which attackers leveraged for injecting malicious steering and acceleration instructions. As bridges between FlexRay and other emerging standards appear in vehicles, rigorous inspection, filtering, sanitization and anomaly detection protections warrant integration to prevent lateral traversal.

Open Challenges & Future Research Opportunities

While sections above summarized various known security mechanisms and leading improvements applicable to FlexRay, considerable gaps remain needing collaborative research across industry, academia and public sector agencies. Open problems cover both foundational science areas and integration challenges translating theoretical crypto defences into tangible vehicular implementations.

On the foundational side, post-quantum cryptography addressing future quantum computing capacity requires another decade at least to mature applicable cryptosystems believed resilient against exponential brute forcing power anticipated. Comparatively little public domain research investigates post-quantum solutions tailored for automotive embedded environments and real-time latency constraints. Long term flexibility for updating cryptosystems across already deployed vehicle fleets also needs consideration given automotive lifecycles exceeding a decade typically.

Developing methods to definitively verify the correctness and absence of exploitable vulnerabilities in large-scale software remains an enormous challenge. Although verification techniques show promise in securing small modules and microcontroller logic, applying these methods across millions of lines of code in vehicle operating systems and layered application software seems like an unattainable goal for the time being (Wang & Liu, 2018). Furthermore, mathematical verification that connects secure hardware elements such as trusted platform modules and boot loaders to complex software is still in its foundational research phase. Advances in this area would greatly benefit security measures, such as FlexRay protections integrated with authenticated boot sequences, which could protect network interface firmware, among other security enhancements.

On the integration side, user experience hurdles around true two-factor authentication need addressing before widespread adoption in vehicles, including challenges like intermittently connected operation, PIN entry while driving, and embedded secure element configurations unaffected by electrical faults or crashes. Secure wireless delivery mechanisms also undergo refinement for resilient over-the-air software update pipelines. Both user authentication and updates criticality magnify as vehicles shift towards self-driving managed predominantly by programming logic. Additionally scalable fleet trust management and policies enforcement architecture requires further development so enterprise-grade public key infrastructure solutions integrate feasibly across manufacturers while avoiding crippling vehicles in the face of backend mis-configurations or temporary certificate authority outages. Finally standardization and tooling for simplified cryptographic keys and identities lifecycle management represents pivotal area still developing. Usability and foolproof processes require considerable improvement before security teams recommend mainstream adoption that remains reliable across entire automotive supply chains.

While FlexRay already provides advanced, robust messaging suitable for autonomous vehicle controllers, active participation in standards advancement, collaborative research resolving open problems, and translation help integrating promising developments together position it for continued vitality securing next generation transportation innovation against looming threat landscapes.

Table 1. Comparison of cryptographic techniques applicable to FlexRay

Technique	Strengths	Limitations
Symmetric Encryption	High speed, low resource requirements	Key distribution and management challenges
Public Key Encryption	Simplified key management	High computational overhead
Hash Functions	Efficient data integrity checking	Does not provide confidentiality
Digital Signatures	Authenticity and integrity	Increased overhead vs simple hashes

Table 2. Intrusion detection approaches comparison

Approach	Detection Methodology	Performance	Limitations
Signature-based	Pattern matching against attack signatures	Low false positives if signatures well-defined	Unable to detect novel, zero-day attacks
Anomaly-based	Identify deviations from normal behaviour	Detect novel attacks	Higher false positives, requires normal models
Machine Learning	Train statistical, AI models on event data	Adaptability to new patterns	Data dependency for accuracy

Table 3. Automotive threat modelling - identifying risks and mitigations

Threat Actor	Motivation	Potential Attacks	Mitigations
Cybercriminals	Financial gain	Sensor spoofing for insurance fraud, Ransomware against vehicle controllers	Encryption, Authentication, Access Controls
Nation States	Strategic disruption	GPS spoofing, Firmware tampering via compromised supplier channel	Hardware protections, Supply chain controls
Insiders	IP Theft, Revenge	Internal reconnaissance and data theft, Logic manipulation	Least privilege policies, Behavioral monitoring

7. CONCLUSION

Realizing Secured Vehicle Networks

Previous sections provided extensive detail across vulnerabilities, defences and enhancements specifically for the FlexRay communications protocol powering connectivity between critical vehicle components. This conclusion summarizes key takeaways from earlier coverage while also connecting broader themes around importance of prioritizing cyber security for emerging autonomous, electric and software-defined automotive systems.

FlexRay Security Summary

FlexRay delivers robust capabilities purposefully designed for safety-critical control systems with determinism, redundancy and reliability exceeding conventional computer networking approaches. Precision synchronization, static scheduling, dual redundant channels, extensive error checking, strong fault containment and built-in authentication support position it well protecting next generation drive-by-wire and vehicle-to-infrastructure automation systems crucial for self-driving realization.

Yet vehicles also increasingly network with external systems across numerous wireless interfaces for practical reasons around remote diagnostics, over-the-air updates, passenger device connectivity and location-based services. Attack surfaces expand as more avenues emerge for external parties to potentially access and manipulate internal vehicle electronics if lacking adequate safeguards.

Therefore securing FlexRay constitutes imperative priority but requires context-specific focus beyond just copying established enterprise IT protections like firewalls, antivirus, password policies and web application scanners. While those find place, automotive environments necessitate additional assurance around firmware integrity, behavioural detection tailored to sensor and actuator patterns, lockdown of debug and access ports, encryption for external communications channels, specialized hardware protections against data bus tampering and denial of service risks, and segmentation of entertainment systems from critically safety domains. Security-informed architecture that contains rather than assumes trust remains crucial starting point as well (Mateus & Königseder, 2014) (Soares et al., 2015) (Sakiz & Sen, 2017).

With these automotive nuances incorporated, recommended FlexRay security controls deliver defence-in-depth protecting integrity and availability through layered mechanisms:

- ✓ Authentication verifying node identities using root of trust hardware protections against spoofing
- ✓ Authorization policies minimizing unnecessary access and resource allocation
- ✓ Cryptography mathematically shielding sensitive vehicle telemetry and commands
- ✓ Redundancy and fault tolerance surviving outliers and random component failures
- ✓ Guard systems filtering inter-network traffic passing between different vehicular zones

- ✓ Anomaly detection identifying statistical deviations from normal sensor and messaging patterns
- ✓ Secure boot mechanisms preventing persistent backdoors or manipulation injection via firmware
- ✓ Monitoring through edge security modules and vehicle intrusion detection systems
- ✓ Over-the-air software update protections preventing infection distribution
- ✓ Physical tamper resistance and evidence mechanisms defending hardware elements
- ✓ Secure fleet management scalable upholding configurations securely

These controls undergo integration starting from initial phases as new automotive systems get architected rather than attempting to bolt on later. Architectural decisions like reduced unnecessary connectivity between units, encrypted storage protections separate from error handling protections, use of hardware elements for root of trust and microcontroller configurations avoiding complete bus accessibility all exemplify foundational precautions benefiting security posture significantly compared to entirely software-based solutions more convenient short term. With regards to FlexRay itself, participation in ongoing protocol standardization also provides opportunities promoting inclusion of enhanced authentication profiles, cryptographic payload protections and formal threat modelling requirements into future iterations or companion specifications (Fadlullah et al., 2013)(Wang & Liu, 2018)(Jadoon et al., 2018)(X-By-Wire Team, n.d.)(A Structured Approach to Anomaly Detection, 2010)(National Institute of Standards and Technology, 2020). Lastly accountability around vehicle cyber risk assessments and incident response planning drives improvement prioritization based on potential safety and financial harms tied to credible real world threats.

Urgency around Automotive Cyber-security

Stepping back, the extensive security considerations around FlexRay speak to overarching urgency required regarding cyber protections for emerging autonomous and electric vehicle innovation. Synergistic advances across vehicular electrification, self-driving programs, intelligent road infrastructure integration and mobility-as-a-service business models promise tremendous societal benefits from reduced emissions to saved lives to increased inclusion. However headlong rushing without applying sufficient caution risks realization of consequences we escaped largely before computers inserted so pervasively. Whether talking future wars targeting infrastructure as weapons systems, ransacking of identifiable location history databases by hackers or stalkers if improperly anonymized, discrimination from algorithmic traffic routing built upon biased data, massive recall costs and legal liability from safety oversights in haste for profits, or simple paralysis of public mobility from things as basic as charging port compatibility fights - examples abound regarding automotive cyber negligence costs. And those just constitute unintended second order effects rather than deliberate criminal or nation state desired harms.

Thankfully, awareness and willingness to invest in vehicular cyber-security keep increasing among manufacturers, vendors, insurers, regulators, infrastructure owners, academic researchers and security practitioners. While more progress is required before matching the scale of risks, encouraging trends emerge. Architecting compartmentalization and principled minimal connectivity into next-generation electric platforms from the outset prevents playing costly catch-up, securing sprawling complexity like the deeply intertwined internal combustion ecosystem necessitated. Cryptography implementations harness modern algorithms and hardware protections that were once lacking or slapdash before (Hung & Hsu, 2018) (Stefan & Groza, 2020). Automakers hire more cyber specialists with software quality, fraud detection and red team penetration testing backgrounds - crucial diversity breaking from traditional mechanical and reliability engineering disciplines. Auto ISAC participation industry-wide drives information sharing and best practices faster based on collective experience rather than isolated trial-and-error. Moreover, policymakers have started appreciating connected and autonomous vehicles as systems requiring oversight similar to seismic building codes, pharmaceutical equipment, FDA approvals, or financial trading safeguards auditing - rather than purely unregulated mobile computing platforms.

Overall, the vehicle transformation era requires commensurate prioritization of cyber-security to match. FlexRay and complementary underlying automotive technologies can responsibly continue fuelling tremendous innovation for decades ahead through continued cross-functional collaboration around design phase threat modelling, runtime vehicular monitoring, emergency response protocols, consumer education and governance advancement.

REFERENCES

Bittl, S. (2014). Attack potential and efficient security enhancement of automotive bus networks using short MACs with rapid key change. In Communication Technologies for Vehicles: 6th International Workshop, Nets4Cars/Nets4Trains/Nets4Aircraft 2014, Offenburg, Germany, May 6-7, 2014. [Springer International Publishing.]. *Proceedings*, 6, 113–125.

Debala Chanu, A., & Sharma, B. (2019). TCP Connection Monitoring System. *International Journal of Computational Intelligence & IoT*, 2(1).

Dunkels, A. (2001). Design and Implementation of the lwIP TCP/IP Stack. Swedish Institute of Computer Science, 2(77).

Ethernet_Frame. Available online: https://en.wikipedia.org/wiki/Ethernet_frame

Fadlullah, Z. M., Nishiyama, H., Kato, N., & Fouda, M. M. (2013). Intrusion Detection System (IDS) for /Combating Attacks Against Cognitive Radio Networks. *IEEE Network*, 31(3), 51–56. DOI: 10.1109/MNET.2013.6523809

FlexRay Consortium. FlexRay Communications System—Protocol Specification—Version 2.1 Revision A. Available online: https://www.google.com.hk/url?sa=t&rct=j&q=&esrc=s&source=web&cd=1&ved=2ahUKEwi--6TIo5nnAhVjL6YKHSzcDb8QFjAAegQIBBAB&url=https%3A%2F%2Fsvn.ipd.kit.edu%2Fnlrp%2Fpublic%2FFlexRay%2FFlexRay%25E2%2584%25A2%2520Protocol%2520Specification%2520V2.1%2520Rev.A.pdf&usg=AOvVaw0snIyyfkFMHWc7KHRLdS82

Herber, C., Richter, A., Wild, T., & Herkersdorf, A. (2015, March). Real-time capable can to avb ethernet gateway using frame aggregation and scheduling. In 2015 Design, Automation & Test in Europe Conference & Exhibition (DATE) (pp. 61-66). IEEE.

Hung, C. W., & Hsu, W. T. (2018). Power Consumption and Calculation Requirement Analysis of AES for WSN IoT. *Sensors (Basel)*, 18(6), 1675. DOI: 10.3390/s18061675 PMID: 29882865

Jadoon, A. K., Wang, L., Li, T., & Zia, M. A. (2018). Lightweight Cryptographic Techniques for Automotive Cybersecurity. *Wireless Communications and Mobile Computing*, 2018(1), 1640167. DOI: 10.1155/2018/1640167

Kim, J. H., Seo, S. H., Nguyen, T., Cheon, B. M., Lee, Y. S., & Jeon, J. W. (2015). Gateway Framework for In-Vehicle Networks based on CAN, FlexRay and Ethernet. *IEEE Transactions on Vehicular Technology*, 64(10), 4472–4486. DOI: 10.1109/TVT.2014.2371470

Lee, T. Y., Lin, I. A., & Liao, R. H. (2020). Design of a FlexRay/Ethernet gateway and security mechanism for in-vehicle networks. *Sensors (Basel)*, 20(3), 641.

Lee, Y. S., Kim, J. H., & Jeon, J. W. (2017). FlexRay and Ethernet AVB Synchronization for High QoS Automotive Gateway. *IEEE Transactions on Vehicular Technology*, 66(7), 5737–5751. DOI: 10.1109/TVT.2016.2636867

Liu, J. J., Zhang, S. B., Sun, W., & Shi, Y. P. (2017). In-Vehicle Network Attacks and Countermeasures: Challenges and Future Directions. *IEEE Network*, 31(5), 55–58. DOI: 10.1109/MNET.2017.1600257

Mateus, K., & Königseder, T. (2014). *Automotive Ethernet* (1st ed.). Cambridge University Press. DOI: 10.1017/CBO9781107414884

Müter, M., Groll, A., & Freiling, F. C. (2010, August). A structured approach to anomaly detection for in-vehicle networks. In *2010 Sixth International Conference on Information Assurance and Security* (pp. 92-98). IEEE.

National Institute of Standards and Technology. Recommendation for Block Cipher Mode of Operation: The CCM Mode for Authentication and Confidentiality. Available online: https://www.nist.gov/publications/recommendation-block-cipher-modes-operation-ccm-mode-authentication-and-confidentiality (accessed on 23 January 2020).

PATAK Engineering FlexRay Controller Documentation. Available online: http://patakengineering.eu/download/FlexRayController.pdf

Puhm, A., Rössler, P., Wimmer, M., Swierczek, R., & Balog, P. (2008, September). Development of a flexible gateway platform for automotive networks. In *2008 IEEE International Conference on Emerging Technologies and Factory Automation* (pp. 456-459). IEEE.

Pullen, D., Anagnostopoulus, N. A., Arul, T., & Katzenbeisser, S. (2020). *"Securing FlexRay-based in-vehicle networks", Microprocessors and Microsystems*. Elsevier.

Sakiz, F., & Sen, S. (2017). A survey of attacks and detection mechanisms on intelligent transportation systems: VANETs and IoV. *Ad Hoc Networks*, 61, 33–50. DOI: 10.1016/j.adhoc.2017.03.006

Sander, O., Hubner, M., Becker, J., & Traub, M. (2008, December). Reducing latency times by accelerated routing mechanisms for an FPGA gateway in the automotive domain. In *2008 International Conference on Field-Programmable Technology* (pp. 97-104). IEEE.

Shane, T., Martin, G., Ciaran, H., Edward, J., Mohan, T., & Liam, K. (2015). Intra-Vehicle Networks: A Review. *IEEE Transactions on Intelligent Transportation Systems*, 16(2), 534–545. DOI: 10.1109/TITS.2014.2320605

Shreejith, S., Mundhenk, P., Ettner, A., Fahmy, A., Steinhorst, S., Lukasiewycz, M., & Chakraborty, S. (2017). VEGa: A High Performance Vehicular Ethernet Gateway on Hybrid FPGA. *IEEE Transactions on Computers*, 66(10), 1790–1803. DOI: 10.1109/TC.2017.2700277

Soares, F. L., Campelo, D. R., Yan, Y., Ruepp, S., Dittmann, L., & Ellegard, L. (2015, March). Reliability in automotive ethernet networks. In 2015 11th International Conference on the Design of Reliable Communication Networks (DRCN) (pp. 85-86). IEEE.

Stefan, P. S., & Groza, B. (2020). *Efficient Physical Layer Key Agreement for FlexRay Networks*. Transactions on Vehicular Technology.

Sumra, I. A., Ahmad, I., & Hasbullah, H., & bin Ab Manan, J. (2011). Classes of attacks in VANET, in 2011 Saudi International Electronics, Communications and Photonics Conference (SIECPC).

Wang, L., & Liu, X. (2018). NOTSA: Novel OBU With Three-Level Security Architecture for Internet of Vehicles. *IEEE Internet of Things Journal*, 5(5), 3548–3558. DOI: 10.1109/JIOT.2018.2800281

Chapter 8
Enhancing User Experiences in Cyber-Physical Systems for Real-Time Feedback and Intelligent Automation

Vishakha Kuwar
https://orcid.org/0000-0003-1764-2475
Centre for Online Learning, Dr. D.Y. Patil Vidyapeeth, India

Amit Yadav
https://orcid.org/0009-0006-3769-4773
Banarsidas Chandiwala Institute of Professional Studies, India

Yatharth Srivastava
https://orcid.org/0009-0003-1134-4722
The LNM Institute of Information Technology, India

Pratibha Bhide
https://orcid.org/0009-0002-2308-8781
MGM University, India

Sonali Gaur
EMPI Business School, India

Shitiz Upreti
Maharishi Markandeshwar, India

ABSTRACT

The hominize of cyber-physical systems (CPS) has create evolution in the user experiences for real-time feedback and intelligent automation. The study explores the advanced sensors and monitoring technologies in cyber-physical systems offer meaning to the user interactions with products, trying massive responsive and adaptable environments. CPS is facilitating the direct feedback and achieve the product usage patterns to discover and implement value-adding features is not

DOI: 10.4018/979-8-3693-5728-6.ch008

explicitly request but would significantly appreciate. The chapter is also discussing these technological advancements. The study illustrating safety and convenience. The chapter is also highlighting -the transformative impact of CPS on product interaction and user satisfaction.

1.INTRODUCTION

Cyber-Physical Systems (CPS) amalgamates sensors and real-time feedback and enhance user experience and intelligent automation across the various industries. Systems plays vital role in different sectors optimize resource management and yield through precise monitoring (Wati K., 2024). CPS leverage advanced level sensors and real-time feedback mechanisms and provide a level of interaction and responsiveness unattainable earlier. The product teams enhance the direct feedback from users but also to gain insights for the products which are used in everyday scenarios. When monitoring user interactions and behavioural patterns, CPS facilitate the continuous improvement of products functionality and overall value from the user point of view (Martinez C.,2023).

The extensive application of fundamental components of cyber-physical systems (CPS) like memory units, sensors, storage devices, and computing units, etc. Notably, the energy and utility industries are expected to drive substantial demand for CPS for smart grid technology. CPS play a vital role in advancing autonomous vehicles and intelligent transportation systems. The increasing focus on boosting electric vehicle sales is anticipated to propel the demand for CPS. The embeds of CPS across these sectors is poised to enhance efficiency, safety, and innovation. The typical Cyber-Physical System framework as visible in the Figure 1 below:

Figure 1. Cyber-Physical System

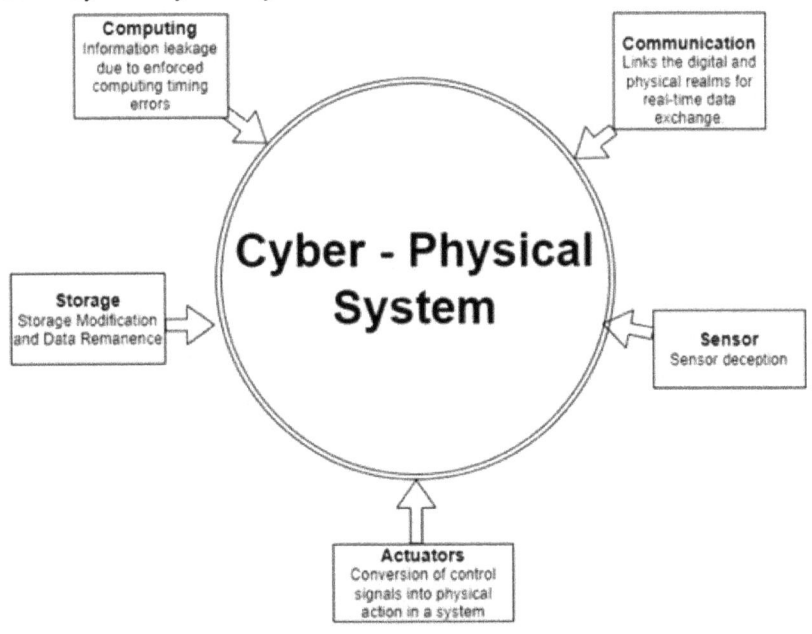

The important advancements offered by CPS ability to implement real-time adjustments based on the recent running data. The surveillance systems are also equipped with heat sensors for the detection of potential fire hazards before they escalate, trigger automatic responses and mitigate risks. The smart home appliances are equipped with user preferences and update setting sensors. The applications such as optimizing lighting and temperature is focused on occupancy and external weather conditions (Paredes C.et al.,2024). These are the capabilities illustrate CPS can enhance safety and convenience issues and user needs adaptation without input.

The impact of CPS extends individual products beyond the entire environments. In smart homes and intelligent infrastructure can orchestrate complex interactions on different devices (Baliyan A.et al.,2023). Cyber Physical systems to create a cohesive and adaptive ecosystem. Integrated systems in a smart home manage the energy consumption efficiently. The real time settings occupancy patterns and external factors. System not only improves the user experience but also contributes to energy conservation and cost savings (Zhang L.et al.,2023). CPS enables a new level of product innovation provides the insights which were previously difficult to obtain. The product developer used user interaction real time data to increase the opportunities. (Nardelli P.,2022). The CPS allows continuous improvement of products and services to ensures the user needs and preferences. The incorporation ability of user feedback and behaviour data to greater customer gratification and loyalty.

The user experiences are climacteric operational efficiency and safety. The responses are automated to detect anomalies or fluctuations. CPS reduces likelihood of human error and timely interventions (Alzubi K.et al.,2022). The capability is explicitly improving the critical environments such as industrial settings and healthcare. The real-time and fast responses are essential for the operational integrity and safety maintenance.

When the technology upgrade continues the potential applications of CPS becomes vast and varied. Everyday conveniences are addressed the complex challenges in industrial and environmental contexts. CPS offers transformation of user experiences and operational effectiveness (Chikurtev D.et al.,2022). The chapter explores innovations, interactions between users and technology, and paving the way for a more responsive and intuitive way.

2. REAL-TIME FEEDBACK AND CYBER-PHYSICAL SYSTEMS (CPS)

Real-time feedback is a component of cyber-physical systems (CPS). Real-time feedback facilitates prompt and adaptive interaction between the user and the environment. CPS incorporates sophisticated sensors and data processing technologies to continuously monitor user behavior and ambient variables. The dissemination of information enables CPS to provide immediate feedback, product features, and system operations (Zhang L. et al., 2023). Smart home systems can modify lighting and climate settings based on the simultaneous presence of people and their preferences. The technology provides a comfortable and efficient living environment without necessitating manual modifications (Upreti K.et al.,2021).

The real-time feedback By providing actionable insights for the products under use, CPS also enhances product development and user satisfaction. The CPS continuous monitoring can collect valuable data on user interactions and enable product developers to identify patterns, preferences, and potential issues. The information facilitates iterative improvements and allows more responsive and user-centric design modifications. The wearable health monitors can track vital signs and activity levels in real-time, offering immediate health insights and recommendations to users while simultaneously providing data to healthcare providers for ongoing care adjustments (Chen J. et al., 2024).

The Vos viewer keywords analysis for Cyber Physical system as seen in the Figure 2 as the analysis used index keywords and authors keywords in which it is clearly visible that the less importance given to the keywords block codes, fault control, reliability, feedback system, physical environment, security etc. Research gaps in cyber-physical systems (CPS) presented significant opportunities for advancement.

One of the major areas is the lack of universal interoperability standards, which is going to complicate integration across diverse platforms. Many CPS are vulnerable to cyberattacks (Singh A. and Jain A., 2018), underscoring the need for adaptive security measures. Data privacy is a critical concern, as extensive data collection in CPS necessitates innovative privacy-preserving techniques.

On top of that, there is a pressing need for research into human-machine interaction to improve usability through better user behavior insights and interface design. The strategies of resilience and disruptions recovery remain underexplored, as do the challenges of scaling algorithms and infrastructure as systems grow. Real-time decision-making capabilities based on dynamic data inputs require further enhancement. Over and above, the blend of emerging technologies like blockchain and edge computing is an area ripe for exploration. Effective lifecycle management practices for maintenance and updates throughout a CPS's lifespan are crucial, and there is an increasing demand for research on how CPS can support sustainability goals. Addressing these gaps can significantly improve the efficiency and effectiveness of CPS in various applications.

Figure 2. Vos viewer Keyword Analysis for Cyber Physical System

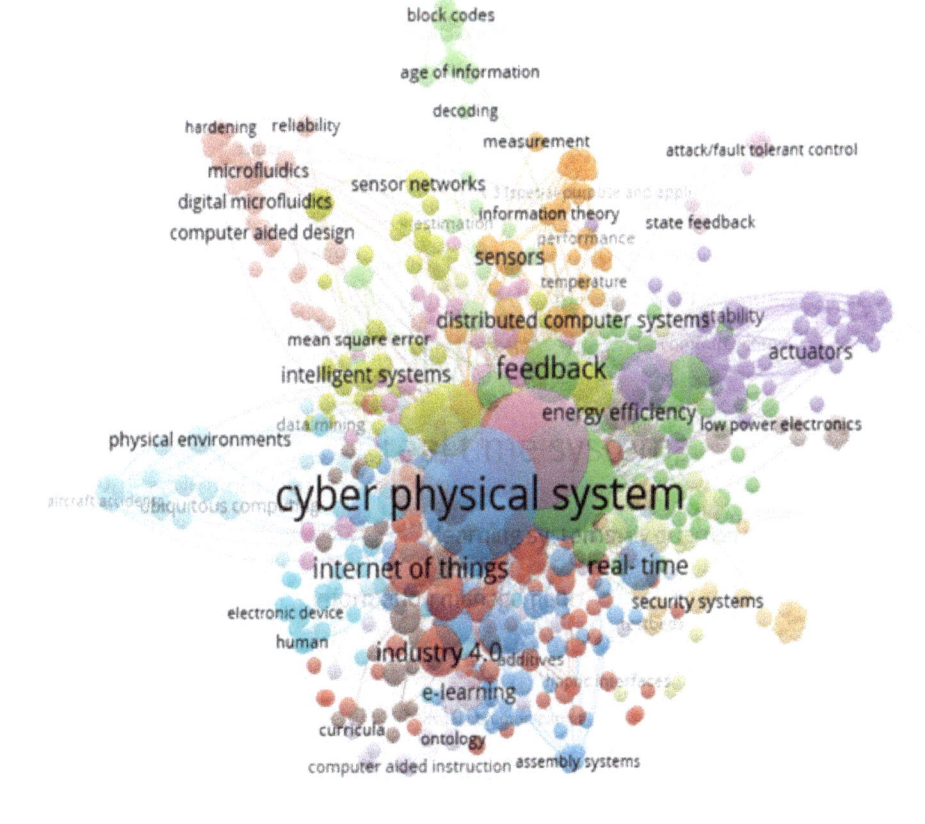

The CPS was able to process and act on real-time feedback, which significantly improves operational efficiency and safety. CPS detected and responded the anomalies in industrial settings, such as machine malfunctions or hazardous conditions. CPS also plays a crucial role in elevating critical issues (Claure R. et al., 2024). According to Das S. and Bhattacharjee S. (2024) added the real-time feedback to CPS making adaptive, smart systems for user needs and environmental changes.

3. INTELLIGENT AUTOMATION AND CYBER-PHYSICAL SYSTEMS (CPS)

Intelligent automation is a capability for operational efficiency and user experience. CPS can automate complex tasks and decision-making processes. Cyber-physical systems (CPS) and intelligent automation are becoming more intertwined and efficiency across various sectors. (Ragunthar T.et al.,2024). Robot controls in cyber-physical systems (CPS) face issues like packet loss and slow responses due to the computer units and physical devices work together. To overcome this study suggests a 2-level fuzzy feedback scheduling method that improve the stability and performance of these systems, even when conditions are uncertain. The Real-time data reliable on sensor information for the computing elements of Autonomous Systems (ASs) is vital for tasks execution that protect human lives and strategic assets. The Internet of Things (IoT) and Cyber-Physical Systems (CPS) the heavily rely on dependable sensor data to functions performance. The sensor based Physically Unclonable Function (PUF) utilization without imposing restrictions on its potential application for cryptographic authentication, identification and generation. There are two key focuses emphasized on the sensor malfunction absence, anticipate predictable responses and for the malfunction of sensor noticeable shift in the responses and indicating the need for disregard and replacement of sensor data information (Claure R.et al.,2024).

The higher education considers advancing digital equality, improving digital fluency, contributing to new ways of teaching complexity, expanding access to learning platforms and preparing next generations to adapt for rapidly changing future of the work conditions (Nazila R.et al.,2024). CPS improve the smart living household operations such as smart water networks, smart grid metering occupancy patterns and environmental conditions. The level of automation enhances convenience for users by proactively managing their environment according to their preferences and routines. The intelligent automation in consumer devices such as smart refrigerators or voice-controlled assistants, anticipate user needs and perform tasks like ordering groceries or adjusting appliance settings based on user habits for a seamless and personalized experience (Sajal K.et al.,2024).

The human-centric smart manufacturing (HSM) is one of the trends towards the integration of human-in-the-loop with technologies, to address challenges of human-machine relationships (Wang B. et al.,2022). HCPS in production, human-centric design, and service are presented with linkage to its enabling technologies and core features/characteristics such as Humachine, Human digital twin and brain robotics.

3. A. Enhancing Robustness through Sensor Technology

Intelligent automation and cyber-physical systems (CPS) require multimodal sensory network to address the problems in posing substantial potential for human health monitoring, human-machine relationships, and human-centric smart manufacturing. The development of flexible porous bimodal pressure sensors (PBPS) enables real-time monitoring of human physiological signals and manufacturing processes and showcasing high sensitivity and durability. The elements to real-time monitor 3D printing manufacturing stress, structure assembling stress and human physiology signals in HCPS indicating the potential applications of PBPS ranging from monitoring human health to facilitating intelligent manufacturing (Li Z.et al., 2023). Industrial CPS also demonstrates the complex deployment of software sensors as well as hardware. The effectiveness of data-centric methodology is demonstrated and represent the two ends of the industrial CPS complexity spectrum (Odyurt U.et al., 2022).

The typical applications of cyber–physical systems (CPSs) are automated and connected vehicles (CAVs) are excellent in measuring the share local information and surroundings with the other vehicles by using wireless networks and multimodal sensors. System is expected to increase safety, efficiency, and capacity of our transportation systems (Yang T. and Lv C.,2021). The adaptive actuator and sensor attack resilient control addresses strategies simplify the control design process and performance. System minimizes the joint effect of estimation communication and errors channel noise on the tracking string behavior and performance.

3.B. Trust and Security Challenges

Intelligent automated industrial process control needs an upper level of connectedness and systems integration than what has traditionally been the case. With such upgradation comes high risk of cyber-attack for Operational Technology (OT) systems like Industrial Control Systems (ICS). The automated process risk taking physical process information into account, to make response decisions based on potential digital and process and physical world consequences (Houmb S.et al., 2023). The models comprise to combine the physical characteristics of the process with the features of the other system layers. The incorporation of intelligent technologies within CPS introduces additional vulnerabilities, making robust cybersecurity frameworks essential to address both cyber and physical threats (Hu S. et al., 2022).

To enhance safety in CPS, a real-time mixed-trust computing framework used for protection and verification. The framework application task consists of an untrusted and a trusted part. The untrusted part permits complex computations supported. The trusted part is implemented by another scheduler within the hypervisor and is thus protected from the untrusted part. If the untrusted part fails to finish the specified time, the trusted part is activated to preserve safety including time span. The framework use of untrusted components for CPS critical functions preserving logical and timing guarantees, even in the presence of malicious attackers. (de Niz D.et al., 2023). Security of cyber-physical systems (CPS) continues to face the new challenges due to the tight integration and operational complexity of the physical components and cyber. To address these challenges, relate to the domain-aware, optimization-based approach to determine an effective defence strategy for CPS (Banik S. et al., 2023).

4. CHALLENGES OF ADVANCED SENSORS IN CYBER-PHYSICAL SYSTEMS (CPS)

CPS education in education is comprehensive and has interdisciplinary approach which emphasis on safety and security in real-world applications and integration of emerging technologies. This education will equip students with the knowledge and skills necessary to thrive in a rapidly evolving technological world while being aware of the trust, safety and security aspects in designing and using the CPS interface (Gaggatur J.,2023). Sensors provide precise measurements and operational in consistently varying conditions. Factors such as calibration, environmental interference, and sensor wear can affect data quality, necessitating rigorous maintenance and calibration practices to ensure dependable operation.

4.A. Anomaly Detection Difficulties

The deep learning advancement is helpful for the detection different anomalies for cyber-physical systems (CPS). The study outlines the challenges of the combination of cyber and physical components, discusses various deep learning models used in CPS applications, and highlights their effectiveness (Luo Y.et al.,2021). The identification of contribution includes a resource constraint, extreme heterogeneity that hampers industry consensus, lack of standardized communication protocols, and different information security priorities between Information Technology (IT) networks and Operational Technology (OT). Potential solutions and/or opportunities for further research are identified to address these selected challenges. (Jeffrey N.et al.,2023).

4.B. Sensor Reliability and Health

The design of a process for evaluating the reliability of Internet-of-Things sensors in a cyber-physical system. The co-simulation framework is designed to enable real-time interaction between real and virtual sensors. The study supporting the experimental evaluation of the co-simulation framework is presented using real and simulated data. The effectiveness for sensor reliability in cyber-physical systems using Internet-of-Things data. (Castaño F.et al., 2019). The technologies capacity is to encourage patient care, upgrade operational efficiency, and facilitate personalized medicine. The vital challenges such as data privacy, interoperability, and the need for robust analytical tools. The sensory reliability highlighted opportunities for innovation such as development of real-time monitoring systems and predictive analytics to support decision-making. Underscore the transformative potential of big data and CPS in healthcare while advocating for continued research to address the associated challenges (Cabello, J.et al.,2020).

4.C. Security Concerns

The security correlated with cyber-physical systems (CPS) is necessary. The unique vulnerabilities that arise from the cyber and physical components which can be exploited by the different types of cyber threats. The key security issues like data integrity, access control, and the potential impact of cyberattacks on physical processes. The more emphasize on the comprehensive security issues and strategies to protect CPS from emerging threats. The study advocates for ongoing research and collaboration to enhance the resilience and security of the critical systems in various industrial applications (Alguliyev R.et al.,2018).

The critical security faced cyber-physical systems (CPS) and proposed the potential solutions. The vulnerabilities inherent in the mesh of cyber and physical components which can lead to significant risks including data breaches and system malfunctions. The complexities of security for diverse applications and developing adaptive security measures reacted to the threats. The need for a multi-layered security approach, incorporating both technological and organizational strategies, to effectively safeguard CPS. The security and resilience of these increasingly interconnected systems (Zhou K.et al.,2017; Singh U.et al.,2023). Human Integration in Cyber-Physical Systems" emphasizes the importance of ergonomic user interfaces in enhancing human merger with cyber-physical systems (CPS). The design principles prioritize usability, safety, and efficiency, the need for user-centred approaches that involve feedback throughout the development process.

5. USER EXPERIENCE AND CYBER-PHYSICAL SYSTEMS (CPS)

User experience (UX) in Cyber-Physical Systems (CPS) is important for enhancing functionality and user satisfaction across various applications, from smart cities to athletic training. User experience and cyber-physical systems (CPS) have recently gained attention in research aimed at improving safety, security, and efficiency across various applications. The dataset of sensor and user-centric design for the effectiveness of CPS (Pavlovic M.,2020).

The fusion of cyber-physical systems (CPS) has revolutionized user experience in more interactive, responsive, and personalized interactions. CPS merged physical devices with digital intelligence by sensors, actuators, and data processing capability, users to engage with systems. Smart home systems CPS is adjusted lighting, temperature, and security settings based on real-time data for user behavior and environmental conditions (Cachada A.et al.,2019). The seamless interaction improves comfort and convenience, transforming routine activities into automated and intuitive experiences.

The impact of CPS on user experience lies in its capability to provide instantaneous feedback. CPS utilizes advanced sensors to monitor user interactions and environmental factors, allowing systems to promptly respond to user requirements (Raisin S. et al., 2020). In a smart manufacturing CPS setup, machinery settings can be automatically adjusted based on real-time sensor data, leading to optimized performance and reduced downtime. Wearable health devices equipped with CPS can offer immediate health insights and alerts, empowering users to make timely decisions regarding their well-being. The real-time adjustment and feedback capacity ensures that users receive instant and pertinent responses to their actions, ultimately enhancing the overall user experience.

Another advantage of CPS is its ability to deliver personalized experiences. By analyzing data from diverse sensors and user interactions, CPS can customize its functions to align with individual preferences and requirements. In the context of smart homes, CPS can learn a user's daily routines and adapt settings such as lighting and climate, accordingly, thereby creating a personalized environment that anticipates and responds to individual preferences. This level of personalization not only enhances user satisfaction but also cultivates a more immersive and intuitive interaction with technology, as systems become more attuned to individual preferences and behaviors (Kathiravelu P. et al., 2019). However, the deployment of CPS also introduces challenges in maintaining a positive user experience. Ensuring the accuracy and reliability of sensor data is crucial for delivering effective and meaningful interactions (Dafflon B.et al.,2021). Inaccurate or inconsistent data can lead to incorrect system responses, diminishing user satisfaction and trust. Additionally, the complexity of CPS systems can sometimes result in a steep learning curve for users, particularly if the systems are not intuitively designed or if users are not familiar with advanced technological interfaces. Addressing these issues requires serious design and user education to ensure that CPS enhances rather than complicates the user experience.

Recent view of value positioning in new industrial production is provided by industrial cyber-physical systems (CPS). Industrial CPS deployment in actual production presents both difficult and satisfying challenges. Following various practical cases of industrial CPS in use, the industrial CPS specifying the components and essential technological characteristics as well as the alignment with previous work such as Industrial revolutions 4.0 principles was explained. Every element and main technological feature is explained together with the variations between conventional industrial systems and industrial CPS. The heterogeneous character of industrial CPS creates difficulties for the development of such systems as well as for long-term sustainability of its design. Industrial CPS research is still under progress; current methods and fresh ideas to solve these sub-challenges are still developing. These understanding will enable industrial practitioners and researchers to build and commercialize industrial CPS. (Hoffmann M. et al., 20211). By addressing these privacy and security challenges, developers can create systems that not only offer advanced functionality and personalization but also uphold the highest standards of data protection, enhancing the overall user experience.

6. CYBER-PHYSICAL SYSTEMS PLATFORM SELECTION

The effective measures taken for the challenges and maximized the calibre of cyber-physical systems (CPS) and where the organizations need robust security strategies. The first layer involves evaluating the existing CPS protection platform to ensure it meets security needs. The Key criteria for selecting a strong CPS protection platform include first is Industry Expertise in which choose platforms with proven experience and a commitment to advancing CPS protection, evidenced by awards and collaborations with manufacturers vulnerabilities (Koutsoukos X.et al.,2017).

Second layer is Deep Visibility Option for solutions both active and passive discovery methods to achieve comprehensive visibility of connected CPS devices. Third layer is broad solution Set looking for vendors that support and ample deployment scenarios and network architectures (Silva E.et al.,2021). Fourth layer is the business outcomes in which centralized management, monitoring, and control, risk management and overall security. Fifth layer is the deployment flexibility option for solutions on-premises or in the cloud, which helps reduce costs and customization based on specific organizational needs.

7. CASE STUDY

The deeper understanding of the potential and capabilities of the user experiences in cyber physical system is seen in wide range of applications. The Industrial Internet of Things (IIoT) network is interconnected various devices which helps to improve the productivity of the processes and the efficiency of the industrial application. The IIoT benefited the predictive maintenance, data analysis in real time, control on quality and smooth supply chain and logistics. In case of Industrial control system is the category of management, regulations and commanding the different industrial operations. The operational efficiency improves in the execution, safety and effective point of view (Tyagi A and Sreenath N.,2021).

The Smart Building and Building Management System are created to monitor, control, optimize and manage the various systems within the infrastructure like HVAC, Fire safety, Security and electricity. When the integration of sensors, software's, controllers use for lighting, power consumption, ventilations and environment sustainability. The Smart Grids technology provides the efficient decision making, real-time monitoring facility and distribution of energy using the digital technology, software and sensor. The Supply Chain Management uses the Cyber Physical System in a Smart Transportations Systems way (Cederbladh J.et al.,2024). The Smart transportation helps in autonomous vehicles, route planning and traffic management effectively.

8. FUTURE TRENDS

The Smart cyber-physical systems (CPS) are suitable for the various applications for manufacturing, healthcare, agriculture and automobiles sector where it comes to the user experience exciting and transformative. CPS ability is driven non-intrusive and convenient for rapid advancements in technology, personalization and integration. (Puliafito A.et al.,2021). Trend is the increased integration of CPS for next-generation interfaces and devices. Augmented reality (AR) and virtual reality (VR) technologies are poised to create more immersive user experiences by digital information blend with the physical world in innovative ways (Yaacoub J.et al.,2020). For example, CPS-powered AR systems could overlay contextual information and interactive elements directly onto users' view of their physical environment. CPS enhancing their ability to interact and manipulate their surroundings (Broo D.et al.,2021).

Cyber-physical systems are dynamic and complicated; so, operational resilience must be addressed in addition to cyber threats. Whatever the reason, systems faults and equipment failures will create disturbance to operations and so affect the company and pose hazards to operational availability and so business continuity. The monitoring and cyber security systems must cover the whole cyber-system to satisfy the demands of the contemporary company since IT systems and ICS are getting more and more linked. Although the effects may be the same, knowing the fundamental cause will help one respond; this calls for constant observation of the cyber as well as the physical parts of the system.

Furthermore, the gathered data across the system must be examined and an answer produced by artificial intelligence technology if one is to address the issue of response speed. For an all-encompassing operational resilience approach, cyber-physical systems thus demand cyber-physical monitoring and automated detection and reaction. The implication of this is that the support teams must be organized such that system owners—information and operational—know how to react and why (Cachada A.et al.,2019). Moreover, the expansion of CPS into new domains and industries will drive further innovation in user experience. From smart cities and autonomous vehicles to healthcare and industrial automation, CPS will increasingly be applied to diverse fields, each with unique requirements and opportunities. This diversification will prompt the development of specialized user interfaces and interaction models tailored to different contexts and user needs, fostering more targeted and effective experiences across various applications (Hoffmann M.et al.,2021). Embracing these emerging trends in CPS is essential for moving toward a smarter, more efficient, and interconnected future.

CPS technology becomes more ubiquitous, there will be an emphasis on creating user-friendly and accessible systems. The challenge of making advanced technology intuitive and easy to use will be a central focus, particularly as these systems integrate into everyday life. Efforts to streamline user interfaces, simplify interactions, and enhance accessibility will ensure that CPS remains inclusive and beneficial to a wide range of users (Lesch V.et al.,2023). By addressing these considerations, the future of CPS will not only advance technological capabilities but also ensure that these advancements contribute positively to the user experience and overall quality of life.

9. CONCLUSION

In this chapter, the era of cyber-physical systems (CPS) user experience which is characterized by interactivity, personalization, and adaptability. Cyber-physical systems and so cyber-physical monitoring are the direction of future. Integration of cyber and physical (IT and ICS) is unavoidable. And it goes beyond industry and manufacturing as well. From smart cities and homes to the cars we drive and travel on, cyber-physical systems are increasingly noticeable in all spheres of human life. It should not distinguish anymore; rather, it should welcome the advantages and benefits that actual cyber-physical systems offer. It also must acknowledge that need the new generation of cyber-physical monitoring technologies using artificial intelligence to make sense of the data and support quick decision-making when time is of the matter.

REFERENCES

Alzubi, K. M., Alaloul, W. S., & Qureshi, A. H. (2022). Applications of cyber-physical systems in construction projects. In *Cyber-Physical Systems in the Construction Sector* (pp. 153–171). CRC Press. DOI: 10.1201/9781003190134-9

Baliyan, A., Kaswan, K. S., Kumar, N., Upreti, K., & Kannan, R. (Eds.). (2023). Cyber Physical Systems: Concepts and Applications.

Banik, S., Ramachandran, T., Bhattacharya, A., & Bopardikar, S. D. (2023). Automated Adversary-in-the-Loop Cyber-Physical Defense Planning. *ACM Transactions on Cyber-Physical Systems*, 7(3), 1–25. DOI: 10.1145/3596222

Broo, D. G., Boman, U., & Törngren, M. (2021). Cyber-physical systems research and education in 2030: Scenarios and strategies. *Journal of Industrial Information Integration*, 21, 100192. DOI: 10.1016/j.jii.2020.100192

Cabello, J. C., Karimipour, H., Jahromi, A. N., Dehghantanha, A., & Parizi, R. M. (2020). Big-data and cyber-physical systems in healthcare: Challenges and opportunities. Handbook of Big Data Privacy, 255-283.

Cachada, A., Barbosa, J., Leitao, P., Deusdado, L., Costa, J., Teixeira, J., . . . Moreira, P. M. (2019). Development of ergonomic user interfaces for the human integration in cyber-physical systems. In 2019 IEEE 28th International Symposium on Industrial Electronics (ISIE) (pp. 1632-1637). IEEE DOI: 10.1109/ISIE.2019.8781101

Castaño, F., Strzelczak, S., Villalonga, A., Haber, R. E., & Kossakowska, J. (2019). Sensor reliability in cyber-physical systems using internet-of-things data: A review and case study. *Remote Sensing (Basel)*, 11(19), 2252. DOI: 10.3390/rs11192252

Cederbladh, J., Eramo, R., Muttillo, V., & Strandberg, P. E. (2024). Experiences and challenges from developing cyber-physical systems in industry-academia collaboration. *Software, Practice & Experience*, 54(6), 1193–1212. DOI: 10.1002/spe.3312

Chen, J., de Mendonça, J. L. V., Ayele, B. S., Bekele, B. N., Jalili, S., Sharma, P., . . . Jeannin, J. B. (2024). Synchronous Programming with Refinement Types. Proceedings of the ACM on Programming Languages, 8(ICFP), 938-972 DOI: 10.1145/3674657

Chikurtev, D., Ivanov, V., Yosifova, V., & Dimitrov, D. (2022). Cyber-physical system for intelligent control of infrared heating. *IFAC-PapersOnLine*, 55(11), 37–41. DOI: 10.1016/j.ifacol.2022.08.045

Claure, R. E. M., Heynssens, J. B., Burke, I., Alam, M., & Cambou, B. (2024). Enhancing cyber-physical systems (CPS) robustness through sensor-pair health indicator. In Autonomous Systems: Sensors, Processing, and Security for Ground, Air, Sea, and Space Vehicles and Infrastructure 2024 (Vol. 13052, pp. 21-28). SPIE

Dafflon, B., Moalla, N., & Ouzrout, Y. (2021). The challenges, approaches, and used techniques of CPS for manufacturing in Industry 4.0: A literature review. *International Journal of Advanced Manufacturing Technology*, 113(7-8), 2395–2412. DOI: 10.1007/s00170-020-06572-4

Das, S. K., & Bhattacharjee, S. (2024). Science of Cyber Physical Security in Smart Living CPS Applications. In 2024 IEEE International Conference on Smart Computing (SMARTCOMP) (pp. 5-5). IEEE de Niz, D., Andersson, B., Klein, M., Lehoczky, J., Vasudevan, A., Kim, H., & Moreno, G. Mixed-Trust Computing: Safe and Secure Real-Time Systems. ACM Transactions on Cyber-Physical Systems DOI: 10.1109/SMARTCOMP61445.2024.00022

Gaggatur, J. S. (2023). *Designing and Implementing a Cyber-Physical Systems (CPS) Education Program for Pre-University Students. In 2023 IEEE Technology & Engineering Management Conference-Asia Pacific (TEMSCON-ASPAC)*. IEEE.

Hoffmann, M. W., Malakuti, S., Grüner, S., Finster, S., Gebhardt, J., Tan, R., Schindler, T., & Gamer, T. (2021). Developing industrial cps: A multi-disciplinary challenge. *Sensors (Basel)*, 21(6), 1991. DOI: 10.3390/s21061991 PMID: 33799891

Houmb, S. H., Iversen, F., Ewald, R., & Færaas, E. (2023, February). Intelligent risk-based cybersecurity protection for industrial systems control-a feasibility study. In International Petroleum Technology Conference (p. D021S014R001). IPTC DOI: 10.2523/IPTC-22795-MS

Hu, S., Yu, S., Li, H., & Piuri, V. (2022). Guest Editorial Special Issue on Security, Privacy, and Trustworthiness in Intelligent Cyber–Physical Systems and Internet of Things. *IEEE Internet of Things Journal*, 9(22), 22044–22047. DOI: 10.1109/JIOT.2022.3207335

Jeffrey, N., Tan, Q., & Villar, J. R. (2023). A review of anomaly detection strategies to detect threats to cyber-physical systems. *Electronics (Basel)*, 12(15), 3283. DOI: 10.3390/electronics12153283

Kathiravelu, P., Van Roy, P., & Veiga, L. (2019). SD-CPS: Software-defined cyber-physical systems. Taming the challenges of CPS with workflows at the edge. *Cluster Computing*, 22(3), 661–677. DOI: 10.1007/s10586-018-2874-8

Koutsoukos, X., Karsai, G., Laszka, A., Neema, H., Potteiger, B., Volgyesi, P., Vorobeychik, Y., & Sztipanovits, J. (2017). SURE: A modeling and simulation integration platform for evaluation of secure and resilient cyber–physical systems. *Proceedings of the IEEE*, 106(1), 93–112. DOI: 10.1109/JPROC.2017.2731741

Lesch, V., Züfle, M., Bauer, A., Iffländer, L., Krupitzer, C., & Kounev, S. (2023). A literature review of IoT and CPS—What they are, and what they are not. *Journal of Systems and Software*, 200, 111631. DOI: 10.1016/j.jss.2023.111631

Li, Z., Wang, S., Ding, W., Chen, Y., Chen, M., Zhang, S., Liu, Z., Yang, W., & Li, Y. (2023). Mechanically robust, flexible hybrid tactile sensor with microstructured sensitive composites for human-cyber-physical systems. *Composites Science and Technology*, 244, 110303. DOI: 10.1016/j.compscitech.2023.110303

Luo, Y., Xiao, Y., Cheng, L., Peng, G., & Yao, D. (2021). Deep learning-based anomaly detection in cyber-physical systems: Progress and opportunities. *ACM Computing Surveys*, 54(5), 1–36. DOI: 10.1145/3453155

Martinez Spessot, C. (2023). Cyber-Physical Automation. In *Springer Handbook of Automation* (pp. 379–404). Springer International Publishing. DOI: 10.1007/978-3-030-96729-1_17

Nardelli, P. H. (2022). *Cyber-physical systems: Theory, methodology, and applications*. John Wiley & Sons. DOI: 10.1002/9781119785194

Odyurt, U., Pimentel, A. D., & Alonso, I. G. (2022). Improving the robustness of industrial Cyber–Physical Systems through machine learning-based performance anomaly identification. *Journal of Systems Architecture*, 131, 102716. DOI: 10.1016/j.sysarc.2022.102716

Paredes, C. M., Martínez Castro, D., González Potes, A., Rey Piedrahita, A., & Ibarra Junquera, V. (2024). Design Procedure for Real-Time Cyber–Physical Systems Tolerant to Cyberattacks. *Symmetry*, 16(6), 684. DOI: 10.3390/sym16060684

Pavlovic, M. (2020). Designing for ambient UX: design framework for managing user experience within cyber-physical systems. L. RAMPINO, I. MARIANI, 39

Ragunthar, T., Kaliappan, S., & Ali, H. M. (2024). Detection of Feedback Control Through Optimization in the Cyber Physical System Through Big Data Analysis and Fuzzy Logic System. In *AI Approaches to Smart and Sustainable Power Systems* (pp. 299–313). IGI Global. DOI: 10.4018/979-8-3693-1586-6.ch016

Raisin, S. N., Jamaludin, J., Rahalim, F. M., Mohamad, F. A. J., & Naeem, B. (2020). Cyber-physical system (CPS) application-a review. REKA ELKOMIKA: Jurnal Pengabdian kepada Masyarakat, 1(2), 52-65

Sajal, K., & Das, S. (2024). Science of Cyber Physical Security in Smart Living CPS Applications. DOI: 10.1109/SMARTCOMP61445.2024.00022

Silva, E. M., & Jardim-Goncalves, R. (2021). Cyber-Physical Systems: A multi-criteria assessment for Internet-of-Things (IoT) systems. *Enterprise Information Systems*, 15(3), 332–351. DOI: 10.1080/17517575.2019.1698060

Singh, A., & Jain, A. (2018, April). Study of cyber-attacks on cyber-physical system. In *Proceedings of 3rd International Conference on Internet of Things and Connected Technologies (ICIoTCT)* (pp. 26-27)

Singh, U. K., Sharma, A., Singh, S. K., Tomar, P. S., Dixit, K., & Upreti, K. (2023). Security and privacy aspect of cyber physical systems. In *Cyber Physical Systems* (pp. 141–164). Chapman and Hall/CRC.

Tyagi, A. K., & Sreenath, N. (2021). Cyber Physical Systems: Analyses, challenges and possible solutions. *Internet of Things and Cyber-Physical Systems*, 1, 22–33. DOI: 10.1016/j.iotcps.2021.12.002

Upreti, K., Syed, M. H., Alam, M. S., Alhudhaif, A., Shuaib, M., & Sharma, A. K. (2021). Generative adversarial networks based cognitive feedback analytics system for integrated cyber-physical system and industrial iot networks.

Wang, B., Zheng, P., Yin, Y., Shih, A., & Wang, L. (2022). Toward human-centric smart manufacturing: A human-cyber-physical systems (HCPS) perspective. *Journal of Manufacturing Systems*, 63, 471–490. DOI: 10.1016/j.jmsy.2022.05.005

Wati, K. M. (2024). Cyber Physical System For Automated Weather Station And Agriculture Node In Smart Farming. Globe: Publikasi Ilmu Teknik, Teknologi Kebumian. *Ilmu Perkapalan*, 2(1), 13–27.

Yaacoub, J. P. A., Salman, O., Noura, H. N., Kaaniche, N., Chehab, A., & Malli, M. (2020). Cyber-physical systems security: Limitations, issues and future trends. *Microprocessors and Microsystems*, 77, 103201. DOI: 10.1016/j.micpro.2020.103201 PMID: 32834204

Yang, T., & Lv, C. (2021). A secure sensor fusion framework for connected and automated vehicles under sensor attacks. *IEEE Internet of Things Journal*, 9(22), 22357–22365. DOI: 10.1109/JIOT.2021.3101502

Zhang, L., Sridhar, K., Liu, M., Lu, P., Chen, X., Kong, F., . . . Lee, I. (2023). Real-time data-predictive attack-recovery for complex cyber-physical systems. In 2023 IEEE 29th Real-Time and Embedded Technology and Applications Symposium (RTAS) (pp. 209-222). IEEE DOI: 10.1109/RTAS58335.2023.00024

Zhou, K., Liu, T., & Liang, L. (2017). Security in cyber-physical systems: Challenges and solutions. *International Journal of Autonomous and Adaptive Communications Systems*, 10(4), 391–408. DOI: 10.1504/IJAACS.2017.088775

Chapter 9
Enhancing Human–Computer Interaction Through Artificial Intelligence and Machine Learning:
A Comprehensive Review

Neha Singh
https://orcid.org/0009-0000-5097-8945
Invertis University, India

Jitendra Nath Shrivastava
Invertis University, India

Gaurav Agarwal
Invertis University, India

Akash Sanghi
https://orcid.org/0000-0002-3532-7981
Invertis University, India

Swati Jha
Invertis University, India

Kamal Upreti
https://orcid.org/0000-0003-0665-530X
Christ University, India

Ramesh Chandra Poonia
https://orcid.org/0000-0001-8054-2405
Christ University, India

Amit Kumar Gupta
KIET Group of Institutions, India

ABSTRACT

The interaction between Human-Cyber-Physical Systems (CPS) has become increasingly critical as CPS technologies permeate various facets of modern life, from smart homes to industrial automation. This highlights the evolving landscape of

DOI: 10.4018/979-8-3693-5728-6.ch009

research aimed at fostering smooth interaction between humans and CPS, stressing the necessity of bridging the gap between users and these intricate systems.Effective interaction with CPS requires a profound understanding of human behaviors, preferences, and cognitive processes. .Furthermore, this emphasizes notable research trends aimed at improving Human-CPS interaction, including the exploration of innovative interaction modalities such as natural language processing, gesture recognition, and brain-computer interfaces.Thanks to Artificial Intelligence (AI) and Machine Learning (ML), computer interactions are changing a lot.

1.INTRODUCTION

This chapter leverages the power of machine learning and artificial intelligence to enable and enhance the productivity of human-computer interaction. Human-computer interaction is a field of emerging basket of technologies which interact with each other and human (Baheti, R. and Gill.H,2011).

Cyber Physical System: A cyber-physical system (CPS) is a network of interconnected computational and physical components that interact to monitor, control, and optimize physical processes. These systems integrate advanced computing, communication, and control technologies to monitor, analyze, and manage physical processes in real-time (Baheti, R. and Gill.H,2011). Let's delve into the details of a cyber system:

It includes:

1. **Cyber Layer**: The cyber layer of a cyber system encompasses the computational and communication infrastructure that processes data, runs algorithms, and facilitates interaction between different system components. Key elements of the cyber layer include:
    - **Computational Units:** These units consist of computing devices such as microprocessors, microcontrollers, and embedded systems that execute software programs and algorithms.
    - **Software Platforms:** Cyber systems rely on software platforms and operating systems to manage computational tasks, handle communication protocols, and provide interfaces for interaction with users or other systems.
    - **Communication Networks:** Networks enable data exchange between different components of the cyber system, allowing seamless coordination and collaboration. Communication technologies include wired and wireless protocols such as Ethernet, Wi-Fi, Bluetooth, and Zigbee.

- **Data Storage and Management**: Cyber systems require storage solutions to store and retrieve data generated by sensors, actuators, and other components. Data management techniques ensure efficient storage, retrieval, and processing of large volumes of data.
2. **Physical Layer**: The physical layer of a cyber system comprises the tangible, real-world elements or processes that are monitored, controlled, or influenced by the system. Physical components interact with the environment and produce data that is captured and processed by the cyber layer. Key elements of the physical layer include:
    - **Sensors:** Sensors find and calculate physical quantities such as temperature, pressure, humidity, motion, or light. They convert physical phenomena into electrical signals that can be processed by computational units in the cyber layer.
    - **Actuators:** Actuators are devices that convert electrical signals from the cyber layer into physical actions or changes in the environment. Examples include motors, valves, switches, and pumps that control movement, flow, or other physical processes.
    - **Physical Processes:** Physical processes refer to the natural or man-made phenomena that the cyber system aims to monitor, control, or optimize. These processes can range from industrial manufacturing and transportation systems to environmental monitoring and smart infrastructure.
3. **Interaction and Control**: The interaction and control mechanisms of a cyber system enable feedback loops, decision-making, and adaptation to achieve desired objectives (Berger, S., Häckel, B., & Häfner, 2021). Interaction between the cyber and physical layers involves sensing physical data, processing it in the cyber layer, and generating appropriate control signals to actuate physical components (Woo, H., Yi, J., and et. al. (2008). Key aspects of interaction and control include:
    - **Feedback Loops**: Feedback loops enable continuous monitoring of physical processes, comparison with desired states or goals, and adjustment of system parameters to maintain desired performance or stability (Berger, S., Häckel, B., & Häfner, 2021).
    - **Control Algorithms**: Control algorithms utilize data from sensors and actuators to make decisions and adjustments in real-time (Bujorianu, M. and Barringer, H. (2009).

- . These algorithms can range from simple PID (Proportional-Integral-Derivative) lers to more complex model-based or adaptive control strategies (Berger, S., Häckel, B., & Häfner, 2021).

- **Adaptation and Learning**: Some cyber systems incorporate machine learning or artificial intelligence techniques to adapt and learn from data, improving performance, efficiency, and autonomy over time (Woo, H., Yi, J., and et. al. (2008).

Applications: Cyber systems find applications across various domains (Baheti, R. and Gill.H,2011)., including:

- **Smart Cities:** Cyber systems enable smart city initiatives by optimizing urban infrastructure, transportation networks, energy grids, and environmental monitoring.
- **Industrial Automation:** In manufacturing and industrial settings, cyber systems enhance automation, process control, predictive maintenance, and quality assurance.
- **Healthcare**: Cyber systems support telemedicine, remote patient monitoring, medical device integration, and personalized healthcare delivery.
- **Transportation:** In transportation systems, cyber systems enable intelligent traffic management, autonomous vehicles, fleet optimization, and logistics.

Challenges and Considerations: Cyber systems face several challenges and considerations, including:

- **Security and Privacy:** Ensuring the security and privacy of data, communication, and control systems against cyber threats and vulnerabilities (Calinescu, R., Camara, J., & Paterson, C., 2019).
- **Reliability and Safety:** Guaranteeing the reliability and safety of cyber-physical systems, particularly in safety-critical applications such as healthcare and transportation (Woo, H., Yi, J., and et. al. (2008).
- **Interoperability and Standards:** Promoting interoperability and compatibility between different cyber-physical systems, devices, and platforms through standardized communication protocols and interfaces.
- **Ethical and Societal Implications:** Addressing ethical, legal, and societal implications of cyber systems, including issues related to data ownership, algorithmic bias, and automation-induced job displacement (Calinescu, R., Camara, J., & Paterson, C., 2019).

2. ROLE OF MACHINE LEARNING (ML) AND ARTIFICIAL INTELLIGENCE(AI) IN HCI ENHANCEMENT

1. The role of Machine Learning (ML) and Artificial Intelligence (AI) in Human-Computer Interaction (HCI) is substantial and continues to evolve with advancements in technology and research (Baheti, R. and Gill.H,2011). Top of FormArtificial Intelligence refers to the simulation of human intelligence in machines that are programmed to Learn, think and perform tasks autonomously. AI encompasses a wide range of techniques, algorithms, and methodologies aimed at enabling computers to mimic various cognitive functions such as pattern recognition, problem-solving, decision-making, and natural language understanding (Z.-H. Zhou,2021). Here are some key systems of AI:

   - **Natural Language Processing (NLP):** NLP enables computers to interpret, comprehend, manipulate and generate human language. NLP techniques include parsing, sentiment analysis, machine translation, and speech recognition, facilitating communication between humans and machines.
   - **Computer Vision:** Computer vision involves the development of algorithms and models that enable computers to object identification and facial recognition, as well as classification, recommendation from images or videos. Applications of computer vision include object detection, image classification, facial recognition, and autonomous navigation.
   - **Machine Learning (ML):** ML is a subset of AI that focuses on building algorithms and statistical models that enable computers to learn from data and make decisions and predictions without being explicitly programmed. ML algorithms can be categorized into supervised learning, unsupervised learning, semi-supervised learning, and reinforcement learning.
   - **Deep Learning:** Deep learning is a subfield of ML that depends on artificial neural networks with multiple layers (deep neural networks). Deep learning models learn hierarchical representations of data, enabling them to achieve state-of-the-art performance in tasks such as image recognition, speech recognition, and natural language processing.
   - **Reinforcement Learning:** Reinforcement learning is a type of ML where an agent learns to make decisions by interacting with an environment and receiving feedback in the form of rewards or penalties or it is a technique that trains software to make decisions to achieve the most optimal results. Reinforcement learning algorithms aim to maximize cumulative reward over time through trial-and-error learning.

Some key feature that can be addressed by these can be:

- **Personalized Interfaces:** Develop AI-powered interfaces that adapt to individual user preferences and behaviors. ML algorithms can analyze user interactions and patterns to tailor the interface in real-time, providing a more intuitive and personalized experience.
- **Predictive Assistance:** Implement AI-driven predictive assistance to anticipate user needs and provide proactive support within CPS. ML models can analyze historical data and user behavior to predict future actions, helping users navigate complex systems more efficiently.
- **Natural Language Processing (NLP):** Integrate NLP capabilities to enable natural and conversational interactions between users and CPS. AI-powered chatbots or voice assistants can understand and respond to user queries, commands, and feedback, enhancing the usability and accessibility of CPS.
- **Gesture Recognition:** Incorporate ML-based gesture recognition technology to enable hands-free interaction with CPS. Users can control and command systems through gestures, improving usability in environments where traditional input methods are impractical or inconvenient.
- **Gesture and Motion Recognition:** ML algorithms can interpret gestures and motions captured by sensors, cameras, or wearables, enabling touchless interaction with computing devices. This technology enhances accessibility and enables new interaction modalities in HCI.
- **Emotion Recognition:** Utilize AI algorithms for emotion recognition to enhance user experience and system responsiveness. ML models can analyze facial expressions, tone of voice, or physiological signals to detect user emotions and adapt system behavior accordingly, fostering a more empathetic interaction.
- **Context-Aware Systems:** Develop context-aware CPS that leverage AI and ML to understand the user's environment and situation. By analyzing contextual cues such as location, time, and activity, systems can dynamically adjust their behavior to better meet user needs and preferences.
- **Adaptive Automation:** Implement adaptive automation techniques that utilize ML to optimize task allocation and decision-making in CPS. Systems can learn from user interactions and performance feedback to autonomously adjust automation levels, balancing user control and system autonomy for optimal efficiency and safety.
- **Continuous Learning:** Enable CPS to continuously learn and improve from user interactions and feedback. Utilize online learning algorithms in ML models to update and refine system behavior over time, ensuring adaptation to evolving user preferences and changing environments.

- **Explainable AI:** Ensure transparency and trustworthiness in AI-powered CPS by incorporating explainable AI techniques. Provide users with insights into how AI algorithms make decisions and recommendations, empowering them to understand and trust the system's behavior.
- **Predictive Modeling:** AI and ML models can predict user actions, intentions, and preferences based on historical data, enabling proactive assistance to CPS and anticipatory user interfaces in CPS. This enhances efficiency and user satisfaction by reducing cognitive load and decision-making effort on CPS ().
- **Recommendation Systems:** AI-powered recommendation systems analyze user data to suggest relevant content, products, or actions, enhancing user engagement and satisfaction. In CPS, recommendation systems help users discover new information and streamline decision-making processes.
- **Affective Computing:** ML techniques enable computers to recognize and respond to human emotions, leading to emotionally intelligent CPS. By detecting emotional cues from facial expressions, voice tone, or physiological signals, computers can adapt their responses to users' emotional states.
- **User Behavior Analysis:** AI and ML algorithms analyze user interactions with interfaces to gain insights into usability issues, user preferences, and engagement patterns. This data-driven approach informs iterative design improvements and enhances the overall user experience in CPS.
- **Accessibility:** AI and ML contribute to improving accessibility in CPS by automating the process of captioning, transcription, and audio description for users with disabilities. Additionally, ML algorithms can adapt interfaces to accommodate diverse user needs, such as font size adjustments for visually impaired users.
- **Gesture and Speech Recognition:** AI and ML techniques power advanced gesture and speech recognition systems, enabling hands-free and voice-controlled interaction with computing devices. These technologies enhance accessibility and facilitate natural interaction modalities in CPS.
- **Autonomous Agents and Intelligent Assistants:** AI-powered autonomous agents and intelligent assistants interact with users to perform tasks, provide information, or offer recommendations. These systems leverage ML algorithms for natural language understanding, decision-making, and learning from user interactions.

Note: Prioritize user-centered design principles throughout the development process of AI and ML-enabled HCI in CPS. Involve end-users in the design and evaluation stages to ensure that systems effectively meet their needs, preferences, and usability requirements. AI and ML play a crucial role in shaping the future of CPS by enabling more natural, adaptive, and intelligent interactions between humans and computers. As these technologies continue to advance, they hold the potential to revolutionize the way we interact with computing systems and enhance the overall user experience.

3. THE IMPACT OF AUTOMATION ON EMPLOYMENT AND SOCIAL IMPLICATIONS ARISE

When integrating Artificial Intelligence (AI) and Machine Learning (ML) into Human-Computer Interaction (HCI), several ethical, legal, data privacy, algorithmic bias, and the impact of automation on employment and social implications arise. These considerations are crucial for ensuring responsible use and fostering public trust in these technologies (Calinescu, R., Camara, J., & Paterson, C., 2019).

Ethical Implications:

a) **Bias and Fairness:** AI systems can perpetuate or amplify biases present in training data, leading to unfair treatment of certain user groups. It's essential to implement fairness-aware algorithms to mitigate these biases and ensure equitable outcomes (Calinescu, R., Camara, J., & Paterson, C., 2019).
b) **Transparency:** Users should understand how AI systems make decisions. Lack of transparency can lead to mistrust, especially in critical applications like healthcare or finance. Employing explainable AI techniques can help users comprehend the rationale behind system actions (Calinescu, R., Camara, J., & Paterson, C., 2019).
c) **Privacy Concerns:** The collection and processing of personal data raise significant privacy issues. Users must be informed about data usage, and robust measures should be in place to protect sensitive information from unauthorized access or misuse (Calinescu, R., Camara, J., & Paterson, C., 2019).
d) **Autonomy and Control:** As AI systems become more autonomous, there is a risk of diminishing human control over critical decisions. It is vital to ensure that users retain the ability to intervene and override automated decisions when necessary (Calinescu, R., Camara, J., & Paterson, C., 2019).

Legal Implications

a) **Data Protection Regulations:** Compliance with laws such as the General Data Protection Regulation (GDPR) is essential when handling personal data. Organizations must establish clear protocols for data collection, storage, and processing to avoid legal repercussions (Calinescu, R., Camara, J., & Paterson, C., 2019).
b) **Accountability:** Determining liability in cases of AI failure or harm is complex. Clear legal frameworks need to be established to define accountability for actions taken by AI systems, particularly in high-stakes environments like autonomous vehicles or medical diagnostics.
c) **Intellectual Property**: The use of AI-generated content raises questions regarding intellectual property rights. Legal standards must evolve to address ownership issues related to creations made by AI systems.

Social Implications

a) **Job Displacement**: The automation of tasks through AI and ML can lead to job displacement in various sectors. Strategies for workforce retraining and upskilling are necessary to mitigate the impact on employment.
b) **Digital Divide**: Access to advanced AI technologies may exacerbate existing inequalities, with marginalized communities potentially missing out on benefits. Efforts should be made to ensure equitable access and inclusion in technology adoption.
c) **Public Trust:** Building trust in AI systems is vital for their widespread acceptance. Engaging with communities, addressing concerns transparently, and demonstrating the benefits of AI can help foster public confidence.
d) **Cultural Impact**: The integration of AI into daily life can alter social interactions and cultural norms. Understanding these shifts is important for designing systems that respect cultural values and promote positive social outcomes.

Data Privacy Protection:

a) **Data Minimization:** Collect only the data necessary for specific tasks to reduce exposure. Implement anonymization techniques to protect user identities.

b) **End-to-End Encryption:** Use encryption for data transmission and storage to safeguard sensitive information from unauthorized access (Calinescu, R., Camara, J., & Paterson, C., 2019).

c) **User Consent Mechanisms:** Develop clear and user-friendly consent processes that inform users about data collection practices and allow them to control their data sharing preferences.

Mitigating Algorithmic Bias:

a) **Diverse Training Data:** Ensure that training datasets are representative of various demographics to minimize biases in AI models. Regularly audit datasets for fairness.

b) **Bias Detection Tools:** Utilize tools and frameworks designed to detect and measure bias in algorithms, allowing for adjustments before deployment.

c) **Inclusive Design Processes:** Involve diverse stakeholders in the design process to gather different perspectives and ensure that the system meets the needs of all user groups.

Addressing Automation's Impact on Employment:

a) **Retraining Programs:** Implement workforce retraining initiatives that equip employees with skills relevant to evolving job markets influenced by automation.

b) **Job Redesign:** Redesign roles to focus on human-centric tasks that leverage uniquely human skills, such as creativity and emotional intelligence, alongside automated systems.

c) **Stakeholder Engagement:** Engage with employees, unions, and industry stakeholders in discussions about automation impacts, ensuring that their voices are heard in decision-making processes.

Continuous Monitoring and Feedback Loops:

a) **Real-Time Monitoring:** Implement systems for continuous monitoring of AI performance and user interactions to identify issues related to bias or privacy breaches promptly (Woo, H., Yi, J., and et. al. (2008).

b) **Feedback Mechanisms:** Establish channels for users to provide feedback on AI behavior, enabling iterative improvements based on real-world usage (Woo, H., Yi, J., and et. al. (2008).

Ethical Governance Frameworks:

a) **Ethics Committees:** Form ethics committees within organizations tasked with overseeing AI implementations, ensuring adherence to ethical standards and practices.
b) **Transparency Reports:** Publish regular transparency reports detailing data usage, algorithm performance, and measures taken to address ethical concerns.
c) Latest Research Insights
d) Recent studies have highlighted various approaches and findings relevant to the ethical implementation of AI and ML in CPS:
e) Privacy-Preserving Machine Learning:
f) Research has explored techniques such as federated learning, where models are trained across decentralized devices without sharing raw data, thus enhancing privacy while still benefiting from collective learning.

By addressing these ethical, legal, and social implications proactively, developers and organizations can create responsible AI systems that enhance Human-Computer Interaction while safeguarding user rights and societal values.

4.BACKGROUND OF AI AND ML IN CPS

The background of Machine Learning (ML) and Artificial Intelligence (AI) in Cyber-Physical Systems (CPS) stems from the convergence of various technological advancements and research efforts. Here's a brief overview of the key factors that have contributed to the integration of AI and ML in CPS:

The proliferation of Internet of Things (IoT)devices (Baheti, R. and Gill.H,2011) and sensor technologies has enabled the collection of vast amounts of data from physical environments. These sensors, embedded in various objects and systems, provide real-time information about the physical world, forming the foundation for CPS. With the increasing volume, velocity, and variety of data generated by CPS, there arose a need for advanced data analytics techniques to extract meaningful insights and enable decision-making. AI and ML algorithms emerged as powerful tools for analyzing large-scale data sets and identifying patterns, trends, and anomalies. The advancement of computational technologies, including parallel processing, distributed computing, and cloud infrastructure, has facilitated the implementation of complex AI and ML algorithms in CPS. High-performance computing resources enable real-time processing of data streams and efficient training of ML models. Control theory, a branch of engineering and mathematics, provides the theoretical foundation for designing and optimizing the control systems in CPS. The integration of AI and ML techniques with control theory enables the development of adaptive, robust, and intelligent control strategies for complex CPS applications. The field of CPS requires interdisciplinary collaboration between computer science, electrical engineering, mechanical engineering, and other domains. AI and ML researchers collaborate with experts in control systems, robotics, cyber security, and other fields to develop integrated solutions that leverage the strengths of each discipline. The growing adoption of CPS in various industry sectors, including manufacturing, transportation, healthcare, energy, and smart cities, has driven the demand for intelligent and autonomous systems. AI and ML technologies offer solutions to optimize operations, improve efficiency, enhance safety, and enable new capabilities in CPS applications. Academic and industrial research institutions have contributed to the advancement of AI and ML techniques specifically tailored for CPS. Researchers explore new algorithms, methodologies, and architectures to address the unique challenges of integrating AI and ML in real-time, safety-critical CPS environments. Standardization bodies and regulatory agencies develop guidelines, protocols, and standards for ensuring interoperability, reliability, and security in CPS. The integration of AI and ML in CPS requires compliance with industry standards and regulations to mitigate risks and ensure system robustness.

5. INTERACTION BETWEEN THE PHYSICAL AND CYBER LAYERS

In a Cyber-Physical System (CPS), sensors and actuators play crucial roles in the interaction between the physical and cyber layers. Here's an overview of how these components interact:

Interaction of Sensors and Actuators in CPS

1. Role of Sensors

Data Collection: Sensors are devices that detect and measure physical phenomena, such as temperature, pressure, humidity, motion, or light. They convert these physical quantities into electrical signals that can be processed by computational units in the cyber layer (Lin, J., Sedigh, 2011).

Real-Time Monitoring: Sensors continuously monitor the physical environment, providing real-time data that is essential for decision-making processes within the CPS (Lin, J., Sedigh, 2011).

2. Role of Actuators

Physical Action: Actuators are devices that convert electrical signals from the cyber layer into physical actions or changes in the environment. Examples include motors, valves, switches, and pumps that control movement, flow, or other physical processes (Woo, H., Yi, J., and et. al. (2008).

Response Mechanism: Actuators respond to commands generated by control algorithms based on sensor data, enabling the system to effect changes in the physical world.

3. Interaction Mechanism

Feedback Loops: The interaction between sensors and actuators is typically governed by feedback loops (Woo, H., Yi, J., and et. al. (2008).

Sensing: Sensors collect data about the current state of a physical process.

Processing: This data is sent to the cyber layer, where it is analyzed using algorithms (including AI and ML) to determine whether any action is needed (Zhang, Y., Yen, I., 2009).

Actuation: If an action is required, control signals are generated and sent to actuators to perform specific tasks (e.g., adjusting temperature or opening a valve).

Continuous Monitoring: The system continuously monitors the results of actuator actions through sensors, allowing for adjustments as needed.

4. Control Algorithms

Control algorithms utilize data from sensors to make decisions about how actuators should respond. These algorithms can range from simple proportional-integral-derivative (PID) controllers to more complex adaptive or model-based control strategies that incorporate machine learning techniques (Berger, S., Häckel, B., & Häfner, 2021).

5. Adaptation and Learning

Some CPS implementations leverage AI and ML to enhance their adaptability. By analyzing historical sensor data and actuator performance, these systems can learn over time to optimize their responses and improve overall efficiency.

Applications

The interaction between sensors and actuators in CPS has numerous applications across various domains:

Smart Cities: Managing traffic lights based on real-time traffic data collected by sensors.

Industrial Automation: Monitoring machinery conditions with sensors and adjusting operations via actuators for optimal performance.

Healthcare: Remote patient monitoring systems that adjust medication delivery based on sensor readings of vital signs.

By effectively integrating sensors and actuators within a CPS framework, these systems can achieve real-time monitoring and control of physical processes, leading to improved efficiency, safety, and responsiveness in various applications (Lin, J., Sedigh, 2011).

6. FOUNDATION OF AI AND ML IN CPS

The foundations of Machine Learning (ML) and Artificial Intelligence (AI)) in Cyber-Physical Systems (CPS) lie at the intersection of several key principles and methodologies. Here are the foundational elements that underpin the integration of AI and ML in CPS:

- **Data Acquisition and Sensing:** CPS relies on sensors and actuators to collect data from the physical world. The foundation of AI and ML in CPS begins with the acquisition of diverse sensor data, including measurements of temperature, pressure, motion, and other environmental parameters. This data forms the basis for learning and decision-making in CPS.
- **Data Representation and Feature Engineering:** AI and ML techniques require data to be represented in a suitable format for analysis and modeling. In CPS, data from sensors may be preprocessed and transformed to extract relevant features that capture important characteristics of the system dynamics. Feature engineering plays a crucial role in preparing data for ML algorithms.
- **Machine Learning Algorithms:** ML algorithms form the core of intelligent decision-making in CPS. These algorithms learn patterns and relationships from data to make predictions, classifications, or decisions. Supervised learning, unsupervised learning, and reinforcement learning are common types of ML techniques used in CPS for tasks such as predictive maintenance, anomaly detection, and optimization.

- **Control Theory and Optimization:** CPS integrates principles from control theory and optimization to regulate and optimize system behavior (Bujorianu, M. and Barringer, H. (2009). Control theory provides the foundation for designing feedback control systems that maintain desired states or trajectories in physical processes (Berger, S., Häckel, B., & Häfner, 2021). Optimization techniques, including linear programming, convex optimization, and evolutionary algorithms, are used to optimize system performance, resource allocation, and scheduling in CPS.
- **Real-Time Processing and Decision-Making:** Many CPS applications require real-time processing of data and decisions to respond to dynamic changes in the environment (Zhang, Y., Yen, I., 2009). AI and ML algorithms must be designed to operate efficiently within the constraints of real-time computing, ensuring timely responses and feedback loops in control systems, autonomous vehicles, smart grids, and other CPS domains.
- **Safety and Reliability:** Safety and reliability are paramount in CPS, where errors or failures can have serious consequences. The foundations of AI and ML in CPS include techniques for ensuring system safety, fault tolerance, and resilience to disturbances. This involves designing algorithms that can detect and recover from faults, as well as incorporating redundancy and fail-safe mechanisms into system architectures.
- **Interdisciplinary Collaboration:** The development of AI and ML solutions for CPS requires interdisciplinary collaboration between researchers and practitioners from various fields, including computer science, electrical engineering, mechanical engineering, and control systems (Bujorianu, M. and Barringer, H. (2009). This collaboration fosters the integration of domain-specific knowledge and expertise into AI and ML algorithms tailored for CPS applications.
- **Ethical and Societal Considerations:** The foundation of AI and ML in CPS includes ethical and societal considerations related to privacy, security, transparency, and fairness. Researchers and developers must address ethical challenges and ensure that AI and ML systems deployed in CPS adhere to ethical principles and legal regulations, promoting trust and accountability in intelligent systems.

By integrating these foundational elements, AI and ML technologies empower CPS with intelligence, autonomy, and adaptability, enabling a wide range of applications across industries such as manufacturing, transportation, healthcare, energy, and smart cities.

6. IMPLEMENTATION OF MACHINE LEARNING (ML) AND ARTIFICIAL INTELLIGENCE (AI) IN CYBER-PHYSICAL SYSTEMS (CPS)

Figure 1. AI-ML Implementation in CPS Environment

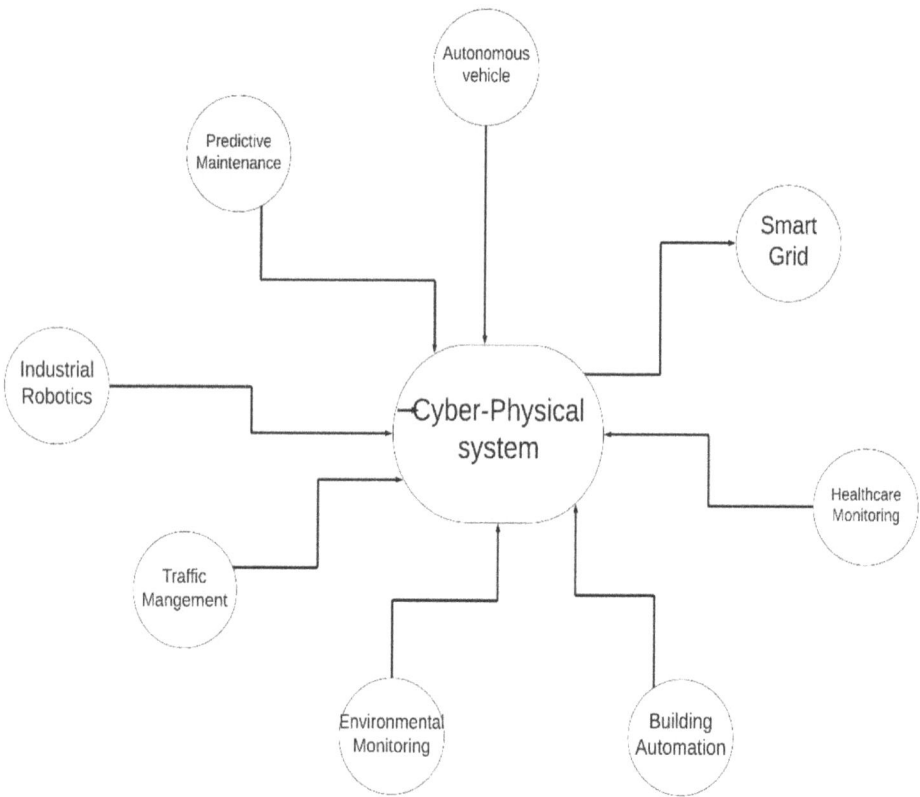

The implementation of Machine Learning (ML) and Artificial Intelligence (AI) in Cyber-Physical Systems (CPS) involves several key steps and considerations to ensure effective integration and deployment (Li, H., 2016). Here's an overview of the implementation process:

- **Problem Definition and Requirements Analysis:** The first step in implementing AI and ML in CPS is to define the problem or task that the system needs to address. This involves understanding the requirements, objectives, and constraints of the application domain, as well as identifying the key performance metrics and success criteria.

- **Data Collection and Preprocessing:** Data collection is crucial for training ML models and making intelligent decisions in CPS. This involves gathering data from sensors, actuators, and other sources in the physical environment. The collected data may need to be preprocessed to remove noise, handle missing values, and extract relevant features for modeling.
- **Model Selection and Training:** Once the data is collected and preprocessed, the next step is to select appropriate ML models for the task at hand. This may involve choosing between supervised, unsupervised, or reinforcement learning techniques based on the nature of the problem and the available data. ML models are then trained using labeled or unlabeled data to learn patterns and relationships.
- **Integration with Control Systems:** In many CPS applications, AI and ML techniques are integrated with traditional control systems to enhance their capabilities (Bujorianu, M. and Barringer, H. (2009). This involves designing feedback loops that incorporate ML predictions or decisions to regulate system behavior in real-time. The integration of AI and ML with control systems requires careful consideration of timing, stability, and robustness (Berger, S., Häckel, B., & Häfner, 2021).
- **Real-Time Processing and Deployment:** Once trained, ML models need to be deployed in CPS environments for real-time processing and decision-making (Zhang, Y., Yen, I., 2009). This may involve deploying models on embedded devices, edge servers, or cloud platforms depending on the latency and resource requirements of the application. Real-time processing ensures timely responses to changes in the physical environment.
- **Monitoring and Feedback:** After deployment, it's important to monitor the performance of AI and ML algorithms in CPS to ensure they continue to meet the desired objectives and requirements (Bujorianu, M. and Barringer, H. (2009). This may involve collecting feedback from sensors, analyzing system behavior, and evaluating model performance over time. Monitoring allows for proactive maintenance and optimization of AI and ML systems.
- **Adaptation and Learning:** AI and ML algorithms in CPS should be capable of adapting to changes in the environment and learning from new data over time. This may involve retraining ML models periodically with updated data to maintain their accuracy and relevance. Adaptive learning enables CPS to continuously improve and optimize their performance.
- **Safety and Security:** Safety and security are paramount in CPS implementations, especially when integrating AI and ML techniques. Robustness against failures, vulnerabilities, and malicious attacks must be considered throughout the design and implementation process. This may involve incorporating re-

dundancy, fault tolerance, and encryption mechanisms to protect CPS from potential threats.
- **Scalability and Interoperability:** CPS implementations should be scalable to accommodate growing data volumes and system complexity over time. Additionally, interoperability between different components and subsystems within CPS is essential for seamless integration and communication. Standards-based approaches and modular architectures facilitate scalability and interoperability in CPS implementations.
- **Validation and Verification:** Finally, it's important to validate and verify the performance of AI and ML implementations in CPS to ensure they meet the specified requirements and objectives. This may involve conducting simulations, testing in controlled environments, and performing field trials to assess system performance and reliability.

By following these steps and considerations, one can effectively implement AI and ML in Cyber-Physical Systems to enhance their capabilities, efficiency, and reliability across various domains and applications.

6. RESULTS OF AI AND ML IN CPS

In conclusion, the integration of Machine Learning (ML) and Artificial Intelligence (AI) in Cyber-Physical Systems (CPS) represents a transformative paradigm shift in the way we design, operate, and interact with complex systems that bridge the physical and digital worlds. Through the fusion of advanced computational techniques with physical infrastructure and sensors, AI and ML empower CPS with intelligence, autonomy, and adaptability, enabling a wide range of applications across industries and domains.

The adoption of AI and ML in CPS offers numerous benefits, including:

- **Improved Efficiency:** AI and ML algorithms optimize system operations, resource allocation, and decision-making processes, leading to increased efficiency and productivity in CPS applications such as manufacturing, transportation, energy management, and healthcare (Zhang, Y., Yen, I., 2009).
- **Enhanced Reliability:** By leveraging real-time data analysis and predictive modeling, AI and ML algorithms enhance the reliability and resilience of CPS by enabling proactive maintenance, fault detection, and recovery strategies, minimizing downtime and maximizing uptime.
- **Greater Automation:** AI and ML enable greater levels of automation in CPS, allowing systems to adapt and respond autonomously to changing envi-

ronmental conditions, user preferences, and operational requirements, reducing the need for manual intervention.
- **Advanced Decision Support:** ML techniques provide valuable insights and recommendations to human operators and decision-makers in CPS, helping them make informed decisions, optimize processes, and mitigate risks based on data-driven analysis and modeling (Bujorianu, M. and Barringer, H. (2009).
- **Innovative Applications:** The integration of AI and ML enables the development of innovative CPS applications and services that were previously unattainable, including autonomous vehicles, smart infrastructure, personalized healthcare systems, and intelligent manufacturing processes.
- However, the adoption of AI and ML in CPS also presents challenges and considerations, including:
- **Safety and Security:** Ensuring the safety, security, and reliability of AI and ML systems in CPS is paramount, as failures or vulnerabilities can have serious consequences. Robustness against cyber threats, adversarial attacks, and system failures must be addressed through rigorous testing, validation, and mitigation strategies.
- **Ethical and Societal Implications:** AI and ML raise ethical and societal concerns related to privacy, fairness, transparency, and accountability, particularly in CPS applications that impact human lives and well-being. Ethical guidelines, regulatory frameworks, and responsible AI practices are essential to address these concerns and foster trust in intelligent CPS.
- **Interdisciplinary Collaboration:** Successful implementation of AI and ML in CPS requires interdisciplinary collaboration between researchers, engineers, domain experts, and stakeholders from diverse fields, including computer science, engineering, physics, and social sciences. Collaboration fosters the integration of domain-specific knowledge and expertise into AI and ML solutions tailored for CPS applications.
- **Continuous Learning and Adaptation:** AI and ML algorithms in CPS must be capable of continuous learning and adaptation to evolving environments, requirements, and user needs. Incremental updates, online learning techniques, and self-tuning mechanisms enable CPS to remain responsive and adaptive over time.

In summary, the integration of AI and ML in CPS holds tremendous promise for advancing the capabilities, efficiency, and reliability of complex cyber-physical systems across various industries and domains. By addressing challenges and considerations related to safety, security, ethics, and collaboration, organizations can harness the full potential of AI and ML to create intelligent CPS that enhance human well-being, drive innovation, and shape the future of technology and society.

REFERENCES

Baheti, R. and Gill, H. (2011). Cyber-physical systems. The Impact of Control Technology, 161-166.

Berger, S., Häckel, B., & Häfner, L. (2021). Organizing self-organizing systems: A terminology, taxonomy, and reference model for enti- ties in cyber-physical production systems. *Information Systems Frontiers*, 23(2), 391–414. DOI: 10.1007/s10796-019-09952-8

Bujorianu, M., & Barringer, H. (2009). An integrated specification logic for cyber-physical systems. In *Proceedings of 14th IEEE International Conference on Engineering of Complex Computer Systems*, 291-300. DOI: 10.1109/ICECCS.2009.36

Calinescu, R., Camara, J., & Paterson, C. (2019). Socio-cyber-phys- ical systems: Models, opportunities, open challenges. In *5th IEEE/ACM International Workshop on Software Engineering for Smart Cyber-Physical Systems (SEsCPS) (pp. 2–6)*. IEEE. https://doi.org/DOI: 10.1109/SEsCPS.2019.00008

Gürdür Broo, D., Boman, U., & Törngren, M. (2021). Cyber-physical systems research and education in 2030: Scenarios and strate- gies. *Journal of Industrial Information Integration*, 21, 100192. DOI: 10.1016/j.jii.2020.100192

Li, D., Li, D., Liu, J., Song, Y., & Ji, Y. (2022). *Backstepping sliding mode control for cyberphysical systems under false data injection attack*. IEEE International.

Li, H. (2016). Introduction to cyber physical systems. Communications for Control in Cyber Physical Systems, 1–8. DOI: 10.1016/B978-0-12-801950-4.00001-9

Lin, J., Sedigh, S., and et.al. (2011). A semantic agent framework for cyber-physical systems.

Woo, H., & Yi, J., (2008). Design and development methodology for resilient cyber-physical systems. In *Proceedings of The 28th International Conference on Distributed Computing Systems Workshop*, 525-528. DOI: 10.1109/ICDCS.Workshops.2008.62

Zhang, Y., & Yen, I., (2009). Optimal adaptive system health monitoring and diagnosis for resource constrained cyber-physical systems. In *Proceedings of the 20th International Symposium on Software Reliability Engineering*, 51-60. DOI: 10.1109/ISSRE.2009.21

Zhou, Z.-H. (2021). *Machine Learning*. Springer Nature. DOI: 10.1007/978-981-15-1967-3

Chapter 10
Harnessing Real-Time Data for Intelligent Decision-Making in Cyber-Physical Systems

T. Premavathi
https://orcid.org/0009-0003-0172-2021
Marwadi University, India

Kumar J. Parmar
https://orcid.org/0000-0002-2502-5680
Marwadi University, India

Rituraj Jain
https://orcid.org/0000-0002-5532-1245
Marwadi University, India

Damodharan Palaniappan
https://orcid.org/0009-0003-0721-3068
Marwadi University, India

Vaishali Vidyasagar Thorat
https://orcid.org/0000-0003-2026-7661
D.Y. Patil School of Engineering and Technology, India

Chetana Vidhyasagar Thorat
Indiana University, Bloomington, USA

ABSTRACT

Cyber physical systems are already transforming different fields, including smart communities and energy systems. These technologies enable CPS to process large volumes of data and come up with insights that enhance processes of taking preventive actions for situations that require quick responses. In addition, we explain how both 5G and edge computing are set to disrupt data handling and transmission as well as outline how both concepts will fit well together in a resource limited environment. Points regarding data quality issues, system architecture, and security

DOI: 10.4018/979-8-3693-5728-6.ch010

threats; resource capacity and the degree of persistence, respectively, are included. For these challenges, we offer solutions like Data management, Modularity or System Decomposition, Security in Layers, and Light-Weight Processing for Improved System Resilience. Lastly, the chapter discusses the current and potential advances in real-time decision making for CPS and the need for CPSs to be interoperable with other CPS.

1. INTRODUCTION

1.1 Brief Definition of Cyber-Physical Systems (CPS)

Cyber-Physical Systems (CPS) transform the cyber and physical systems into an integrated system in which cyber components interact with the physical components to monitor or control other systems. CPS refer to the integration of physical substrates, be they machinery, vehicles, structures, or even whole infrastructures with computational or communication substrates (Rajkumar, 2023). These interactions create complex systems that are capable of inputs from the physical world, processing, making decision and self-correcting in the event of change. CPS thus has at its core systems and networks of sensors that communicate information from the physical environment to a computational environment. It allows physical systems to adapt to changes in a surrounding environment; it increases systems' production capability/decreases human interference, and generally increases the robustness of a system. CPS are core components of solutions like self-driving automobiles, power utility, industries, hospitals, and above all the smart city. As physical and computational systems become interconnected at the core of automation, CPS are becoming ubiquitous in the development of urban environment, manufacturing, transportation, and many other spheres (Quincozes et al., 2022) (Hoffmann et al., 2021). In Table 1, it will be possible to present brief information on the major components of a CPS and their purpose. Figure 1 described the system architecture of the cyber physical systems. Accordingly, excited techniques like edge computing, 5G, the AI-identification of abnormal data, and unique cybersecurity models play a critical role in addressing major issues in CPS namely data explosion, delay, distorted data, and cyber threats. These innovations guarantee that the CPS handles wanted actual-time data efficiently and safely in the modern purposes. Figure. 1 refers Overview of Cyber Physical Systems.

Table 1. Key Components of Cyber-Physical Systems (CPS)

Component	Function
Sensors	Collect real-time data from the physical environment
Actuators	Perform actions based on decisions made by CPS
Embedded Systems	Interface between physical components and computational systems
Communication Networks	Facilitate data exchange between components
Decision-Making Engine	Analyzes data and makes real-time decisions

Figure 1. Overview of Cyber Physical Systems

Components of Cyber-Physical Systems

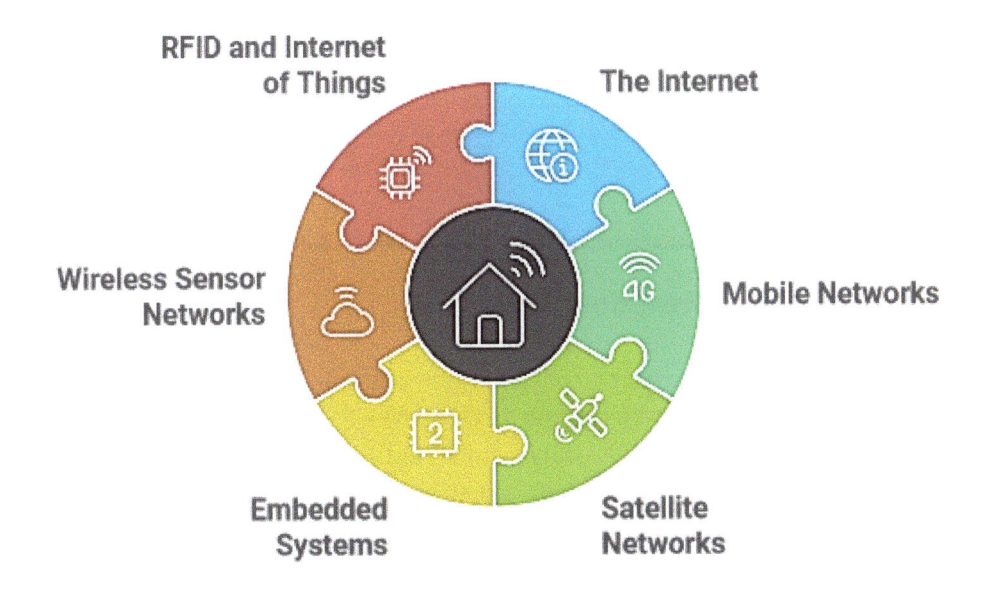

- RFID and Internet of Things
- The Internet
- Wireless Sensor Networks
- Mobile Networks
- Embedded Systems
- Satellite Networks

Figure 2 shows outline the system architecture of the layers of Cyber-Physical Systems (CPS): At the Physical Layer, there's sensors, IoT devices, and actuators, Data Processing Layer – edge computing, communication networks, and real-time filtering, and Decision-Making Layer – AI/ML models, Predictive Analytics. A cross-cutting Security and Privacy Layer guarantees protecting and maintain the data and system stability.

Figure 2. Basic Architecture of Cyber Physical Systems

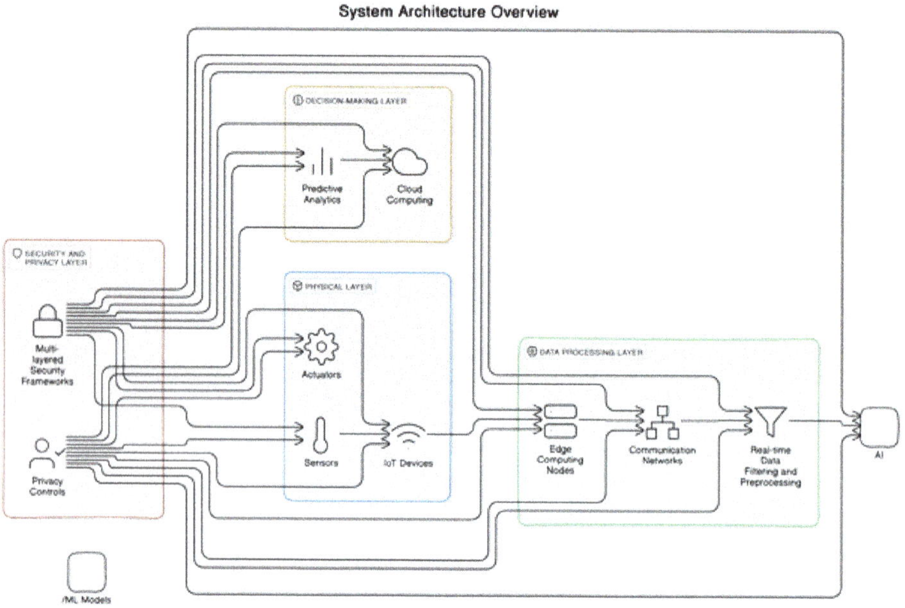

1.2 Importance of Real-time Data

The real-time data is the enabler of all the knowledge processing and operational feedback the CPS requires. When implemented, the CPS can operate fast utilizing real-time data which is accrued, analyzed, and utilized in due course. Most CPS applications like smart city traffic management, industrial robot systems, and healthcare monitoring systems demand real-time data (Zhang & Huang, 2024), (Halder et al., 2024), (Andersen, 2023). Without accurate, up-to-date information, systems would fail to react to changes in the operational environment adequately, reducing access to their benefits. The real-time data enable smart traffic system adapt traffic signals by congestion type, thus reducing journeys time and emissions. 'Smart' devices can be used to notify medical care givers of trends of worrying vital signs in real time. Business intelligence leverages the inputs from a system to predict future occurrences in an organisation and envision ways to enhance the performance of the system (Shaik et al., 2023).

1.3 Challenges in Decision-making for CPS

Real-time data integration with CPS decisions proves a challenge even were beneficial. A CPS system is expected to deal with huge amounts of data. Real-time big data platforms must work in parallel allowing no delays from millions of sensors and devices. This requires real time data architectures to filter, analyse and react to data. In mission-critical use cases such as self-driving cars and medical life-and-death situations, latency significantly impairs CPS functionality (Kankanamge et al., 2024), (Chen et al., 2023), (Aboualola, 2023). Incorrect and unreliable information for decision-making is the other problem that is closely related to this issue. Sensors are misconfigured, or the network is down, or there is a cyberattack; erroneous data enter real-time systems, creating unsafe deductions. CPS system integrity rests with the capability to identify and handle bad data. Also, the decision making of CPS is often elaborate and often must oscillate between automation and manual control. Therefore, when some issues are critical, it may not be appropriate to make totally autonomous decisions since they maybe best decided by humans (Chamberlain et al., 2020; De-Arteaga et al., 2020).

Further, cps decision-making is hampered by security as well as the privacy aspects. Security and privacy concerns are significant barriers to CPS decision-making. Such information that should be safeguarded includes; personal health records or city-wide monitoring feeds in these systems. The frameworks for decision making as a significant element of risk management should incorporate cybersecurity to protect information from hostile entities' intrusion and interference with functions. Hence, the CPS design should accommodate present time computation while at the same time considering data security and privacy (Quincozes et al., 2022) (Keshk, 2021) (Kanwade et al., 2023)). Finally, the distributed CPS's decision-making over multiple interacting systems makes it even more challenging. These environments demand that the subsystems must cooperate and time their decisions with one another since failures or slippages can generate system failures. Finally, effective distributed decision making is always a work in progress in the CPS field. In Table 2 we have synthesized the information from the current paper by listing key challenges to real-time decision making and the possible solutions to these challenges which give a broad perspective of the topic under discussion.

Table 2. Challenges in CPS Decision-Making and Solutions

Challenge	Description	Solution
Data Overload	Excessive data from sensors can create bottlenecks	Use distributed data processing and edge computing
Faulty or Inaccurate Data	Incorrect data can lead to poor decision-making	Implement error-detection and data validation mechanisms
Latency	Delays in data processing can hinder real-time decisions	Use low-latency communication protocols and edge computing
Security Vulnerabilities	Data breaches can compromise system integrity	Employ multi-layered security frameworks

1.4 Emergence of Intelligent Decision-Making in CPS

One must appreciate that Decision making with incorporation of AI & ML has been one of the most significant changes in the CPS. These technologies enable CPS to move from the level of relying upon predetermined set of operational rules to the level where systems learn from data, decide based on the feedback received from the surrounding environment and make complex decisions by themselves. Intelligent CPS gathers delicate information about the physical world, processes it, and derives decisions promptly (Akhuseyinoglu & Joshi, 2020). AI/ML models in CPS can be said used when there exists stochastic/fully dynamic environment in which rule-based systems cannot be employed. A self-driving car that contains an AI engine can maneuver around any object or path and even switch from a lesser path to a much better one. In smart grid systems, AI can control electric power distribution and mitigate blackouts by processing real time energy consumption data (Franki et al., 2023). CPS can respond and adapt to current events, as well as predict events in the future and then take preventive measures with the help of Artificial Intelligence and real time data. Predictive maintenance on the other hand, is the use of real time sensor data to determine when a machine is likely going to fail and schedule the maintenance to be done on the same before it fails (Proactive Measures to Avoid Failure, 2021). Getting intelligence in our decision making brings improvement in the productivity cost and stability of the systems we are using. However, bringing AI and real-time data into the CPS decision also introduce many new issues with this. For AI models to be effective and perform well they require good data to train and update continually (Upreti et al., 2024). For these models, data without error or bias is needed to prevent poor decision outcomes when using these models. AI-based decision making can also be complex to understand or justify, which can be especially problematic in decision-critical zones where explainability is important, for example in healthcare sector (Han et al., 2023).

2. REAL-TIME DATA ACQUISITION IN CPS

Based on the real time data stated in the previous section, data acquisition is the primary process of data integration in Cyber Physical Systems (CPS). The capacity with which CPS can make right decisions at the right time and at the right times depends on the capability of the CPS to collect real-time data of the physical world. Such data are collected from various sensors and IoT devices and must be channelled through preset protocols for processing and transmission to computational platforms that can analyze and utilize the data for decision making purposes. This section examines the practice of acquiring real-time data, the technologies that support the practice, and the issues involved with the management of such data. Table 1 summarize the main elements for Real-time Decision-making in CPS

2.1 Sensors and IoT Devices

CPS real time data acquisition is through sensors and IoT devices. These components interact with the physical environment to provide data acquired from actual systems and context. Temperature, pressure, motion, humidity and light sensors exist. It is also important to note that the smart city's infrastructure can contain sensors that can track traffic or air quality or energy utilization. In industrial CPS, sensors monitor the performance of the machinery to enable prediction and reduce the duration of the machinery breakdowns (Chaudhuri & Kahyaoğlu, 2023), (Liu & Yang, 2022). CPS is made possible through IoT devices to gather more detailed and dispersed data collecting opportunities. Numerous sensors are integrated into IoT devices, and these devices communicate via wireless links with one another as well as with data hubs or computing on the edge. This integration allows CPS to acquire status data from the various subsystems with the ability to give a real-time system status. Smart homes can include IoT devices for checking energy use, security systems and health monitoring gadget in people's homes, thus improving their efficiency and, security (Rejeb et al., 2024). However, the volume of IoT devices and sensors integrated in CPS causes new challenges relating to data management. Proper data captured, analyzed, and transferred without overloading the system is not easy when millions of devices are submitting data. Environmental sensors, for instance, lack consistency in measurement and, therefore, require accurate data to enhance decision making. To make this guide helpful to a diverse range of readers familiar with the concept of sensors and IoT devices for real-time data acquisition, the main point of this article is to provide a brief overview of the diverse types of sensors and IoT devices available and their usage. Real-time data acquisition of the different types of sensors and IoT devices are explained below in the table 3.

Table 3. Real-Time Data Acquisition Technologies

Type of Sensor/Device	Measured Parameter	Application
Temperature Sensor	Temperature	Industrial Automation, HVAC Systems
Pressure Sensor	Pressure	Autonomous Vehicles, Smart Grids
Motion Sensor	Movement	Surveillance, Robotics
Light Sensor	Light Intensity	Smart Lighting Systems, Environmental Monitoring

2.2 Data Collection Frameworks

CPS utilize strong data collecting frameworks with a view of managing tremendously big sensing and IoT data in real time. These frameworks consist of system hardware and software used for collecting, processing and transmitting real-time data. It deals with the actual data collected at the sensor level and the preliminary processing of the collected data to eliminate redundancy information's. Preprocessing is here relevant since otherwise passing and analyzing raw, unfiltered data would be counterproductive and slow down the result (He, 2022). Communication networks convey information to a central process unit or, in some cases, an edge computing node. Earlier centralised setup models input and analyze data at cloud or centralized server (Upreti et al., 2023). This can lead to delays which is a big no if timely decision-making is required. And therefore, edge computing—data processing where it is needed, i.e., at the network edge—has appeared. Edge computing enhances the rate of operation and decisions made across the device and a centralized system by limiting data exchange between them (Dave et al., 2021). In smart cities, edge computing can analyze traffic data at the intersection level to change the traffic lights in the same level depending on the density present. There is less congested traffic and time is saved wherever employed. In industrial CPS, Edge computing provides real-time monitoring and controlling of the industrial equipment to enhance system proficiency and minimize response to any irregularities.

2.3 Challenges in Real-time Data Acquisition

Despite the CPS real-time data collecting technologies are developing at a very high rate, there are some issues that need to be addressed to make it run smoothly. One problem is lag time or the amount of time that elapses between data collection and action. Minor delay can be fatal in time-sensitive CPS applications such as self-driving cars and medical application systems. Efficient communication channels, and other structures to convey and manner data swiftly is also required (Peros et al., 2020). Another problem is data dependability is another issue. Sensors' and IoT-based measurement data can be impacted by external conditions, hardware damage, and hacking. CPS safety and efficiency may be in jeopardy due to poor decisions made because of misinterpreted or misrepresented information. To address this shortfall, CPS designers must rely on redundancy to have many of the same sensors that send identical data to check the correctness, and error detection to find and rectify errors. The ability to expand when real-time gathered data increases or the demand for data collection increases is vital. Increasing the level of complexity of CPS and integration of new devices to the network more data are produced and the system should not be overloaded. This requires efficient data processing frameworks and flexible structures in an organization's communicating system. Due to the increased data transfer speed and low latency of the 5G network, CPS's scalability is expected to be improved and real-time data acquisition is expected to be improved as well (Nejković et al., 2019; Quintana et al., 2018). Finally, the privacy and security of the CPS real-time data taking into consideration. Data for smart cities, healthcare, and other CPS applications is often private and requires security (Upreti et al., 2023). Again, protection of data from all forms of risks during collection, transmission, and storage is not easy. Threats that are associated with cyber element can be managed and contain through including encryption, secure communication method, and access in data collecting.

2.4 Synchronizing Data Acquisition with Decision-making

That is, the acquisition of real-time data is tightly integrated with decision-making mechanisms in the CPS. Even if the business uses sophisticated algorithms to make its decisions, it will not work well if it does not get the data in time and if it is inaccurate. Real time data is collected from various sensors and devices that are placed across the CPS and provides real-time feedback to the decision-making mechanism of the CPS. A good example of this synchronization is self-driving cars whereby data from cameras, radar, and LiDAR sensors is continually collected and processed to enable the car to make decisions on steering, acceleration and braking in a very short span of time. For this reason, the vehicle must possess an effective decision maker to interpreted this data effectively in real-time to avoid end up in a wrong position, traffic junctions, and most importantly guarantee the passengers safety. Likewise, in smart grids, information on demand for energy and offer of energy is reported in actual time to avoid blackouts and modify the flow of energy in the grid.

3. PROCESSING AND ANALYZING REAL-TIME DATA

Cyber-Physical Systems or abbreviated as CPS is transforming different industries through, real-time analysis and control and smart operations. Self-driving cars and drones rely on sensors to detect objects and paths on the route while smart grids and management predict consumer consumption to determine how to distribute energy. In industrial automation, CPS decides to make certain that production does not stall due to the breakdowns. Transport sector data may be useful for traffic control and making it safer whereas healthcare sector data may be useful for tracking patients' health and act consequently. Security and lightning, as well as home appliances, work depending on the information about the environment in smart homes. A common is CPS in precision farming where crops are favorably monitored, with optimized irrigation employing the technology. Smart watches monitor the user's health parameters in real time, and robotic systems aid surgeons with information feedback. Self-propelled underwater vehicle inspects and manage aquatic ecosystems, smart rail controls train operation and consumption. CPS in smart buildings regulates, lighting and HVAC systems to provide comfort and at the same time, conserving energy. In all the applications, edge and cloud computing improve analyses and response times. Pertaining to CPS, the real-time information needs to be processed and analyzed in real-time to drive performance and decision making. In figure 3 below, some of the use cases of CPS in which real time data processing is essential have been illustrated. Self-driving cars have embedded environments managing sensors to drive, while intelligent power systems managing energy usage and demands. Automated industrial systems involve the use of algorithms and heuristics on machine data to foretell breaks down and enable smooth running. In smart cities, systems enhance traffic conditions, while health care organizations assess signs to support unexpected incidents. Smart homes use sensed data to control provisions in the home, while smart agricultural use sensed data to monitor resources used in farming; drones and autonomous vehicles use environmental data to get work done. In all these systems, edge and cloud computing improve data analysis, so that prompt decisions are made. The second most important aspect in CPS is to process and analyze real-time data acquired from sensors and IoT devices for making intelligent decisions. As seen in the earlier parts of this paper, real-time data is the foundation of all decision making in CPS. However, the utility of this data does not end with data collection, but rather, with the speed with which the data collected can be analyzed to make sound business decisions. This section deals with the foundational elements of data processing infrastructures, real time analysis in CPS and the use of edge and cloud computing in processing of massive volumes of real time data for actionable purposes (Viswas et al., 2023).

Figure 3. List of the Some of the Use Cases of CPS in Which Real-Time Data Processing is Required

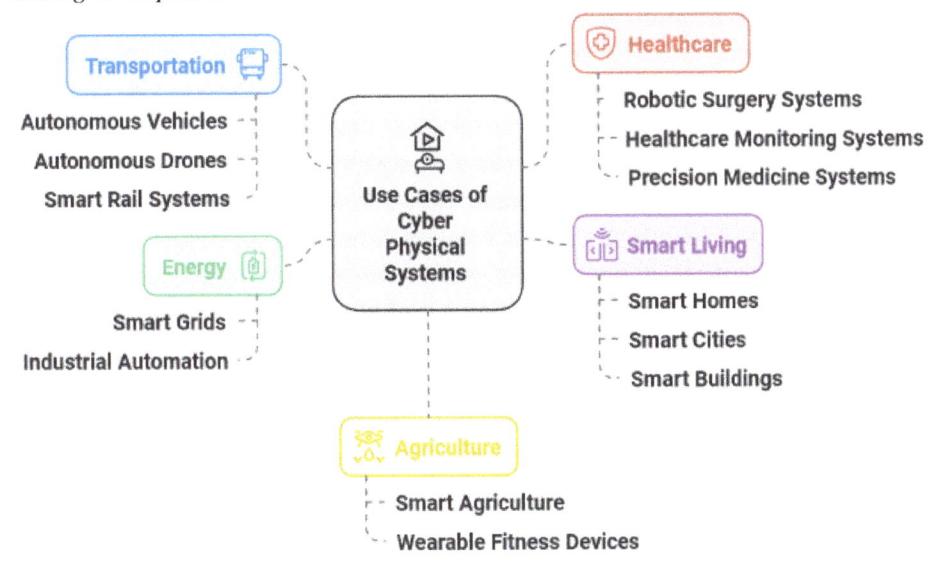

3.1 Data Processing Architectures

CPS performance, therefore, relies on a real-time data processing architecture. CPS produce large amounts of information from devices and sensors which need to be processed for timely decision making. Conventional data analytics structures moved data to a central server or the cloud for processing and analysis. The latency and bandwidth of this model are problematic (Surya & Rajam, 2023), (Nguyen et al., 2021). As it will be seen more in details, even small processing latency time can lead to system failure or to emergence of dangerous situations in most of the CPS applications, which require real time response, such as, autonomous automobile, or manufacture automation, etc. Today's CPS employ distributed and decentralised processing structures to counter these issues. Decentralised systems help to decrease loading on the server and time delay due to data processing at many nodes. Edge computing takes place where data is, usually at the device or at the sensor level. The concept of edge computing reduces the processing time, network congestion, scalability of system and capacity which are all constraints (Fu-xing et al., 2021).

Vehicles moving through a smart city may have some data read by the cameras or sensors mounted along arteries to control traffic signal lights in real-time without necessarily linking up to a main framework. Smart accessories also enable registration of signs of a patient and notify healthcare providers about unfavourable trends to initiate necessary urgent actions without additional data processing at the specialized center. This means decentralisation of decision making when rapid local decisions are needed.

3.2 Real-time Data Analytics

Rather than conducted in batches, real-time analytics deals with the continuous processing of data in real time. Through CPS, decision making must be able to adapt to the change in environmental factors over time. Real-time analytics directly processes real-time data to produce timely results while the traditional data analytics processes real-time data to find out patterns needed in the future. This skill is important for optimized CPS operations in smart grids, automated manufacturing, and autonomous transportation (Chen et al., 2023). CPS real-time data analysis employs stream processing, event driven systems and machine learning. Event processing identifies deviations, patterns or occurrences in the real time processing of streams of data (Angelopoulos et al., 2019). Real-time stream processing capability can track electricity consumption in smart grids so that the distribution of loads is controlled efficiently while avoiding power blackouts.

On the other hand, evented systems rely on triggers from real-time data steam rather on pure events. These systems operate based on stimulus response mechanisms where some pre-determined stimuli elicit responses. For instance, an industrial equipment might stop functioning if it regulates a temperature beyond a certain level to avoid overheating (Chen et al., 2023). This technology minimises human contact since the responses are prompt and dynamic to the messages received. Present CPS data analysis for real-time data is evidently involving the use of machine learning algorithms. These models can categorize the data and make some form of prediction or decision making. At runtime, occurrences can be predicted, processes adapted, or anomalies that need remediation identified by ML models. An ML model in a smart city may process traffic data in real time and then issue recommendations for avoiding traffic jams to cars, which will effectively eliminate congestion.

3.3 Role of Edge and Cloud Computing

CPS real-time data processing and analysis depend on edge and cloud computing. Edge computing handles the data closer to its source, at the device edge or at the neighbouring edge nodes edge nodes. While CPS can be centralized or distributed, local decision making is much better suited for lowering latency and is much more important for time critical applications such as autonomous cars and industrial robotics. Coping of data on the edge is appropriate when the amount of data is large and cannot be easily transferred to a centralized system. Following the analysis of such data on a local level, CPS can provide an instantaneous response that would not be possible when such massive amounts of data must be transmitted to the cloud for analysis. Sensors attached to manufacturing equipment in a smart factory can use edge computing for its ability to measure performance and identify problematic situations in near-real time, thereby avoiding costly downtime or other output errors.

However, cloud computing has profound computation capability and storage capability, not suitable for real-time data analyses, yet for huge amount data. Advanced, computer resource demanding operations which do not require real time responses in CPS, can be managed by cloud computing. Edge computing gives quick, on-site decisions, while the cloud could collect data from multiple edges, process the information to understand long-term patterns and present the results with valuable optimizations. The twin domain of edge and cloud computing enables CPS to balance real-time decision making and data aggregation. A smart city can use one or both of so-called 'edge computing' to control traffic in real-time use and cloud computing to top optimize transportation and planning of city architecture. CPS can make local decision in a short time and analysing data for all strategic change with the help of both technologies. As highlighted in table 4 below, edge and cloud computing in CPS has some similarities.

Table 4. Comparison Between Edge and Cloud Computing for CPS

Feature	Edge Computing	Cloud Computing
Processing Location	Near the data source (locally)	Centralized (remote servers)
Latency	Low	Higher (dependent on network speed)
Scalability	Limited by local hardware	Highly scalable (virtually unlimited resources)
Best Suited For	Time-sensitive applications (e.g., autonomous vehicles)	Complex, large-scale data analysis (e.g., smart city planning)

3.4 Challenges in Real-time Data Processing

However, data processing of real-time data within CPS remains very challenging even with highly developed data processing architectures and technologies. From sensors and IoT technology, the size of generated data and its rate are significant issues. Individual CPS systems can generate large streams of data that can overwhelm even advanced processing systems as systems grow more complex. CPS designers need to develop methods that can facilitate enlarged data quantities to be processed at an equivalent or higher rate while being more accurate than previously ambient systems (Quincozes et al., 2022, Papageorgiou, 2015). Another significant concern is accuracy and reliability of the data processing of the huge amount of information. Discrimination Process 2: Real-time data are messy because of the presence of noise, partial data, and mistakes. This issue can be solved by deploying mistake detection and repair systems, as well as filtering and pre-processing the data before analysis. In the areas where microsecond delays can lead to system failure latency still becomes an issue. The CPS research mainly concentrates on how the systems coupled with processors, how the communicational networks, and how the algorithms can be minimised that causes latency. Finally, many important application scenarios like healthcare and smart city require real-time complex data sensitive and private. When engaging in data processing, personal and organizational data privacy and security must be observed. The several aspects of CPS security include encryption of data streams, protection of edge and cloud nodes, and the presentation of robust access control methodologies to safeguard real-time data.

4. DECISION-MAKING IN CYBER-PHYSICAL SYSTEMS

Having explored the stages of real-time data acquisition, processing, and analysis, we now focus on the goal of these efforts: decision-making. In Cyber-Physical Systems (CPS), decision-making is understood as the approach defining how information is processed to manage physical systems in real time. CPS entails the need for an efficient decision-making mechanism with the key informatics being extracted and processed from the data streams to enhance CPS performance, safety, as well as to accommodate the changing operational requirements. This section describes the decision-making approaches in CPS, the rule-based systems, machine learning decision models as well as the decentralized decision-making models. This also captures the fact that decisions are made within a real-time operation and thus CPS must be designed to feedback and learn as a system.

4.1 Rule-based Decision-making

The formation of the decisions at APS follows conventional decision-making and rule-takers systems. These systems make decisions based on programming that has he/she defined the input, the process, and the output. Hypothesis are conditional statement that state something should be done if some conditions are fulfilled (Pathak et al., 2023). Rule-based systems are easy to implement, predictable, and accurate, making them a go-to CPS system in general, but indispensable in smaller cases where a limited number of possible decisions can be made. Analyzing the patterns and applying the control rules of a particular area where the sensors have been placed, sensor-based lighting and temperatures can also be controlled in smart building management. An example of Smart Control Scenario is when a sensor triggers a rule that switches off the light or reduces heating in an area that is empty. In industrial automation, processes and equipment may be halted for a failed part or an unfavorable environment detected by a sensor. Uncertain and intricate conditions demonstrate the vulnerability of applying rule-based systems. The number of sensors and possible situation adopting the sensor can exponentially grow up the decision-making rules so that the decision-making itself follows great complexity and rule conflict. Since rule-based systems cannot change with dynamic environment situations, they should not be used in environments that need learning and growth of decision-making authorities.

4.2 Machine Learning-driven Decision-making

A few CPS decision-making frameworks are adopting techniques of ML to counter rule-based system limitations. It is a methodology that allows systems to learn from data, recognize patterns, and make predictions without the use explicit rules. The use of artificial intelligence in decision-making processes can change with increasing complexity and unpredictability of CPS. CPS can use data to come up with better behaviour by training other machine learning models to make such decisions. In smart transport systems, the process of machine learning entails that an algorithm can estimate congestion by analyzing and provide information that can be used in redirecting vehicles to areas where congestion is low. In the field of industrial CPS, the application of ML models helps to analyse the data of machine sensors and make predictions on future equipment breakdowns to reduce cases of unexpected downtime (Filios et al., 2022), (Kumar & Raubal, 2021).

This is probably one of the biggest advantages to using ML for decision making – data uncertainty and variabilities are more easily identified. This report about ML models, can easily train on patterns from noisy or incomplete data and are more suitable in the real world than rule-based systems. According to ML models, as more data is collected, the models can refine their decision-making thus making it easier for the CPS to adjust in new situations and increase its levels of efficiency. However, there are also difficulties of using CPS machine learning. Training an ML model requires large amounts of good data, while real-time systems need to process this data as quickly as possible to make decisions. Infinitely, certain safety-critical applications that demand displays of transparency could be hampered by the many ML algorithms that do not have explainable judgements. CPS designers consider explainable, accurate and secure machine learning algorithms as their primary goal. The following are the benefits of implementing the use of artificial intelligence and machine learning in context to the CPS as presented in table 5 below.

Table 5. Key Advantages of AI and Machine Learning in CPS Decision-Making

Advantage	Description
Adaptability	Can adjust to new data and conditions without manual updates
Predictive Capabilities	Can forecast future trends and optimize decisions proactively
Handling Complex Environments	Excels in dynamic and unpredictable scenarios
Reduced Human Intervention	Enables automation of decision-making with minimal oversight

4.3 Decentralized Decision-making

Other approaches adopted includes decentralised decision making whereby different components or sub-systems of CPS make local decisions based on some facts. Centralised decision-making systems on the other hand relay all information to a central controller for analysis and decision making. Applications of a large scale require decentralisation since centralised approaches can lead to bottlenecks and latency, although enabling decentralised approaches. Local devices or nodes are decision makers that employ only local data for decentralized decision making. In a smart city the traffic signals fitted with local sensors can change frequency according to the actual traffic situation in any corner rather than a fixed hub (Wang et al., 2024). This enhances quick action towards local conditions such as rush hour or an accident to enhance traffic flow.

Flawed roles of decision-making are decentralised minimising chances of a single point of failure hence enhancing system reliability. This is because in the system one node can come off without affecting the functioning of the other nodes. In these cases, it is easy to add new devices or nodes in the system without stressing the central processor, an important aspect of scalability. But here I will say that decentralised systems require very good communication to avoid contradicting local decisions. Micro-grids must communicate to co-ordinate distributed generation and consumption in a smart grid to avoid a blackout and sudden loads. Such coordination means intricate computations and the protocols for exchanging information enough to ensure coherent decision making among various nodes in parallel.

4.4 Real-time Decision-making Challenges

Real-time decisions in CPS judgements are an enormous challenge since these aspects include speed, accuracy, and flexibility. A major problem is the ability to make judgements with low latency, especially when delay may have negative consequences for safety or system performance. Self-driving cars need to decide how to steer, when to apply the brakes or accelerate in a matter of milliseconds to prevent an accident. The previous sections discussed fast communication networks, highly optimized ones as well as fast computation architectures such as edge computing which mimic the time taken in arriving at decisions (Zeng et al., 2023).

Another challenge is making accurate decisions where data can be noisy or in many cases insufficient. There are environmental factors, faulty devices, or cyberattacks that can cause the sensor data stream to become corrupt. For this reason, decision-making frameworks need to identify and counteract these errors so that undesirable decisions will be prevented. Some ways are redundancy where many sensors check data or data validation algorithms that filter out anomalies will do this. Flexible is vital when the environments are dynamic. Values in smart city may vary for traffic, weather, and energy usage and this may require the CPS to make decisions in real time. As mentioned before machine learning models are perfect for this, however, they require data to remain relevant and up to date. Recurrent feedbacks in CPS are capable of perceiving prior decisions with intention of enhancing the outcomes of the decisions made.

4.5 Continuing Feedback and Self Improvement

The issue of feedback is central in most CPS decision making frameworks. Systems also contain the ability of decision making and code of conduct modifications. Feedback control systems can measure outcome of decisions in real time and feed back into decision making process. Through learning, CPS can improve performance with slight variance with the surrounding environment.

SMART GRID can use feedback to observe how much energy is consumed and changes the flow of electricity in its network. The technology can switch supply to different component if it feels that some component is drawing more energy. About driving modes in autonomously driven vehicles, sensors covering speed and distance from the leading car, distance as well as road conditions can enhance real time driving behaviour and increase safety and efficiency. In the ML-aided decision-making processes, feedback helps the existing models to update and adapt to new data and improve the prediction. CPS can get ahead of patterns, identify when things deviate, and make improved decisions with new data infusing models. Feedback loops in real-time systems require fast data processing and data analysis architecture to cater to an incoming information as early as possible with little or no latency.

5. USES OF REAL-TIME DATA IN CPS

Real-time decision making in Cyber-Physical Systems (CPS) is the incorporation of several intricate parts resulting in a system that can actively accept data and navigate physical processes. As noted in the previous sections, CPS rely on data acquisition processing and analysis of real-time data to make decision on what improvements, changes, repairs, designs or replacement of subparts, etc that need to be made to facilitate improved performance and safety as well as improved operational efficiency. The actualization process is complex and needs system architectures, software frameworks, communication protocol, and algorithm models to see that the right decision is made right and on time (Karaaslan et al., 2021), (Liu et al., 2020), (Mashayekhi & Heravi, 2020), (LaSorda et al., 2020). This section considers the concerns of deploying real-time decision making in CPS such as the system architecture, middleware, operational communication protocols and the use of RTOS and an emphasis on testing and validation.

5.1 System Design for Real Time Decision

System architecture determines who and how receives, stores, processes, analyzes, protects, and disseminates data in any CPS. To perform well in real-time decision-making style the architecture used must be low on latency and high in dependability. System architecture in CPS typically has three layers: physical, data processing and decision making (Konstantopoulos et al., 2020), (Frysak, 2020). In the physical layer, there are activities by the sensor, actuators and the other associated hardware components which encapsulate physical things. While certain actuators effect system decisions, others aggregate corresponding environmental information as sensors. Real-time processing depends greatly on a reliable, physically high frequency transport layer. The raw data gathered in the physical layer is refined, and conditioned by the data processing layer. Data collected by edge nodes are processed locally to minimize latency while more complex analysis is done at the cloud servers. The speed of data analysis which rapidly converts datum into information to provide timely decision support is required. In many cases decision is made in real time based on the data which has been processed in the decision-making layer. Adaptable foresight, instance learning techniques, application of rules and decentralized decision-making approaches are typical of this layer. To make the system responsive to new information and changes in the environment the last layer, the decision-making layer, must be real-time. Each of these three tiers of data needs to be incorporated into CPS that will be able to make real time decisions. This enables proper organization of design and bestows prompt movement of data from one layer to the next and fast decision making.

5.2 Middleware for CPS

Communications between the CPS components require the use of middleware. Middleware is the centralized communication between the OS and applications software and determines both the content and the time relation of the messages that are exchanged. Middleware enables the encapsulated interaction of sensors, processors, and actuators in CPS systems across platforms and geographical distances (Noureldin & Fadel, 2021). Middleware must deal with very large amounts of data, high data rates and very low latency for real-time decision making. These functionalities are provided with middleware frameworks including ROS, DDS, and ZeroMQ for CPS. These frameworks help in maintaining the identically synchronized real time data by controlling the remote component's interaction. It is responsible for enabling interaction between individually acting sensors, cameras, navigation systems, and control units in autonomous vehicle systems. Middleware enables the vehicle to make fast rational deciding about accelerating, braking, and steering because it delivers current information to all the components. Middleware is also responsible for the timing and synchronization of the tasks so that choices are made in an orderly basis.

5.3 Communication Protocols

Real-time CPS decision-making in systems with numerous sensors, devices, and nodes in a large area needs proper communication. That is why, the standard communication protocols of data exchange between the components of CPS should reflect the aspects of high speed, reliability and scalability. Based on CPS, the data transmission that occurs most frequently and in real time is done through the MQTT, CoAP, and OPC UA protocols (Elattar, 2019), (Burg et al., 2018), (Cai & Qi, 2017), (Jawhar, 2017), (Huang & Zhang, 2017), (Liu, 2017), (Shih, 2016). These protocols include security connectivity to the device, high data and a low level of latency. Smart grid networks enable power production plants, substations and consumer appliances through communication protocols to continuously feed real time information to the network. These protocols facilitate the real time generation, distribution, load management and demand response by sharing energy utilization, power availability, and grid stability.

5.4 Real-time Operating Systems (RTOS)

Real-time operating systems are important to the CPS real-time decisions. RTOS are intended to control the hardware resources to enable the tasks to run at moderate and well-timed rates, unlike those found in general purpose operating systems.

CPS does require this due to the consequences of uncontrolled decision-making delays even at elementary level. RTOS are necessary in embedded systems such as industrial automation, autonomous car and health system. They achieve strict deadlines for time-sensitive measurement of sensor data, signal processing and decision implementation (Chen et al., 2018). RTOS also pre-empt tasks, allowing the system to dedicate time to critical processes without its progress being hindered by less important procedures. The real-time operating system enables the robot in an industrial robotic system to respond to sensor inputs and to move without delays. The RTOS enhances robot accuracy and precision by controlling the execution of tasks and division of resources. Safety and reliability of CPS are directly dependent on predictability.

5.5 Testing and Validation

Another one is applicable to real-time decision-making in CPS where testing and validation are employed to guarantee proper system performance under various conditions. Testing reveals the latency, imprecision, and correspondence problems, while validation verifies that the system satisfies the requirements of an application. Simulation type environments provide multiple scenarios and settings to practice real time decision-making frameworks.

6. CHALLENGES AND SOLUTIONS IN REAL-TIME DECISION-MAKING FOR CPS

Although the cointegration of real-time decision-making into CPS may enrich the field, it also poses several difficulties, this review explained how the realities of integrating real-time data, communication and security all need to be managed to fully unlock the CPS potential as discussed in previous sections. This section analyses the potential difficulties associated with implementing the proposed real-time decision-making in CPS, and discusses how to overcome these challenges, based on data quality and management, system complexity, security threats, limited availability of resources, and constant learning.

6.1 Data Quality and Management

This paper identified four main factors which CPS real-time decision-making is likely to experience whenever the data is being acquired: The quality of data which includes data accuracy, data timeliness and data relevance can all have cumulative impacts on decision making. CPS are typically in the business of amassing massive quantities of data from sensors, IoT devices, and databases. However, the absence of data, or the presence of noisy or missing, wrong data, results in wrong decision making. This lead to data quality problems, for which CPS requires efficient management of data. This is in form of data governance frameworks which stipulate the procedures for accruing, storing and processing the data. Routine ways of data validation and its cleansing can fix several mistakes and we could use only the clean data for the decision making. Sophisticated methods in harmonic analysis such as anomaly detection algorithms remove luck and noise making decision-making system data superior. Everyone knows that when data processing takes place at the edge, data filtration and analysis which occur locally help reduce data quantity to be transferred to main servers, which in turn improves data quality. These deburdens communication networks and pass only the stringent data for processing thus enabling accurate real time decisions in the process.

6.2 System Complexity

Different and often complex CPS designs intervene with the possibility of real-time decision making. It may also reveal that systems can have complex interaction patterns when they grow and scale. This complexity has implications for system dependability, data flow management as well as timely decision support. Specifically, as issues arise with simplified CPS design, utilizing a modular design approach must be incorporated when developing CPS. Due to this, developers can pinpoint and earmark lessens which have a negative impact on the whole system while they remain functional. Modular structure allows for a faster integration of roll changes/improvements, maintenance, new technologies and methodologies without having to redesign the entire structure. Practical approaches such as model based system engineering (MBSE) can to some extent reduce complexity through organization of system design and analysis. This increases efficiency through the ability to assess all the components of CPS and their interactions with MBSE, observe constant and variable factors that slow teams down, and consider ways to enhance the CPS system by determining localized weak points through simulated prototyping before a CPS is has been deployed. Performance and reliability of real-time decision-making is enhanced by this proactive method.

6.3 Security Vulnerabilities

Security threats are realized as more CPS are interfaced with each other. Real-time decision-making systems are most at risk because the repercussions for either data accuracy or system availability in the event of a cyber-attack are grave. CPS security is required in the transportation sector, healthcare and industrial automation sectors. The threats that the CPS might face in its operation can only be reduced when an integrated durable/soft hardware security mechanism has been adopted. This has a bearing with the coding, audits and authorization to allow only permitted personnel to access your system. Real-time systems also an detect a cyber-attack stimulus for immediate response to the recognized aberrations. CPS security requires all stakeholders to be aware of the vulnerability hence tooking up a cybersecurity culture is important. It may enable staff and users accept the security practices and possibly notice or even perhaps avoid undesirable incidents. There are several ways through which organisations can enhance the resilience of CPS: Proactive security.

7. FUTURE DIRECTIONS FOR REAL-TIME DECISION-MAKING IN CPS

With new development in Cyber-Physical Systems (CPS), the realm of real-time decision-making stands on the verge of entering a new paradigm. As with any field of technology, developments in decision-making methodologies and frameworks will be constantly pressed forward by growing and changing needs for smarter systems and better ways of operating. This section discusses trends in real-time decision making in CPS, specifically in the aspects of Artificial Intelligence and machine learning, the potential of 5G and edge computing, development in data analytics, the issue of security and privacies, and standardization and interoperability.

7.1 Integration of Artificial Intelligence and Machine Learning

Real-time CPS decision-making utilizing AI and advanced forms of ML appear to have considerable potential. As mentioned, ML models enable the computer to find its way and make conclusions from past data; this redefines decision-making. CPS enhancement is feasible in the future through the integration of more complex algorithms into the AI and ML architecture. For CPS, deep learning can reveal subtle features in millions of data points, enhancing predictions and CPS In self-driving cars, AI can process data of the traffic situation through multiple sensors and cameras, to manage complex emerging traffic conditions safely. CPS are most benefitted from a subsector of ML known as reinforcement learning which enables CPS to engage in decision making when responding to decision made earlier. AI also leads to a low intervention of human decides in the CPS to take an action autonomously. The ability of AI algorithms in analyzing various elements in the real-time enables the systems to make real-time decisions based on the relevant facts enhancing critical response times.

7.2 Potential of 5G and Edge Computing

One will expect real-time decisions in a CPS to be transformed by Ultra-reliable low-latency communication (URLLC) from the 5G system. CPS performance and effectiveness will increase due to 5G networks' characteristics, such as the high data transfer rate and low latency. 5G means systems can manage more details from more devices simultaneously, supporting more extensive and comprehensive CPS. Real-time decision-making is enhanced by both edge computing and 5G. Edge computing shortens data analysis and action time since data is computed at the network periphery or edge (source). This is beneficial to industrial automation and smart city development which require fast response. In manufacturing, edge devices may analyze sensor data in real time and learn that it contains deviations from the normal flow of production, thus avoiding interruptions. Real-time analytics and edge computing make CPS effective in the places where the bandwidth is limited. Hence, more edge processing means that less data must be sent to centralized servers to alleviate the communication network or to ensure timely action on important data. This architecture enhances the CPS performance, which in turn, may enable the CPS to run under many operating conditions.

7.3 Advancements in Data Analytics

The advancement in data analytical techniques will also influence CPS real-time decision making. In the aspect of data analysis, it may become a challenge to sustain the processing of the huge volume and complicated nature of CPS data using the traditional data analysis methods. Concerning levels, advanced analytics methods like a predictive and prescriptive analytics technique will become more meaningful. Predictive analytics enables CPS to know in advance what action to take using historical data and statistical models. In smart manufacturing, the use of Predictive Analytics can predict when the manufacturing equipment may develop a fault and thus plan for maintenance to minimize any time when the equipment is not usable.

7.4 Importance of Security and Privacy

Increasing interconnectedness and data orientation of CPS make data security and privacy more relevant. As a result, future real-time decision-making frameworks need to be secure to reduce incidences of cyber-attacks and unauthorized access. This is important because CPS are frequently used in sectors of critical national infrastructure where breaches can have consequences. Particularly when the CPS architecture is to support real-time decision-making functions, security measures must be incorporated at each level of the system's hierarchy. This option includes encryption of data transmitted over the network; secure authentication; and threat identification. Another is that decentralised decision-making frameworks help to avoid centralised control systems with high risks to security.

8. CONCLUSION

Real-time decision-making is essential in Cyber-Physical Systems (CPS) due to the propagation of innovation in cyber space technology and globalization. CPS revealed that real-time data utilisation enhances utility, effectiveness, and reaction in real-time in different sectors in this chapter. Real time information ensures that CPS responds to the current environment, maximizes resource utilization processes, and security. Of particular interest are artificial intelligence and machine learning for the purpose of making prompt decisions. These tools allow CPS to analyze enormous quantities of information and make conclusions to prevent certain scenarios and respond more quickly to critical situations in facilities. Moreover, 5G and edge computing will evolve the way data is processed and transmitted when integrated and used in resource-scarce environments.

The process of getting to real-time decision is not smooth. For CPS to realise its full possibility, the aspects such as data quality, system sophistication, threat to security, constraints on resources, and the necessity of constant evolution require consideration. Modularity, multilevel security, good data management, and lightweight computations can be used to enhance system robustness and longevity to meet organizational goals. Further improvements in data analysis tools and new technologies will make further improvements in real time decision-making applications possible. This focus will push different systems to align consequently achieving their targets of fulfilling CPS objectives. These features make real-time decision making as one of the revolutionary factors in CPS, which new directions can have an impact on society. Prevention strategies that incorporate the most advanced solutions with modern effective practices will allow establishing enhanced, secure and effective systems that positively affect people's lives and enable the sustainable development. Sound decisions made at appropriate time will help define the complex interconnections of the environment as we move through the digital world. The future is never dull, as there are hurdles and giant opportunities for learning, coming together, and making the world a better place.

REFERENCES

Aboualola, M., Abualsaud, K., Khattab, T., Zorba, N., & Hassanein, H. S. (2023, January 1). Edge Technologies for Disaster Management: A Survey of Social Media and Artificial Intelligence Integration. *IEEE Access : Practical Innovations, Open Solutions*, 11, 73782–73802. DOI: 10.1109/ACCESS.2023.3293035

Akhuseyinoglu, N B., & Joshi, J. (2020, September 1). A constraint and risk-aware approach to attribute-based access control for cyber-physical systems. Elsevier BV, 96, 101802-101802. DOI: 10.1016/j.cose.2020.101802

Andersen, A. H. D., Zhang, Z., Hørsholt, S., Ritschel, T., & Jørgensen, J. B. (2023, January 1). Software principles and concepts applied in the implementation of cyber-physical systems for real-time advanced process control. Cornell University. https://doi.org//arxiv.2302.13595DOI: 10.48550

Angelopoulos, A., Michailidis, E. T., Νομικός, N., Trakadas, P., Hatziefremidis, A., Voliotis, S., & Zahariadis, T. (2019, December 23). Tackling Faults in the Industry 4.0 Era—A Survey of Machine-Learning Solutions and Key Aspects. *Sensors (Basel)*, 20(1), 109–109. DOI: 10.3390/s20010109 PMID: 31878065

Burg, A., Chattopadhyay, A., & Lam, K. (2018, January 1). Wireless Communication and Security Issues for Cyber–Physical Systems and the Internet-of-Things. *Proceedings of the IEEE*, 106(1), 38–60. DOI: 10.1109/JPROC.2017.2780172

Cai, Y., & Qi, D. (2017, July 1). Physical control framework and protocol design for cyber-physical control system. *International Journal of Distributed Sensor Networks*, 13(7), 155014771772269–155014771772269. DOI: 10.1177/1550147717722692

Chamberlain, L B., Davis, L E., Stanley, M., & Gattoni, B. (2020, May 1). Automated Decision Systems for Cybersecurity and Infrastructure Security. DOI: 10.1109/SPW50608.2020.00048

Chaudhuri, A., & Kahyaoğlu, S B. (2023, March 8). CYBERSECURITY ASSURANCE IN SMART CITIES: A RISK MANAGEMENT PERSPECTIVE. Taylor & Francis, 67(4), 1-22. DOI: 10.1080/07366981.2023.2165293

Chen, C., Hasan, M., & Mohan, S. (2018, December 10). Securing Real-Time Internet-of-Things. *Sensors (Basel)*, 18(12), 4356–4356. DOI: 10.3390/s18124356 PMID: 30544673

Chen, W., Milošević, Z., Rabhi, F., & Berry, A. J. (2023, January 1). Real-Time Analytics: Concepts, Architectures, and ML/AI Considerations. *IEEE Access : Practical Innovations, Open Solutions*, 11, 71634–71657. DOI: 10.1109/ACCESS.2023.3295694

Dave, R., Seliya, N., & Siddiqui, N. (2021, October 29). The Benefits of Edge Computing in Healthcare, Smart Cities, and IoT. *The Benefits of Edge Computing in Healthcare, Smart Cities, and IoT.*, 9(1), 23–34. DOI: 10.12691/jcsa-9-1-3

De-Arteaga, M., Fogliato, R., & Chouldechova, A. (2020, April 21). A Case for Humans-in-the-Loop: Decisions in the Presence of Erroneous Algorithmic Scores. DOI: 10.1145/3313831.3376638

Elattar, M. (2019, August 7). Reliable Communications within Cyber-Physical Systems Using the Internet (RC4CPS). Springer Nature. DOI: 10.1007/978-3-662-59793-4

Filios, G., Katsidimas, I., Nikoletseas, S., Panagiotou, S H., & Raptis, T P. (2022, December 9). Agnostic learning for packing machine stoppage prediction in smart factories., 3(3), 793-807. DOI: 10.52953/LEDZ3942

Franki, V., Majnarić, D., & Višković, A. (2023, January 18). A Comprehensive Review of Artificial Intelligence (AI) Companies in the Power Sector. *Energies*, 16(3), 1077–1077. DOI: 10.3390/en16031077

Frysak, J. (2020, January 1). S-BPM Diagrams as Decision Aids in a Decision Based Framework for CPS Development. Springer Science+Business Media, 23-32. DOI: 10.1007/978-3-030-64351-5_2

Fu-xing, L., Li, L., & Peng, Y. (2021, December 3). Research on Digital Twin and Collaborative Cloud and Edge Computing Applied in Operations and Maintenance in Wind Turbines of Wind Power Farm. IOS Press. DOI: 10.3233/ATDE210263

Halder, S., Afsari, K., & Akanmu, A. (2024, February 10). A Robotic Cyber-Physical System for Automated Reality Capture and Visualization in Construction Progress Monitoring. Cornell University. https://doi.org/DOI: 10.48550/arXiv.2402

Han, Y., Chen, J., Dou, M., Wang, J., & Feng, K. (2023, May 20). The Impact of Artificial Intelligence on the Financial Services Industry. *The Impact of Artificial Intelligence on the Financial Services Industry.*, 2(3), 83–85. DOI: 10.54097/ajmss.v2i3.8741

He, Z. W. (2022, August 21). Research on the Civic Policy Model and Reform Innovation of Intelligent Sensor Technology Course. *Journal of Sensors*, 2022, 1–8. DOI: 10.1155/2022/2499421

Hoffmann, M., Malakuti, S., Grüner, S., Finster, S., Gebhardt, J., Tan, R., Schindler, T., & Gamer, T. (2021, March 11). Developing Industrial CPS: A Multi-Disciplinary Challenge. *Sensors (Basel)*, 21(6), 1991–1991. DOI: 10.3390/s21061991 PMID: 33799891

Huang, J., & Zhang, L. (2017, January 1). Research and challenges of CPS. American Institute of Physics. DOI: 10.1063/1.4992851

Jawhar, I., Al-Jaroodi, J., Noura, H., & Mohamed, N. (2017, June 1). Networking and Communication in Cyber Physical Systems. DOI: 10.1109/ICDCSW.2017.31

Kankanamge, M. W., Hasan, S. M., Shahid, A. R., & Yang, N. (2024, July 2).. . *Large Language Model Integrated Healthcare Cyber-Physical Systems Architecture.*, 1540-1541, 1540–1541. Advance online publication. DOI: 10.1109/COMPSAC61105.2024.00228

Kanwade, A. B., Sardey, M. P., Panwar, S. A., Gajare, M. P., Chaudhari, M. N., & Upreti, K. (2023). Combined weighted feature extraction and deep learning approach for chronic obstructive pulmonary disease classification using electromyography. *International Journal of Information Technology : an Official Journal of Bharati Vidyapeeth's Institute of Computer Applications and Management*, 16(3), 1485–1494. DOI: 10.1007/s41870-023-01498-y

Karaaslan, E., Bağcı, U., & Çatbaş, F. N. (2021, June 25). A Novel Decision Support System for Long-Term Management of Bridge Networks. *Applied Sciences (Basel, Switzerland)*, 11(13), 5928–5928. DOI: 10.3390/app11135928

Keshk, M., Turnbull, B., Sitnikova, E., Vatsalan, D., & Moustafa, N. (2021, January 1). Privacy-Preserving Schemes for Safeguarding Heterogeneous Data Sources in Cyber-Physical Systems. *IEEE Access : Practical Innovations, Open Solutions*, 9, 55077–55097. DOI: 10.1109/ACCESS.2021.3069737

Konstantopoulos, G. C., Alexandridis, A. T., & Papageorgiou, P. (2020, May 1). Towards the Integration of Modern Power Systems into a Cyber–Physical Framework. *Energies*, 13(9), 2169–2169. DOI: 10.3390/en13092169

Kumar, N., & Raubal, M. (2021, November 1). Applications of deep learning in congestion detection, prediction and alleviation: A survey. Elsevier BV, 133, 103432-103432. DOI: 10.1016/j.trc.2021.103432

LaSorda, M M., Borky, J M., & Sega, R M. (2020, March 1). Model-Based Systems Architecting with Decision Quantification for Cybersecurity, Cost, and Performance. DOI: 10.1109/AERO47225.2020.9172283

Liu, L., Lu, S., Ren, Z., Wu, B., Yao, Y., Zhang, Q., & Shi, W. (2020, December 10). Computing Systems for Autonomous Driving: State of the Art and Challenges. *IEEE Internet of Things Journal*, 8(8), 6469–6486. DOI: 10.1109/JIOT.2020.3043716

Liu, Y., Peng, Y., Wang, B., Yao, S., & Liu, Z. (2017, January 1). Review on cyber-physical systems. *Institute of Electrical and Electronics Engineers*, 4(1), 27–40. DOI: 10.1109/JAS.2017.7510349

Liu, Y., & Yang, K. (2022, September 12). Communication, sensing, computing and energy harvesting in smart cities. *IET Smart Cities*, 4(4), 265–274. DOI: 10.1049/smc2.12041

Mashayekhi, A N., & Heravi, G. (2020, August 1). A decision-making framework opted for smart building's equipment based on energy consumption and cost trade-off using BIM and MIS. Elsevier BV, 32, 101653-101653. DOI: 10.1016/j.jobe.2020.101653

Nejković, V., Visa, A., Tošić, M., Petrović, N., Valkama, M., Koivisto, M., Talvitie, J., Rančić, S., Grzonka, D., Tchórzewski, J., Kuonen, P., & Gortázar, F. (2019, January 1). Big Data in 5G Distributed Applications. Springer Science+Business Media, 138-162. DOI: 10.1007/978-3-030-16272-6_5

Nguyen, D. C., Cheng, P., Ding, M., López-Pérez, D., Pathirana, P. N., Li, J., Seneviratne, A., Li, Y., & Poor, H. V. (2021, January 1). Enabling AI in Future Wireless Networks: A Data Life Cycle Perspective. *IEEE Communications Surveys and Tutorials*, 23(1), 553–595. DOI: 10.1109/COMST.2020.3024783

Noureldin, H. F., & Fadel, M. (2021, June 1). Rationalizing Resource Utilization in Cloud Computing Using Coalition Formation Strategy. *Science Publications*, 17(6), 539–555. DOI: 10.3844/jcssp.2021.539.555

Papageorgiou, A., Cheng, B., & Kovács, E. (2015, November 1). Real-time data reduction at the network edge of Internet-of-Things systems. DOI: 10.1109/CNSM.2015.7367373

Pathak, S., Srivastava, K B L., & Dewangan, R L. (2023, January 14). Decision styles and their association with heuristic cue and decision-making rules. Taylor & Francis, 10(1). DOI: 10.1080/23311908.2023.2166307

Peros, S., Delbruel, S., Michiels, S., Joosen, W., & Hughes, D. (2020, June 1). Simplifying CPS Application Development through Fine-grained, Automatic Timeout Predictions. *Association for Computing Machinery*, 1(3), 1–30. DOI: 10.1145/3385960

Proactive Measures to Avoid Failure. (2021, April 21)., 309-320. DOI: 10.1002/9781119615606.ch17

Quincozes, S. E., Mossé, D., Passos, D., Albuquerque, C., Ochi, L. S., & Santos, V F D. (2022, March 1). On the Performance of GRASP-Based Feature Selection for CPS Intrusion Detection. *IEEE Transactions on Network and Service Management*, 19(1), 614–626. DOI: 10.1109/TNSM.2021.3088763

Quintana, I., Tsiopoulos, A., Lema, M. A., Sardis, F., Sequeira, L., Arias, J., Raman, A., Azam, A., & Döhler, M. (2018, December 1). The Making of 5G: Building an End-to-End 5G-Enabled System. *Institute of Electrical and Electronics Engineers*, 2(4), 88–96. DOI: 10.1109/MCOMSTD.2018.1800024

Rajkumar, R. (2023, September 24). Cyber-physical systems. https://dl.acm.org/doi/10.1145/1837274.1837461

Rejeb, A., Rejeb, K., Appolloni, A., Jagtap, S., Iranmanesh, M., Alghamdi, S., Alhasawi, Y., & Kayıkçı, Y. (2024, January 1). Unleashing the power of internet of things and blockchain: A comprehensive analysis and future directions. Elsevier BV, 4, 1-18. DOI: 10.1016/j.iotcps.2023.06.003

Shaik, T., Tao, X., Higgins, N., Li, L., Gururajan, R., Zhou, X., & Acharya, U R. (2023, January 5). Remote patient monitoring using artificial intelligence: Current state, applications, and challenges. Wiley, 13(2). DOI: 10.1002/widm.1485

Shih, C., Chou, J., Reijers, N., & Kuo, T. (2016, December 1). Designing CPS/IoT applications for smart buildings and cities. *IET Cyber-Physical Systems*, 1(1), 3–12. DOI: 10.1049/iet-cps.2016.0025

Surya, K. C. N., & Rajam, V. M. A. (2023, January 2). Novel Approaches for Resource Management Across Edge Servers. *Springer Nature*, 11(1), 20–30. DOI: 10.1007/s44227-022-00007-0

Upreti, K., Arora, S., Sharma, A. K., Pandey, A. K., Sharma, K. K., & Dayal, M. (2024). Wave Height Forecasting Over Ocean of Things Based on Machine Learning Techniques: An Application for Ocean Renewable Energy Generation. *IEEE Journal of Oceanic Engineering*, 49(2), 1–16. DOI: 10.1109/JOE.2023.3314090

Upreti, K., Peng, S., Kshirsagar, P. R., Chakrabarti, P., Al-Alshaikh, H. A., Sharma, A. K., & Poonia, R. C. (2023). A multi-model unified disease diagnosis framework for cyber healthcare using IoMT- cloud computing networks. *Journal of Discrete Mathematical Sciences and Cryptography*, 26(6), 1819–1834. DOI: 10.47974/JDMSC-1831

Upreti, K., Vats, P., Srinivasan, A., Sagar, K. V. D., Mahaveerakannan, R., & Babu, G. C. (2023). Detection of Banking Financial Frauds Using Hyper-Parameter Tuning of DL in Cloud Computing Environment. *International Journal of Cooperative Information Systems*, 2350024. Advance online publication. DOI: 10.1142/S0218843023500247

Viswas, A., Dabla, P. K., Gupta, S., Yadav, M., Tanwar, A., Upreti, K., & Koner, B. C. (2023). SCN1A Genetic Alterations and Oxidative Stress in Idiopathic Generalized Epilepsy Patients: A Causative Analysis in Refractory Cases. *Indian Journal of Clinical Biochemistry*. Advance online publication. DOI: 10.1007/s12291-023-01164-x

Wang, X., Jerome, Z., Wang, Z., Zhang, C., Shen, S., Kumar, V., Bai, F., Krajewski, P. E., Deneau, D., Ahmad, J., Jones, R., Piotrowicz, G., & Liu, H. (2024, February 20). Traffic light optimization with low penetration rate vehicle trajectory data. *Nature Communications*, 15(1), 1306. Advance online publication. DOI: 10.1038/s41467-024-45427-4 PMID: 38378680

Zeng, L., Wang, W., Feng, D., Zhang, X., & Chen, X. (2023, January 1). A3D: Adaptive, Accurate, and Autonomous Navigation for Edge-Assisted Drones. Cornell University. https://doi.org/DOI: 10.48550/arXiv.2307

Zhang, Y., & Huang, H. (2024, February 23). Photolithography Control System: A Case Study For Cyber-Physical System. Cornell University. https://doi.org//arxiv.2402.15693DOI: 10.48550

Chapter 11
Leveraging Big Data and Advanced Technologies for Enhanced Sustainability in Healthcare:
An IPO Model Approach

Mary Metilda Jayaraj
https://orcid.org/0009-0002-4725-7295
Dayananda Sagar College of Engineering, India

Subbulakshmi Somu
https://orcid.org/0000-0002-3744-1075
Dayananda Sagar College of Arts, Science, and Commerce, India

ABSTRACT

The convergence of big data and technology is transforming the collection, processing, and application of information in the healthcare industry. It is now crucial to comprehend and manage the enormous amounts of data that are constantly produced from various sources, including wearable technology, medical imaging, and electronic health records (EHRs). This chapter examines the five main characteristics of big data that healthcare companies may use to improve sustainability in patient care and operational efficiency, i.e. Volume, velocity, variety, Veracity and Variability. These elements are integrated into a comprehensive IPO (Input, Process, Output) model, illustrating how they function as input and process components that leads to substantiality. Based on earlier literature and real world scenario, the chapter

DOI: 10.4018/979-8-3693-5728-6.ch011

highlights the importance of cutting-edge technologies in managing the complexity of healthcare data and improving delivery and operational effectiveness.

INTRODUCTION

The contemporary healthcare sector relies more on big data and advanced technologies. Data is generated in huge quantity on a daily basis from various sources in the healthcare sector, including electronic health records (EHRs), medical imaging, genetic information, and wearable sensors. Effectively managing and using this data has the potential to significantly enhance the quality of patient care, streamline operations, and enable evidence-based decision-making. This chapter examines the use of big data and emerging technologies to improve sustainability in the healthcare sector via the implementation of a complete Input-Process-Output (IPO) model. The IPO model provides a structured framework for understanding the flow of data inside healthcare systems. The input component of the model comprises several types of data, including structured, semi-structured, and unstructured data, which are collected. Examples of such data include patient records, laboratory results, medical images, and data obtained from wearable devices. The process component encompasses the techniques and technologies used to standardize, analyze, and interpret the data, including data cleansing, validation, and interoperability frameworks. The output component emphasizes the practical insights obtained from the data, which may be used to influence clinical choices, enhance operations, and ultimately enhance patient outcomes.

The chapter explores the five dimensions of big data in healthcare, namely Volume, Velocity, Variety, Veracity, and Variability. Every dimension poses a distinct difficulty and opportunity when it comes to handling healthcare data. Volume refers to the large amount of data created, which requires storage solutions that can be easily expanded and strong techniques for combining different sets of data. Velocity refers to the pace at which data is produced and processed, highlighting the need of instantaneous analytics and decision-making. Diversity emphasizes the many types of data, requiring sophisticated methods to manage organized, semi-structured, and unstructured data. Veracity is dedicated to assuring the precision and dependability of data, which is crucial for efficient patient care and research. Variability pertains to the ever-changing characteristics of data, which are impacted by variables such as patient demographics and seasonal patterns.

This chapter demonstrates the interplay of these factors inside the IPO model to generate enduring results in the healthcare sector. We illustrate the actual use of data analysis insights in healthcare delivery by providing examples, showing how these insights may lead to quantifiable changes. The primary objective is to provide a complete framework that healthcare organizations may use to improve their data management procedures, resulting in more sustainable and efficient healthcare systems.

LITERATURE REVIEW

Big Data in HealthCare

The healthcare is progressively adopting a data-driven approach as it transforms to a patient-centric paradigm. The use of big data analytics has the potential to provide excellent patient care at cost-effective rates. However, healthcare organizations have not completely capitalized on this promise owing to a range of difficulties. Advanced data analytics tools have the capability to not only forecast results but also provide the most optimal solutions. This may be beneficial for healthcare organizations in enhancing their operations and decision-making processes (Mehta et al., 2020 & Khan et al., 2022). The main determinants for the use of big data in healthcare may be categorized into two groups: external variables (such as government policies, economic conditions, social issues, technological advancements, and market demands) and endogenous elements (including the drive for development and the ability for development). The potential development models for big data applications in healthcare are founded on the invention of value derived from big data and the motives of various stakeholders, such as government-led, industry-driven, socially-initiated, and mixed development models (Dai et al., 2020).

The use of big data analytics in healthcare has a significant effect providing revolutionary prospects for precise diagnosis, personalized therapy, and anticipatory healthcare (Ali et al., 2023& Rayan et al., 2022). The technology has the capacity to reduce healthcare expenses, avert illness epidemics, and enhance the quality of life for patients. Healthcare organizations have a significant problem in developing practical applications based on Big Data Analytics (BDA). Future research should prioritize investigating the correlation between digitization and resource management, while also striving to create more effective data-sharing mechanisms. Utilizing the potential of big data analytics will be essential for predicting and tracking epidemics (Cozzoli et al., 2022). The use of big data in healthcare offers a greater number of advantages and prospects compared to disadvantages and risks. The three primary characteristics of big data, namely volume, velocity, and diversity, provide significant advantages but may also give rise to infrastructure complications, subpar data quality, and difficulty in data aggregation. However, the potential benefits of using big data in healthcare are extensive, including both financial advancements and improvements in the quality of service. The primary challenges are privacy issues over personal data and a dearth of motivation for healthcare organizations to use big data solutions (Dias et al., 2020).

Dimensions of Bigdata in Healthcare:

The five dimensions of big data are: Volume, Velocity, Variety, Veracity, and Variability. Each of these factors plays a vital part in overseeing the immense quantities of healthcare data produced on a daily basis, processed and using it efficiently. Volume is the huge amount of data produced from diverse sources such as genetics, wearable sensors, medical imaging, and electronic health records (EHRs). Velocity is the rate at which data is produced, gathered, handled, and examined. It underscores the importance of managing data in real-time or near-real-time to provide timely insights and facilitate swift decision-making. Variety encompasses the assortment of data types that healthcare organizations need to handle, which include unstructured, semi-structured, and structured data. Veracity refers to the precision and dependability of healthcare data. Ensuring the accuracy and reliability of data is of utmost importance, as any mistakes, inconsistencies, or missing information may have a substantial influence on patient care and decision-making processes. The temporal variations in data caused by patient demographics, seasonal trends, and changes in data quality is given by variability.

Volume

The term 'Volume' in the context of the healthcare business refers to the vast amount of data generated, including sensor data from wearable devices, medical imaging, genetic data, and electronic health records (EHRs). The rise in healthcare data, along with the adoption of electronic health records (EHRs) and the increased use of genetic and medical imaging data, has made it crucial to have dependable and scalable storage solutions. Smith and Brown (2023) argue that using advanced methods of data integration is crucial for merging diverse datasets and enabling comprehensive analysis. The vast number of datasets presents both benefits and problems (Smith & Brown 2023). Data warehousing and cloud storage solutions are becoming essential in this context. Despite the challenges, the process of combining several data sources, such as wearable sensor data, imaging data, and electronic health records (EHRs), remains complex.

Johnson and Lee (2022) argue that modern data processing frameworks like Hadoop and Spark offer effective solutions for managing large healthcare datasets. These frameworks support real-time analytics and decision-making. A study by Williams & Davis(2021) shows that the growing volume of data requires new approaches to data storage and management. Allocating resources to scalable storage technologies is essential for organizations for successfully handling increasing amounts of data. Utilizing essential data processing methods such as data extraction and machine learning algorithms are essential for extracting important insights from large datasets. According to Thompson & Garcia (2020), the quantity of healthcare data has both positive and negative aspects. On one hand, it provides a wealth of information that may improve patient outcomes and make hospital operations more efficient. Nevertheless, managing and examining this vast amount of data poses several challenges. Furthermore, it is essential to use efficient data processing techniques to handle the vast quantities of data while maintaining both precision and swiftness. Technologies such as distributed computing and in-memory processing are becoming important in this context. Martin and Patel (2019) note that efficient data processing techniques, such as cloud computing and parallel processing, are essential for managing large volumes of data and extracting valuable insights.

Velocity

This dimension will examine the speed at which medical data is generated, collected, and analyzed. This encompasses rapid changes to patient information and live data feeds from monitoring devices. The following paragraphs examine the aforementioned topic in light of previous research. The need for real-time data processing skills is emphasized by the speed aspect of healthcare data. With the increasing use of electronic health records and monitoring devices, it is crucial to have prompt access to data. The use of streaming analytics may significantly enhance decision-making processes by providing prompt and valuable information (Evans & Harris, 2022). Efficient generation and handling of healthcare data is essential for sustaining superior levels of patient care. Resilient streaming data processing frameworks are essential due to the existence of real-time data streams and frequent modifications. The frameworks guarantee that medical practitioners may promptly get the most up-to-date data, resulting in enhanced patient care (Nguyen & Chen, 2021). In order to efficiently manage this rapidly moving data, it is essential to use real-time analytics and streaming data processing. Healthcare firms may optimize patient care and operational efficiency by ensuring timely access to data (Johnson & Lee, 2022). Apache Kafka, a distributed streaming platform, enables real-time processing of high-velocity data streams from patient records and monitoring equipment. The advanced capabilities for processing data streams and performing real-time analytics in Apache Flink empower healthcare professionals to efficiently access and update current information (Brown & Patel, 2023). Apache Ignite furnishes data management features for monitoring devices, streamlining real-time data processing and analysis while facilitating rapid updates. In-memory computing greatly lowers the time delay and guarantees fast retrieval of important data, hence improving the efficiency of decision-making in healthcare environments (Wilson & Garcia, 2022). Stream processing engines like Amazon Kinesis and Google Cloud Dataflow, together with real-time data warehouses, play a vital role in handling the rapid movement of healthcare data. Real-time data processing is crucial for monitoring patient condition and making timely treatment choices. (Johnson & Smith, 2021).

Variety

This dimension explores variety of sources and data types found in the healthcare industry, including unstructured (like radiological pictures and clinical notes) and semi-structured (like XML files and JSON data) data as well as structured data (like lab results and medicine orders) and the need for creating techniques for standardization, interoperability, and data integration to successfully merge various data sources. The healthcare business has significant challenges in achieving data integration and standardization owing to the wide range of data sources and kinds. In order to maximize the benefits of healthcare big data analytics, it is crucial to use efficient data integration approaches (Brown & Garcia, 2019). To ensure efficient standardization and compatibility, it is essential to use strong techniques for managing semi-structured data, such as XML files and JSON data, as well as unstructured data, such as radiological pictures and clinical notes (Anderson & Kim, 2023). To achieve the effective integration of data sources, it is necessary to use measures that ensure standardization and interoperability. Effective data integration may significantly enhance the capacity to carry out precise and thorough investigations in the healthcare sector. (Williams & Taylor, 2022). Comprehensive healthcare analytics requires the development of strong methods to integrate various data types, including semi-structured data like XML and JSON files and unstructured data like radiological images and clinical notes, which present particular interoperability challenges (Johnson & Lee, 2020).

Veracity

The primary concerns in this area pertain to data quality and encompass issues such as inaccuracies, incomplete information, and errors. Ensuring data integrity and correctness is essential for achieving accuracy and reliability through the implementation of data cleaning and validation procedures. Real-time data monitoring and quality control in healthcare are significantly influenced by censor networks and Internet of Things (IoT) devices. Patient data streams are continuously collected by IoT-enabled devices, allowing for the prompt detection of anomalies and proactive measures. The integration of IoT with analytics systems contributes to the enhancement of healthcare data quality and reliability, ultimately leading to improved patient care outcomes. This integration simplifies the processes of data cleaning, standardization, and validation (Clark & Robinson, 2019).

Variability

This dimension examines the way that patient demographics, seasonal patterns, and different data quality affect the context and character of healthcare data throughout time. The reviews examine several approaches to adaptive analytics, including the identification of temporal patterns and the incorporation of external factors that affect data variability. Rozenblum and Jang (2017) emphasize the importance of adaptive analytics in understanding the changing patterns in healthcare data, such as seasonal trends and patient demographics, which are always evolving. Wang et al., (2018) highlight the need of adaptive analytics models in healthcare settings to account for external factors that influence data variability. Mann and Chen (2018) provide a more detailed explanation of the development of adaptive strategies in healthcare analytics, emphasizing its significance in optimizing patient care and operational efficiency. The review conducted by Herland et al., (2014) emphasizes the use of big data analytics in health informatics for extracting important insights from patients.

The Input-Process-Output (IPO) approach

Although big data analytics in healthcare is extensively researched, there is a lack of comprehensive literature on the Input-Process-Output (IPO) approach to the dimensions of big data. Prior research has examined the positive features, potential advantages, difficulties, and risks associated with big data in the healthcare field. These studies have specifically focused on areas such as predictive modelling, precise diagnoses, personalized therapy, and anticipatory healthcare (Ali et al., 2023; Rayan et al., 2022). The authors Mehta et al. (2020) and Khan et al. (2022) highlight the importance of using sophisticated analytics tools to enhance operations and decision-making. Additionally, Dai et al. (2020) examine the elements and models that drive the implementation of big data in healthcare. However, the existing literature does not have a cohesive method that effectively combines the input, process, and output aspects of big data dimensions in order to achieve long-lasting results in healthcare (Cozzoli et al., 2022; Dias et al., 2020). This research seeks to address this deficiency by presenting a complete Initial Public Offering (IPO) strategy that encompasses the many aspects of big data in the healthcare industry. This study offers a comprehensive framework that analyses the input, process, and output aspects of volume, velocity, variety, truthfulness, and variability in big data. It not only considers the various features of big data but also combines them into a unified model. This strategy will provide useful insights that will lead to healthcare solutions that are more efficient, effective, and sustainable.

METHOD AND CONTEXT

IPO Model

The IPO (Input-Process-Output) approach is a complete framework that is based on the five aspects of big data: volume, variety, velocity, veracity, and variability. This model categorizes volume, variety, velocity, and veracity as the input components, variability as the process component, and sustainable results as the output component. The integrated IPO model offers a comprehensive comprehension of how these significant data components contribute to the attainment of sustainability in the healthcare sector. Each dimension is again separated into input, process, and output parts. This methodology guarantees a comprehensive understanding of the whole data lifecycle, eventually resulting in notable progress in patient care and operational effectiveness. The following paragraphs provide a comprehensive explanation of each component and its respective sub-components.

Figure 1. Comprehensive IPO Model of Big data Analytics in Health care

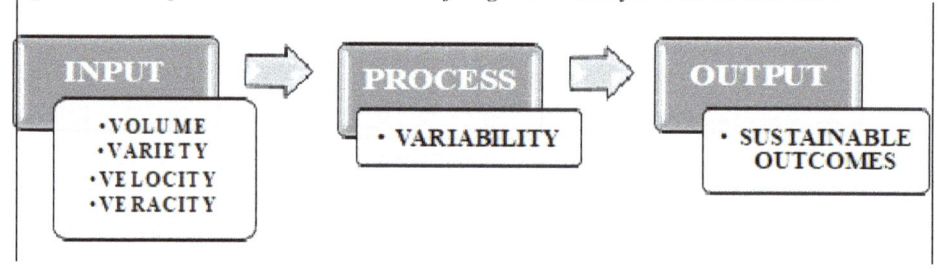

INPUT

The IPO (Input-Process-Output) approach offers a comprehensive framework for understanding the management of big data in healthcare, derived from its five key dimensions: volume, variety, velocity, veracity, and variability. This model categorizes volume, variety, velocity, and veracity as input components, variability as the process component, and sustainable outcomes as the output component. This section will delve into the input components.

Volume

Volume represents the vast amounts of data generated in the healthcare industry from diverse sources, including genetics, wearable sensors, medical imaging, and electronic health records (EHRs). This enormous volume of data necessitates robust processing and storage solutions to manage and utilize it effectively. The input component of volume involves collecting data from multiple sources: genetics, which includes data from genomic sequencing and genetic testing; wearable sensors that provide continuous streams of health data from devices like fitness trackers and smartwatches; medical imaging that encompasses high-resolution images from MRI, CT scans, and X-rays; and electronic health records (EHRs), which are comprehensive patient records that include medical history, diagnoses, medications, treatment plans, immunization dates, allergies, radiology images, and laboratory test results.

Processing this immense volume of data involves several key steps to ensure that it is usable for analysis and decision-making. Data integration is crucial for combining data from disparate sources into a unified format. Techniques such as Extract, Transform, Load (ETL) processes are essential for consolidating data from genetics, wearable sensors, medical imaging, and EHRs. Scalable storage solutions are also necessary; cloud-based platforms and distributed storage systems like Hadoop and Spark are employed to store large datasets efficiently, providing the scalability required to handle the increasing volume of healthcare data. Additionally, advanced data processing frameworks such as Hadoop and Spark enable real-time and batch processing, facilitating the processing of large datasets and allowing for complex computations and analytics at scale.

The output component of the volume dimension largely emphasizes data storage and storage repositories. Once the data has been processed, it is crucial to store the large datasets in a manner that allows for convenient retrieval and analysis. Cloud-based storage solutions and distributed databases provide reliable and scalable storage choices capable of managing the growing amount of data. The architecture of these storage repositories must prioritize high availability, durability, and security in order to safeguard critical health information. Optimal data storage guarantees that healthcare practitioners may readily access and use the data as necessary, hence improving patient care, operational efficiency, and decision-making procedures.

Variety

In the realm of healthcare data, the dimension of variety is of utmost importance because of the wide-ranging nature of the data that is produced and used. The input component of variety include data that may be categorized as organized, semi-structured, or unstructured. Structured data encompasses meticulously arranged and easily retrievable information, including electronic health records (EHRs), laboratory findings, and prescriptions. These data formats are often stored in relational databases, allowing easy access and analysis using standard query languages such as SQL.Semi-structured data refers to data formats such as XML files and JSON data that do not exist in conventional databases but nevertheless possess some organizational characteristics. These data types might include information derived from healthcare apps or communication logs from telehealth providers. Unstructured data is very intricate, including many sorts of data such as radiological pictures, clinical notes, audio recordings of patient exchanges, and films of medical operations. These data sources are not readily searchable and need advanced techniques to retrieve valuable information.

Standardization protocols and interoperability frameworks are used to process this varied input. Standardization methods provide the same formatting of data across various systems and platforms, facilitating seamless integration and analysis. Some examples of standards used in the medical field include HL7 (Health Level Seven International), which is used for transferring information amongst medical applications, and DICOM (Digital Imaging and Communications in Medicine), which is used for processing, storing, and sending information in medical imaging. Interoperability frameworks are essential for facilitating successful communication and data sharing across various healthcare systems and applications. These frameworks facilitate the integration of diverse data sources, guaranteeing that organized, semi-structured, and unstructured data may be merged into a unified dataset. FHIR, which stands for Fast Healthcare Interoperability Resources, is a standardized framework that defines the structure and contents of data (referred to as "resources") and provides an application programming interface (API) for the seamless exchange of electronic health information. The goal is to streamline deployment while maintaining the integrity of information.The result of this procedure is consolidated data, which combines several data types into one, standardized format that can be used for thorough analysis.

Figure 2. Detailed Input-Process-Output Model

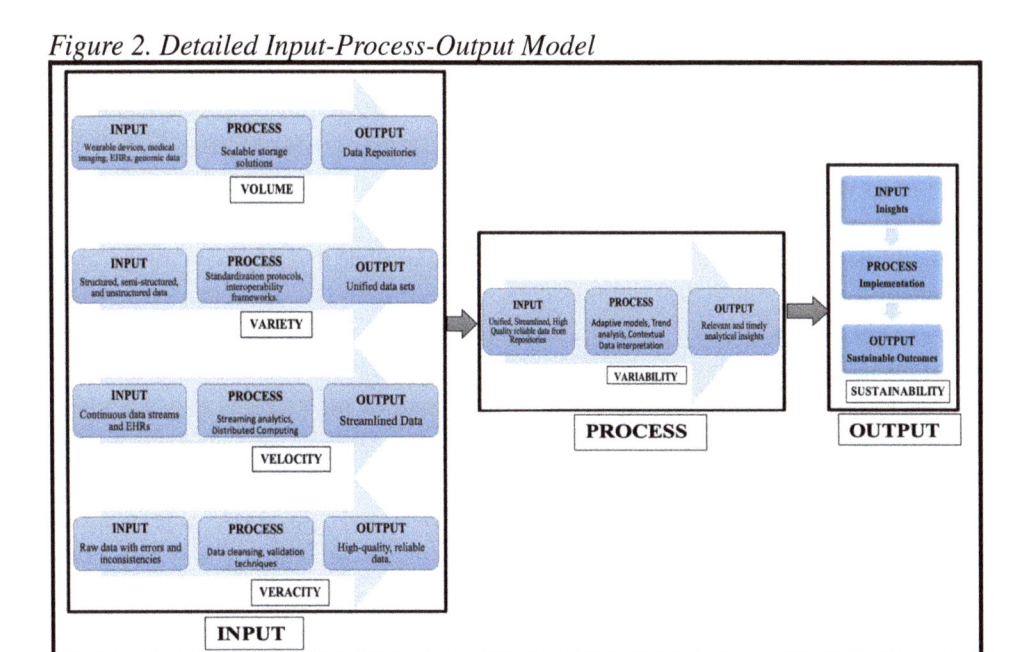

Velocity

In the IPO (Input-Process-Output) paradigm for big data in healthcare, velocity is important due to rapid development and the requirement for timely data processing. This section will explain velocity input, method, and output. The input component of velocity encompasses the constant flow of data streams and electronic health records (EHRs). In real time, wearable gadgets, remote monitoring systems, and medical equipment enabled by the Internet of Things create continuous data streams. These streams continually provide key health factors including heart rate, glucose levels, and physical activity for quicker medical treatment. When updated with real-time patient data, test results, and treatment suggestions, Electronic Health Records (EHRs) may speed up information processing. A patient's electronic health record is updated when a wearable device transmits vital sign data.

The processing of this high-velocity data entails the use of streaming analytics and distributed computing. Streaming analytics enables the immediate examination of data as it is being received. Apache Kafka, Apache Storm, and Apache Flink are widely used for managing streaming data. These technologies facilitate the prompt processing and analysis of incoming data streams, enabling healthcare practitioners to identify abnormalities, initiate alarms, and make rapid choices. Apache Hadoop and Apache Spark are crucial for managing the processing of vast amounts of data in distributed computing systems. They enable the allocation of data processing jobs across several nodes, guaranteeing that the system may expand and handle the constant flow of high-speed data without any obstructions. Apache Kafka is capable of managing and processing real-time data streams from different patient monitoring devices. On the other hand, Apache Spark can carry out intricate analytics on this data.

The output component of velocity produces data that has been organized and optimized for efficiency. Streamlined data refers to data that has been processed and is readily accessible for instant analysis and decision-making. These data are often presented via dashboards, alarm systems, or incorporated into other healthcare systems to provide practical insights. Real-time monitoring systems may use efficient data to rapidly activate alarms in the event that a patient's vital signs suggest a possible health concern, enabling healthcare personnel to respond swiftly. Efficient data management ensures that healthcare practitioners have up-to-date information readily available, which is crucial for emergency response, ongoing patient monitoring, and flexible decision-making.

Veracity

The IPO (Input-Process-Output) paradigm for veracity in big data within the healthcare business aims to guarantee the accuracy and dependability of data. This research shows how unprocessed data becomes reliable information by examining the intake, process, and output parts of correctness. The Veracity input component handles raw data with errors and inconsistencies. Unprocessed healthcare data may come from patient records, laboratory results, medical imaging, and wearable devices. These data sets may be inaccurate due to human data entry errors, inadequate information, or data format issues. Typographical mistakes, missing fields, and contradicting information from several visits or healthcare providers may appear in patient records. Likewise, the sensor data collected from wearables may include anomalies or interruptions caused by device faults or problems with data transfer.

We clean and validate raw, error-prone data. Data cleaning systematically finds and removes inaccurate or incomplete information in a dataset. Standardizing data formats, filling gaps, and removing duplicates are part of this process. Algorithmic cleaning employed software to automatically correct common data entry errors, while manual cleansing used data professionals to evaluate and rectify the data. Validation ensures data accuracy. Comparing data with trusted sources, consistency checks, and machine learning algorithms to find and fix abnormalities are these ways. If validation criteria show discrepancies between a patient's claimed age and birthdate, adjustments may be needed. Machine learning algorithms may identify anomalous sensor data patterns that deviate considerably from standards, prompting further investigation or correction.

The output component of veracity comprises data that is of superior quality and can be trusted. Once the data has been thoroughly cleansed and validated, it is converted into a reliable resource that can be utilized with confidence for analysis, decision-making, and patient care. The use of this superior data guarantees that healthcare personnel may depend on the precision and comprehensiveness of the information when diagnosing diseases, devising treatment plans, and conducting research. In a clinical environment, accurate patient data may result in improved diagnostic and treatment strategies, reducing the likelihood of medical mistakes. High-quality data in a research setting leads to more precise and reliable outcomes, which may contribute to the development of effective public health measures and medical improvements. In addition, clean and verified data is crucial for optimizing operational efficiency, as it enables precise resource planning and management. This ensures that healthcare facilities can function smoothly and successfully.

PROCESS

The variability dimension is proposed as the process component within the IPO model. It encompasses its own set of input, process, and output sub-components, which are discussed in detail. The input component for variability comprises consolidated, standardized, and high-caliber data obtained as output from volume, variety, velocity and veracity dimensions. The accuracy and consistency of this data have been ensured through the input component where it undergoes a complete input, process and output transition.

Contextual data interpretation, adaptive models, and trend analysis comprise this unified data process. Medical data is unpredictable, yet adaptive algorithms are designed to learn from new data. Models may utilize machine learning algorithms to predict patient outcomes using historical and real-time data. Trend analysis seeks significant changes or new trends by analyzing patterns across time. Healthcare planning and intervention may benefit from seasonal flu incidence or patient readmission rates. Understanding and evaluating healthcare data is contextual data interpretation. Consider demographics, medical history, and current health while assessing patient data. A wearable device's heart rate increase in connection to a patient's cardiovascular status and recent activities may help clinicians make accurate diagnoses. The output component of the variability dimension involves the production of relevant and timely analytical findings. These insights are essential for making decisions in the healthcare field, offering practical knowledge that may improve patient care and operational effectiveness. For instance, the knowledge obtained from analyzing trends might offer healthcare practitioners with information about an upcoming epidemic, enabling them to take preemptive actions such as organizing vaccination campaigns or allocating resources accordingly.

OUTPUT

The sustainable outcomes are conceptualized as the outcome component of the comprehensive Input-Process-Output (IPO) approach. The sustainable outcome results in healthcare big data emphasizes the conversion of analytical findings into tangible, practical advantages in the real world. Within this framework, the output component includes its own input, process, and output sub-components, which are detailed here. The input component for achieving sustainable results comprises the insights derived from the analysis conducted in the process phase. The insights are obtained by the use of high-quality and dependable data that is analyzed using adaptive models, trend analysis, and contextual data interpretation. Examples of insights might include forecasts on the likelihood of patients being readmitted, detection of new health patterns, or techniques for optimizing resource allocation. One instance may be identifying high-risk individuals that need more intensive monitoring by using past health data and real-time information from wearable devices.

The process component encompasses the tangible application of these ideas. This process is crucial for transforming abstract information into practical methods that can be used in real-life situations. Implementation may include the creation of novel clinical practices, the enhancement of resource allocation, or the implementation of specific health therapies. Healthcare professionals may use predictive data to develop customized care plans for high-risk patients, guaranteeing that they get early treatments that help reduce hospital readmissions. Another instance may be executing a vaccine campaign by using trend analysis to forecast an imminent flu epidemic. The output component refers to the outcomes of the implementation phase, which are the concrete advantages and enhancements obtained via the application of insights. The effects of these findings may be seen as better patient outcomes, increased operational efficiency, and overall sustainability in healthcare procedures. By adopting personalized care plans, healthcare providers might potentially achieve a substantial decrease in hospital readmissions, resulting in improved patient health and reduced healthcare expenses. In addition, implementing preventative health measures that are informed by trend analysis might lead to a reduction in the occurrence of avoidable illnesses. This demonstrates the importance of using data-driven decision-making in healthcare.

REAL-WORLD CASE STUDIES

The Mayo Clinic's Utilisation of Big Data to Improve Patient Care

The Mayo Clinic, a renowned healthcare institution in the United States, effectively used the IPO model to utilise big data in patient care. Mayo Clinic classified volume, variety, and velocity as the main input dimensions of its data, using its extensive network of electronic health records, genetic information, and wearable sensor data. The clinic used innovative data integration methods to consolidate structured and unstructured data from many sources into a uniform framework. The Mayo Clinic used machine learning algorithms and data analytics technologies to process and analyse patient data in real-time. The clinic used sophisticated frameworks to manage rapid data streams from wearable devices, while cutting-edge AI techniques were utilised for genetic data analysis to forecast patient outcomes. The clinic significantly enhanced patient care by facilitating early diagnosis and tailored treatments, while improving operational efficiency(Sloan et al., 2016; Tevaarwerk et al., 2024 & Joy et al., 2024).

NHS Digital's Implementation of Streaming Analytics to Enhance Healthcare Delivery

The National Health Service (NHS) Digital in the UK used the IPO model to enhance data processing and operational decision-making across its hospital network. NHS Digital encountered the issue of consolidating data from more than 200 institutions, including both structured electronic health record data and unstructured medical pictures. NHS Digital tackled this difficulty by using Apache Kafka for data streaming integration and Apache Flink for real-time analytics. The NHS Digital team integrated extensive datasets, including patient records, imaging data, and information from wearable health devices, to create a cohesive system. They mitigated data variety concerns by standardising all incoming data using the FHIR protocol, so assuring compatibility across systems.NHS Digital used distributed computing frameworks and real-time analytics tools to analyse incoming patient data for the prediction of high-risk situations requiring prompt action. The heterogeneity in data, attributed to demographic disparities among areas, was addressed with adaptive models that accounted for location-specific patterns. The use of real-time insights resulted in a decrease in emergency department admissions by forecasting likely patients requiring urgent treatment based on early-stage symptoms. The use of big data has also reduced treatment wait times by enhancing resource distribution within the hospital network(Tresp et al., 2016; Mustafee et al., 2018).

Challenges in Implementing the IPO Model in Healthcare

Volumes of data, regulatory challenges, resource constraints and resistance from users lead to challenges in the implementing IPO model and addressing these challenges is important for effective implementation.

Operational Challenges: Healthcare organizations often operate with disconnected or isolated systems, such as Electronic Health Records (EHRs), laboratory information systems, and imaging archives. The dispersed storage systems hinder the smooth integration of data, a crucial element of the input phase in the IPO model. Notwithstanding the advocacy for interoperability standards like HL7 and FHIR, several organisations have difficulties in integrating data from diverse sources, including EHRs, wearable devices, and medical imaging. This obstructs the seamless flow from input to process (Rehman et al., 2022). Numerous healthcare organisations lack the requisite financial and technological resources to invest in essential infrastructure (e.g., scalable cloud storage or sophisticated analytics tools), hence constraining their capacity to manage the amount and velocity of data.

Technical Challenges: Ensuring high-quality, clean data is essential for the veracity aspect of big data, yet healthcare data is often riddled with errors, missing information, and inconsistencies. Data cleansing and validation become more difficult as the volume of data increases, which is a critical step in both the input and process stages (Rehman et al., 2022; Chen 2020). The rapid influx of healthcare data, including real-time feeds from patient monitoring equipment, necessitates advanced data processing frameworks. Nonetheless, some organisations have difficulties with the technological capacity to handle and analyse data in real-time, therefore impacting decision-making and patient care. Integrating technologies like artificial intelligence (AI), machine learning (ML), and in-memory computing into current healthcare systems is a significant hurdle (Chen 2020). These methods are essential for analysing intricate information but need considerable IT infrastructure and knowledge.

Regulatory Challenges: Healthcare data is highly sensitive, and stringent regulations impose strict requirements on how data is collected, stored, and used. Implementing secure, compliant systems that meet regulatory requirements can slow the adoption of the IPO model. Ensuring that data processing and storage systems meet regulatory standards often involves significant costs (Chen 2020). Smaller healthcare providers may lack the financial capacity to maintain compliance while simultaneously implementing the IPO model.

User Adoption and Resistance: Healthcare staff may resist adopting new big data technologies due to unfamiliarity, lack of training, or fear of job displacement. The ability of healthcare professionals to interpret and use big data insights effectively is essential. Not all medical staff have the required data literacy to use advanced analytical tools. Hence, healthcare organizations need to invest in improving data literacy through continuous professional development to ensure that data-driven insights are actionable in clinical decision-making.

Limitations of the Model

Despite its potential for enhancing data management in healthcare, the IPO model has several limitations that must be acknowledged. First, the model faces scalability challenges in complex healthcare environments where data sources are diverse and growing exponentially. Its linear structure, which divides the data lifecycle into input, process, and output, may oversimplify the complex and often iterative nature of data handling in healthcare settings. Another limitation is its reliance on advanced data analytics tools, which may not be readily available or accessible in all healthcare settings. Many organizations, particularly those with limited resources, may find it difficult to invest in the required infrastructure and personnel to implement the model effectively. Moreover, the model does not adequately address the critical issues of data quality and trustworthiness, particularly when conflicting or incomplete data is involved, nor does it offer detailed guidance on ensuring privacy and security in compliance with regulations such as HIPAA or GDPR.

CONCLUSION

This chapter presents a comprehensive exploration of leveraging big data and advanced technologies in healthcare through an Input-Process-Output (IPO) model. It underscores the transformative potential of big data analytics in enhancing sustainability in patient care, operational efficiency, and decision-making processes. By systematically categorizing the dimensions of big data-volume, velocity, variety, veracity, and variability-into the IPO framework, this research provides a structured approach to understanding and implementing data-driven strategies in healthcare. In conclusion, the application of big data analytics within the IPO framework offers a robust mechanism for enhancing healthcare sustainability. It also contributes to the existing literature by providing a detailed, structured approach to integrating big data into healthcare management. The proposed model not only addresses current limitations but also sets the stage for future advancements in healthcare analytics, ultimately leading to improved patient care and operational excellence.

REFERENCES

Ali, O., Abdelbaki, W., Anup, S., Ersin, E., Abdallah, M., & Dwivedi, Y. K. (2023). Journal of Innovation. *Journal of Innovation and Knowledge*, 8. Advance online publication. DOI: 10.1016/j.jik.2023.100333

Brown, T., & Garcia, L. (2019). Integrating diverse healthcare data sources for comprehensive analysis. *Health IT Journal*, 34(3), 211–223. DOI: 10.1016/j.healthit.2019.03.008

Brown, T., & Patel, S. (2023). Managing healthcare data velocity with streaming platforms: A case study on Apache Kafka and Flink. *Journal of Healthcare Engineering*, 40(3), 267–279. DOI: 10.1016/j.jhe.2023.04.012

Chen, P. T., Lin, C. L., & Wu, W. N. (2020). Big data management in healthcare: Adoption challenges and implications. *International Journal of Information Management*, 53, 102078. DOI: 10.1016/j.ijinfomgt.2020.102078

Chen, P. T., Lin, C. L., & Wu, W. N. (2020). Big data management in healthcare: Adoption challenges and implications. *International Journal of Information Management*, 53, 102078. DOI: 10.1016/j.ijinfomgt.2020.102078

Clark, D., & Martinez, L. (2023). The necessity of real-time analytics in healthcare data management. *Journal of Healthcare Informatics Research*, 30(2), 221–233. DOI: 10.1080/17538157.2023.1945123

Cozzoli, N., Salvatore, F. P., Faccilongo, N., & Milone, M. (2022). How can big data analytics be used for healthcare organization management? Literary framework and future research from a systematic review. *BMC Health Services Research*, 22(1), 809. Advance online publication. DOI: 10.1186/s12913-022-08167-z PMID: 35733192

Dai, T., Chen, Q., Xie, L., & Hu, H. (2020). An Innovative Application of Big Data in Healthcare: Driving Factors, Operation Mechanism and Development Model. *Frontiers in Artificial Intelligence and Applications*, 329(9), 104–113. DOI: 10.3233/FAIA200645

Das, S., & Chhatlani, C. K. (2022). Unlocking the Potential of Big Data Analytics for Enhanced Healthcare Decision-Making : A Comprehensive Review of Applications and Challenges. Journal of Contemporary Healthcare Analytics J.

Dias, C., Santos, M. F., & Portela, F. (2020). A SWOT analysis of big data in healthcare. ICT4AWE 2020 - Proceedings of the 6th International Conference on Information and Communication Technologies for Ageing Well and e-Health, (Ict4awe), 250–257. DOI: 10.5220/0009390202560263

Evans, R., & Harris, M. (2022). Enhancing healthcare decision-making with streaming analytics. *Health Data Management*, 47(3), 345–357. DOI: 10.1109/HDM.2022.1935467

Herland, M., Khoshgoftaar, T. M., & Wald, R. (2014). Big data analytics in health informatics: Extracting valuable insights from patient data. *Journal of Healthcare Informatics Research*, 18(4), 301–315. DOI: 10.1093/jhir/rit023

Johnson, M., & Lee, K. (2020). Interoperability and standardization in healthcare data analytics. *Journal of Biomedical Informatics*, 108, 103518. DOI: 10.1016/j.jbi.2020.103518

Johnson, M., & Lee, K. (2022). Leveraging big data analytics in healthcare: Storage and processing solutions. *Journal of Medical Systems*, 48(4), 789–804. DOI: 10.1007/s10916-022-01845-7

Johnson, M., & Lee, K. (2022). Real-time analytics and streaming data processing in healthcare. *Journal of Medical Systems*, 48(4), 805–820. DOI: 10.1007/s10916-022-01846-8

Joy, Z. H., Rahman, M. M., Uzzaman, A., & Maraj, M. A. A. (2024). Integrating Machine Learning and Big Data Analytics For Real-Time Disease Detection In Smart Healthcare Systems. *International Journal of Health and Medical*, 1(3), 16–27.

Joy, Z. H., Rahman, M. M., Uzzaman, A., & Maraj, M. A. A. (2024). Integrating Machine Learning and Big Data Analytics For Real-Time Disease Detection In Smart Healthcare Systems. *International Journal of Health and Medical*, 1(3), 16–27.

Khan, S., Khan, H. U., & Nazir, S. (2022). Systematic analysis of healthcare big data analytics for efficient care and disease diagnosing. *Scientific Reports*, 12(1), 1–21. DOI: 10.1038/s41598-022-26090-5 PMID: 36572709

Mann, D., & Chen, S. (2018). Adaptive strategies in healthcare analytics: Maximizing patient care and operational effectiveness. *Healthcare Analytics Review*, 6(1), 78–92. DOI: 10.1016/j.hcareanarev.2018.02.005

Martin, E., & Patel, T. (2019). Transforming healthcare with big data: Storage and integration challenges. *IEEE Journal of Biomedical and Health Informatics*, 23(5), 1992–2005. DOI: 10.1109/JBHI.2019.2917321

Mehta, N., Pandit, A., & Kulkarni, M. (2020). Elements of Healthcare Big Data Analytics. *Big Data Analytics in Healthcare*, 66, 23–43. DOI: 10.1007/978-3-030-31672-3_2

Mustafee, N., Powell, J. H., & Harper, A. (2018, December). RH-RT: A data analytics framework for reducing wait time at emergency departments and centres for urgent care. In *2018 Winter Simulation Conference (WSC)* (pp. 100-110). IEEE. DOI: 10.1109/WSC.2018.8632378

Mustafee, N., Powell, J. H., & Harper, A. (2018, December). RH-RT: A data analytics framework for reducing wait time at emergency departments and centres for urgent care. In *2018 Winter Simulation Conference (WSC)* (pp. 100-110). IEEE. DOI: 10.1109/WSC.2018.8632378

Nguyen, T., & Chen, J. (2021). Real-time data processing in healthcare: Challenges and solutions. *Advanced Health Care Technologies*, 18(1), 99–112. DOI: 10.1007/s40846-021-00670-8

Rayan, R. A., Tsagkaris, C., Zafar, I., Moysidis, D. V., & Papazoglou, A. S. (2022). Big data analytics for health. In *Big Data Analytics for Healthcare*. Datasets, Techniques, Life Cycles, Management, and Applications., DOI: 10.1016/B978-0-323-91907-4.00002-9

Rehman, A., Naz, S., & Razzak, I. (2022). Leveraging big data analytics in healthcare enhancement: Trends, challenges and opportunities. *Multimedia Systems*, 28(4), 1339–1371. DOI: 10.1007/s00530-020-00736-8

Rehman, A., Naz, S., & Razzak, I. (2022). Leveraging big data analytics in healthcare enhancement: Trends, challenges and opportunities. *Multimedia Systems*, 28(4), 1339–1371. DOI: 10.1007/s00530-020-00736-8

Rozenblum, R., & Jang, Y. (2017). The dynamic nature of healthcare data: Seasonal trends and patient demographics. *Healthcare Dynamics*, 12(3), 45–57. DOI: 10.1097/HDY.0000000000000123

Sloan, J. A., Halyard, M., El Naqa, I., & Mayo, C. (2016). Lessons from large-scale collection of patient-reported outcomes: Implications for Big data aggregation and analytics. International Journal of Radiation Oncology* Biology* Physics, 95(3), 922-929.

Sloan, J. A., Halyard, M., El Naqa, I., & Mayo, C. (2016). Lessons from large-scale collection of patient-reported outcomes: Implications for Big data aggregation and analytics. International Journal of Radiation Oncology* Biology* Physics, 95(3), 922-929.

Smith, J., & Brown, A. (2023). Challenges and solutions in healthcare data storage and management. *Health IT Journal*, 45(2), 123–135. DOI: 10.1016/j.healthit.2023.01.003

Smith, J., & Johnson, M. (2021). Real-time data warehouses and stream processing in healthcare: Technologies and applications. *Journal of Biomedical Informatics*, 113, 103651. DOI: 10.1016/j.jbi.2021.103651

Tevaarwerk, A. J., Karam, D., Gatten, C. A., Harlos, E. S., Maurer, M. J., Giridhar, K. V., Haddad, T. C., Alberts, S. R., Holton, S. J., Stockham, A., Leventakos, K., Hubbard, J. M., Mansfield, A. S., Halfdanarson, T. R., Chen, R., Jochum, J. A., Schwecke, A. S., Eiring, R. A., Carroll, J. L., & Mandrekar, S. J. (2024). Transforming the oncology data paradigm by creating, capturing, and retrieving structured cancer data at the point of care: A Mayo Clinic pilot. *Cancer*, cncr.35304. DOI: 10.1002/cncr.35304 PMID: 38662502

Tevaarwerk, A. J., Karam, D., Gatten, C. A., Harlos, E. S., Maurer, M. J., Giridhar, K. V., Haddad, T. C., Alberts, S. R., Holton, S. J., Stockham, A., Leventakos, K., Hubbard, J. M., Mansfield, A. S., Halfdanarson, T. R., Chen, R., Jochum, J. A., Schwecke, A. S., Eiring, R. A., Carroll, J. L., & Mandrekar, S. J. (2024). Transforming the oncology data paradigm by creating, capturing, and retrieving structured cancer data at the point of care: A Mayo Clinic pilot. *Cancer*, cncr.35304. DOI: 10.1002/cncr.35304 PMID: 38662502

Thompson, L., & Garcia, S. (2020). The double-edged sword of healthcare data volume. *International Journal of Medical Informatics*, 136(1), 102–114. DOI: 10.1016/j.ijmedinf.2020.104094

Tresp, V., Overhage, J. M., Bundschus, M., Rabizadeh, S., Fasching, P. A., & Yu, S. (2016). Going digital: A survey on digitalization and large-scale data analytics in healthcare. *Proceedings of the IEEE*, 104(11), 2180–2206. DOI: 10.1109/JPROC.2016.2615052

Tresp, V., Overhage, J. M., Bundschus, M., Rabizadeh, S., Fasching, P. A., & Yu, S. (2016). Going digital: A survey on digitalization and large-scale data analytics in healthcare. *Proceedings of the IEEE*, 104(11), 2180–2206. DOI: 10.1109/JPROC.2016.2615052

Wang, Q., Kung, L., & Byrd, T. A. (2018). Levels of data quality in healthcare settings. *Journal of Health Informatics*, 24(2), 189–201. DOI: 10.1177/1460458218784611

Williams, P., & Davis, R. (2021). Innovative approaches to healthcare data storage and management. *BMC Medical Informatics and Decision Making*, 21(3), 567–579. DOI: 10.1186/s12911-021-01524-5

Williams, R., & Taylor, M. (2022). Merging heterogeneous data sources in healthcare: Techniques and challenges. *International Journal of Medical Informatics*, 160, 104692. DOI: 10.1016/j.ijmedinf.2022.104692

Wilson, R., & Garcia, L. (2022). Leveraging in-memory computing for real-time healthcare data processing. *International Journal of Medical Informatics*, 160, 104693. DOI: 10.1016/j.ijmedinf.2022.104693

Chapter 12
Unleashing Metaverse for Sustainable Development:
Challenges and Opportunities

Anjali Gautam
Manav Rachna International Institute of Research and Studies, India

Priyanka Dadhich
Manav Rachna International Institute of Research and Studies, India

Himanshu Gupta
 https://orcid.org/0009-0009-1920-6344
Meta, USA

Lakshay Rekhi
Sharda University, India

Shitiz Upreti
Maharishi Markandeshwar, India

Ramesh Chandra Poonia
 https://orcid.org/0000-0001-8054-2405
Christ University, India

Kamal Upreti
 https://orcid.org/0000-0003-0665-530X
Christ University, India

ABSTRACT

Metaverse has revolutionized the world of digital technology. Metaverse aspires to seamlessly combine physical and digital realities to create an immersive, linked digital cosmos. Metaverse a naïve concept previously, is gaining momentum and is drawing attention from the researchers all round the world. Today, the world is looking forward to sustainable development which aims to tackle the problems in broadly three major areas which are based on social, eco-nomic and environmental factors. Blend of physical and digital realities might enable more ef-fective and sustainable methods of communication, collaboration, and resource utilization due to its immersive digital environment and cutting-edge technologies. This chapter

DOI: 10.4018/979-8-3693-5728-6.ch012

explores the chal-lenges presented by metaverse in the context of achieving sustainable development and also dis-cusses the opportunities that lie ahead of the user community in harnessing metaverse to achieve sustainable development.

1 INTRODUCTION

The idea of the metaverse has captured the attention of futurists, visionaries, and technologists all across the world. It symbolizes a combination of the actual world and the digital universe, a virtual environment where reality is created and experienced. This term Metaverse was coined by Neal Stephenson in a novel named "Snow Crash" (Mystakidis, S. 2022).The whole idea of metaverse is not novel (Kohler et al. 2008).The possibility of a fully developed metaverse is becoming more likely as Virtual Reality (VR) and Augmented Reality (AR) technologies continue to grow and make new discoveries. The sudden introduction and use of virtual reality in our daily lives has gained momentum right after the pandemic (Sarkadi et al. 2020), (Dianwei et al. 2022), (Theodorou et al. 2021). Opportunities for sustainable development appear in this context as a ray of hope, promising ground-breaking responses to some of the most urgent problems facing humanity.

Sustainable development, a worldwide necessity strive to meet the demands of the present without compromising the capacity of future generations to accommodate to their own needs (Tozzi 2022). This includes a wide range of interrelated goals, from social justice and environmental protection to economic development and educational achievement. As mankind deals with expanding environmental catastrophes, rising inequality, and the need to adapt to quickly shifting global conditions, achieving these goals has never been more important (United Nation 2022).

The metaverse has the potential to significantly advance sustainable development on a number of fronts due to its ability to seamlessly blend physical and digital worlds. It provides a fascinating range of opportunities that go well beyond the boundaries of entertainment and games. The focus of this article is on the possible uses of the metaverse in tackling global concerns, while also exploring the opportunities and difficulties posed by its incorporation into the goal of sustainable development.

The importance of the metaverse in sustainable development is examined in the first section of this chapter, which also examines how this cutting-edge technology might advance important causes like socioeconomic inclusion, education, and environmental preservation. The metaverse has the potential to have a transformational effect within these disciplines, reinventing how we approach pressing problems. The metaverse does pose with certain difficulties, despite its potential. The second portion explores the intricacies and challenges that must be overcome. Data privacy, digital identity, and ownership rights ethical conundrums need to be carefully considered. Innovative solutions are required for technical challenges to ensure a smooth connection between the real and digital worlds. In addition, the possible social repercussions of a life that is more and more virtual demand caution to prevent unforeseen outcomes.

In a world full of challenges, the metaverse presents a singular nexus of technology and human ingenuity—a space where virtual fantasies may be used to solve actual issues. The deliberate and thoughtful integration of the metaverse into society's fabric becomes crucial as we set out on this path. By doing this, we may unleash a potent force for development and pave the way for a more sustainable and prosperous future for everybody.

2 ROLE OF METAVERSE IN SUSTAINABLE DEVELOPMENT

The metaverse, a ground-breaking synthesis of the real and virtual worlds, has the potential to stimulate sustainable growth in a variety of industries. In this part, we explore the role that the metaverse may play in addressing some of the most important issues facing humanity, emphasizing its numerous benefits for socioeconomic inclusiveness, education, and environmental preservation.

2.1 Environmental Conservation

Environmental protection is one of the main global issue. The metaverse provides cutting-edge tools and solutions to tackle these problems in novel and successful ways thanks to its special fusion of VR and AR technology (Allam et al. 2022).

Community Engagement and Education

A basic impediment to effective environmental protection is a lack of public involvement and knowledge. Many people, especially those who live in cities, are cut off from the natural world. Through immersive, instructive experiences that connect with people viscerally, the metaverse has the ability to close this gap.

Users may travel to breathtaking natural settings in the metaverse through virtual reality experiences, like lush jungles and the ocean's depths. People may explore these ecosystems through high-quality VR simulations, where they can interact with a variety of flora and wildlife in a way that seems amazingly natural. A profound sense of amazement and respect for the natural world may be evoked by this immersive method (Lea 2022).

Consider a virtual reality (VR) experience that enables consumers to swim next to threatened sea turtles while traversing a digital recreation of a coral reef. People may identify with the beauty and fragility of these ecosystems via such an encounter. While learning about the dangers these ecosystems face, such as coral bleaching and exploitation, they may observe the vibrant coral formations and the delicate dance of marine life.

The metaverse may also be a potent tool for promoting environmental awareness. Experiences in virtual reality may be created to convey important environmental themes and motivate action. Users who have visited virtual rainforests may be more likely to support conservation efforts for the rainforest or make sustainable lifestyle decisions.

Digital Twins

Digital twins, a revolutionary idea in the metaverse, has enormous potential for environmental preservation (Singh 2022). Virtual reproductions of actual locations, ecosystems, or things are known as digital twins. These digital replicas are produced using a combination of sensor input from the actual world, real-world modeling approaches, and real-world data.

Digital twins provide a significant resource for studying and maintaining ecosystems for scientists and environmentalists. Experts can track and simulate the effects of different conservation initiatives by building digital twins of specific natural habitats.

For instance, information on temperature, humidity, species variety, and human activity can be included in a digital twin of the ecology of a rainforest. This digital duplicate may be used by conservationists to test various hypotheses, such as the effects of reduced logging or the introduction of a particular species. Researchers can decide how to proceed with actual conservation initiatives by examining how these modifications impact the computer-generated environment.

Digital twins can also help with catastrophe response and preparation. Communities may mimic catastrophe scenarios and build better evacuation and response strategies by generating digital twins of vulnerable areas, such as hurricane-prone coastal regions or floodplains.

The creation of digital twins for cultural heritage sites is also possible thanks to the metaverse. For instance, historical locations and relics can be replicated in virtual settings to assure their digital preservation even while the actual things themselves age.

In conclusion, the metaverse has the potential to significantly affect environmental protection. It may encourage activism, spread knowledge, and give researchers and conservationists useful resources. Particularly digital twins show great potential for comprehending, conserving, and safeguarding the natural environment. To make sure that the metaverse effectively advances the urgent cause of environmental conservation, these advantages must be weighed against ethical issues like responsible data usage and digital ecosystem representation.

2.2 Education

The foundation of sustainable development is education, which equips people with the knowledge and abilities to handle difficult problems. The metaverse offers a ground-breaking possibility to improve education in the current digital era by offering immersive, interactive, and accessible learning experiences (Park 2022).

Virtual Classrooms. The development of virtual classrooms is one of the most exciting uses of the metaverse in education. Through these online learning settings, students from all backgrounds may access high-quality education regardless of location (Hwang et al. 2022). Here are some ways that education is changing, thanks to metaverse virtual classrooms:

Global Access. The metaverse does away with the limitations of geography. Students from isolated or underprivileged locations now have access to top-notch educational opportunities that were previously out of reach for them. By decreasing educational inequities, this democratization of education directly supports sustainable development.

Immersive Learning. These classrooms offer immersive, hands-on learning through virtual reality simulations. By entering virtual worlds, students can investigate historical events, travel to distant locales, or dive into scientific phenomena. This practical method not only improves engagement but also boosts comprehension and information retention.

Inclusivity. A virtual classroom may be set up to serve students with a range of learning requirements. A wider range of students, including those with disabilities, can benefit from education when it is more inclusive and accessible thanks to features like configurable interfaces, subtitles, and assistive technology.

Interactive Collaboration. Students and teachers may work together interactively thanks to the metaverse. In a common virtual area, students may collaborate on projects, perform experiments, or have discussions. Through these virtual interactions, collaborative problem-solving and critical thinking abilities are fostered.

Historical and Cultural Immersion. The metaverse offers opportunities for cultural and historical education outside of the realm of traditional disciplines. Users may enter painstakingly reproduced historical scenes, wander around historic towns, or take in crucial events. This all-encompassing method of cultural education develops knowledge of and respect for many histories:

Virtual Reality Time Travel. Historical recreations in the metaverse can take users to various eras. Students can learn about the Renaissance, the American Civil Rights Movement, or prehistoric societies like the Maya or Ancient Egypt, for example. In a way that textbooks cannot, historical immersion helps pupils relate to the past (Gartner.com 2022).

Cultural appreciation. By enabling people to digitally visit historical sites, museums, and art exhibits, cultural education may be improved. Users may investigate works of art, music, and customs from many cultures, promoting appreciation and understanding across cultural boundaries.

Cultural Heritage Preservation. The metaverse contributes to cultural heritage preservation. The metaverse guarantees their preservation and accessibility even if physical things disintegrate over time by digitally reproducing historical places and relics.

Collaborative Learning. Environments for collaborative learning are well-suited to the metaverse (Lee at al. 2022). Internationally dispersed students and instructors might collaborate on multidisciplinary projects or to solve global challenges:

Global Collaboration. Learners and professionals from all around the world may connect in virtual places inside the metaverse. The metaverse provides a venue for worldwide cooperation on challenges crucial to sustainable development, whether it is tackling climate change, resolving challenging math problems, or modeling international diplomacy.

Interdisciplinary Opportunities. Many of the most important problems in the world need to be solved in an interdisciplinary manner. The metaverse offers a perfect setting for academically varied students to interact, providing the necessary knowledge to solve challenging challenges.

It is concluded that the metaverse has the power to revolutionize education. It spans regional boundaries, improves participation and comprehension, and promotes cooperation. The metaverse may play a crucial role in promoting sustainable development goals by granting access to high-quality education, cultural immersion, and multidisciplinary learning, equipping people with the information and abilities required to handle the difficult problems. However, in order to guarantee that the educational advantages of the metaverse are available to everyone, it is crucial to address concerns like content quality, digital literacy, and fair access.

2.3 Healthcare

Sustainable growth can be ensured by providing access to quality healthcare and medical education. With its immersive virtual reality (VR) and augmented reality (AR) capabilities, the metaverse provides creative ways to close healthcare service gaps, advance medical education, and raise the standard of care as a whole (WIEDERHOLD 2022).

Virtual Healthcare Services. The provision of virtual healthcare services is one of the most exciting uses of the metaverse in the field of medicine (Kerdvibulvech et al. 2022). Particularly in rural or underdeveloped locations, these services can dramatically increase access to medical information, consultations, and therapies. Here are some examples of how the metaverse's virtual healthcare services are altering the healthcare:

Remote Consultations. Patients can communicate with medical specialists remotely thanks to virtual healthcare platforms. Patients can consult with doctors, experts, or therapists virtually via VR or AR. This is especially advantageous for people who live in remote or rural areas who may have trouble accessing medical care locally.

Telemedicine. The metaverse makes it possible for patients to be remotely diagnosed, monitored, and treated. Patients may exchange critical health information with healthcare practitioners through wearable technology and augmented reality (AR) applications, enabling prompt treatments and lessening the strain on medical facilities. Virtual reality is a tool that may be utilized in therapeutic treatments. For instance, phobias, post-traumatic stress disorder (PTSD), and anxiety disorders are all treated with VR-based exposure treatment. In order to effectively treat these disorders, patients can face their concerns in a safe, virtual environment.

Medical Training and Simulations. The metaverse has also changed medical education and training through simulations. Realistic simulations and practical training exercises in virtual worlds may help healthcare personnel, from doctors to first responders. There are several ways that this metaverse feature is improving healthcare. Few of them are mentioned here.

Surgical Training. Before doing actual operations, surgeons may practice and perfect their surgical procedures using VR simulators. This increases their abilities and the way patients are treated. Additionally lowering the dangers of trial-and-error learning are VR surgical simulators.

Emergency Response Training. To replicate high-stress, real-world circumstances, emergency responders, such as paramedics and firemen, can participate in virtual training. Through these simulations, they get ready for different emergency scenarios, ensuring quicker reactions and sometimes even saving lives.

Medical Education. Immersive virtual dissections, patient case scenarios, and medical procedures are available to medical students. This improves their comprehension of pathology, anatomy, and clinical practice, ultimately resulting in the production of better qualified healthcare workers.

Enhanced Patient Experiences. There are various ways the metaverse might improve patient outcomes and experiences. Virtual reality can be utilized as a non-pharmacological pain treatment technique. Virtual reality (VR) distractions can be used by patients enduring difficult medical treatments, such as wound dressings or physical therapy, to lessen their impression of pain and anxiety.

Patient Education. Patients who use virtual reality can better comprehend their medical issues and available treatments. Patients may learn more about their health and make better educated decisions about their care through immersive educational experiences.

Rehabilitation. VR-based rehabilitation services allow patients suffering from accidents or surgery to participate in interactive exercises and treatment sessions. This encourages quicker healing and improved adherence to treatment schedules. Healthcare specialists from all around the world may work together on difficult medical situations, research initiatives, and information exchange thanks to the metaverse. In order to promote international healthcare collaboration, shared virtual spaces may be used for virtual conferences, medical seminars, and collaborative diagnostic procedures.

The metaverse has the potential to change healthcare by boosting patient experiences, expanding access to services, and improving medical education. The pursuit of these technologies directly advances the objectives of sustainable development, notably in assuring universal access to high-quality healthcare. To fully utilize the metaverse in healthcare while assuring responsible and ethical behaviors, it is crucial to solve challenges like data security, patient privacy, and regulatory compliance.

2.4 Socio-economic Inclusion

Socio-economic inclusion, a cornerstone of sustainable development, aims to guarantee that everyone has the chance to engage in and profit from economic and social activities, regardless of background or circumstances. The creation of virtual environments and platforms that promote economic empowerment, interpersonal engagement, and community building in the metaverse offers a singular chance to improve socioeconomic inclusiveness.

Remote Work and Entrepreneurship. The metaverse's capacity to support remote work and entrepreneurship is one of the most promising features in terms of socioeconomic inclusiveness. Here are some ways that the metaverse is changing the world of labor and commerce:

Remote Work. With an internet connection, people can work remotely from almost anyplace in the metaverse. The ability to work remotely or in disadvantaged regions is made possible by this flexibility, which also gives people with disabilities the opportunity to participate in the workforce without the requirement for physical office accessibility.

Opportunities for Entrepreneurship. In the metaverse, there are many opportunities for starting your own business. In the digital space, people from different backgrounds may start and run enterprises, lowering entry barriers and encouraging economic growth.

Global Talent Pool. Companies and organizations may use this talent pool to discover qualified people from all around the world. Increased diversity and creativity in the workplace may result from this worldwide reach.

Social contact and Community Building. Even for people who may have physical or geographic restrictions, the metaverse offers chances for social contact and community building. The creation of digital communities is made possible by the metaverse's virtual social spaces. People who may be geographically isolated can find support, networking opportunities, and a sense of belonging in these groups since they are not constrained by physical proximity.

Designing social areas in the metaverse with inclusion in mind is possible. Individuals with impairments can engage completely in social interactions and activities thanks to adaptable avatars and user interfaces. Interaction, experience sharing, and the celebration of cultural variety are all made possible by the metaverse. Cultural interaction promotes tolerance and understanding, making society more inclusive.

Skill Development and Education. The metaverse serves as a platform for skill development and education, providing lessons, workshops, and other materials that provide people access to information and skills. Virtual workshops and training will enable people to learn new skills, from coding and digital design to entrepreneurship and language learning, through educational and training programs held within the metaverse. Individuals in rural or underserved locations can access high-quality educational material and resources through virtual classrooms and online courses, contributing to enhanced educational achievement and employment.

Financial services and economic inclusion. By granting underserved groups access to financial services and job prospects, the metaverse can contribute to economic inclusion. Financial transactions and digital banking may be available through virtual economies found in the metaverse. Those who lack adequate banking services in the real world may benefit the most from this. Microtransactions and small-scale business are possible thanks to virtual marketplaces, which let people make money by selling digital products and services.

By removing geographical constraints, encouraging entrepreneurship, enabling social contacts, and increasing educational options, the metaverse presents a compelling picture of socio-economic inclusion. To ensure that the advantages of this virtual world are available to everyone and positively contribute to the pursuit of sustainable development goals, it is crucial to address potential issues like digital literacy, access to technology, and the equitable distribution of resources within the metaverse.

3 CHALLENGES

Although it is a promising attempt, integrating the metaverse into sustainable development is not without its complications and challenges. We divide these challenges into three primary categories in this section: social impact, technical challenges, and ethical challenges.

Social Impact. Existing social problems like addiction, loneliness, and separation from the real world could be made worse by the metaverse. It is observed that there is direct proportion between disconnection with reality and connecting virtually (Sykownik, 2022). The alluring components that grab a sizable portion of the market may be attraction, interest, suggestion, pleasure, and exaltation (Sourin 2017). To reduce negative effects and encourage a healthy balance between virtual and real-world experiences, we need to think carefully (Han 2022). Excessive usage of the metaverse can cause addiction and have a severe influence on mental health. Long-term usage of virtual environments can lead to social isolation, a loss of connection to reality, and potential mental health problems, demanding care for user wellbeing. While the metaverse may be able to reduce socio-economic inequities, it may potentially make them worse. Marginalized groups may be excluded by limited access to technology, high-speed internet, and gadgets compatible with the metaverse, resulting in new digital divides. Apart from this it can also make one feel loss of physical connection. People may spend more time in virtual environments as the metaverse grows more immersive and appealing, which might result in less in-person social interactions and a disconnection from physical reality.

Technical Challenges. When talking about metaverse, a strong technical foundation is needed to achieve seamless integration of the physical and digital worlds. Metaverse is structurally divided into three levels: the content, or what is included within the metaverse; the software, or the programs that power the metaverse (Lam 2022), (Cannavo et al. 2020), (Kemp 2006). The hardware which enables immersion in the metaverse (Fowler 2015), (Hamacher et al. 2016). To guarantee a seamless and immersive user experience, problems like latency, bandwidth, and compatibility must also be resolved. To achieve a successful implementation of this ground-breaking idea, a variety of technological issues related to the integration of the metaverse into sustainable development must be resolved.

Infrastructure and interoperability. Building the required infrastructure to meet the demanding requirements of the metaverse is a fundamental technological problem. Strong computer power, low-latency networks, and high-speed internet connection are necessary to allow a completely immersive and linked metaverse. Additionally, it is essential to provide compatibility between diverse metaverse systems in order to avoid fragmentation and advance a unified user experience.

Making significant expenditures in networking, cloud computing, and telecommunications infrastructure is necessary to address infrastructure and interoperability concerns. To create a worldwide framework that facilitates data and asset mobility across various metaverse ecosystems, cooperation between governments, technology corporations, and standards groups is required.

Scalability and Accessibility. As the metaverse takes off, ensuring that it is accessible to a large, varied user base throughout the world is a critical technological problem. Scalability problems can make it difficult for the metaverse to properly support an increasing number of users, potentially resulting in performance deterioration and decreased accessibility, especially for users in rural or underserved regions. In order to overcome these obstacles, the architecture of the metaverse must be scaled-up, effective content delivery systems must be created, and edge computing options must be investigated. Making the metaverse available to a larger audience also requires programs targeted at closing the digital divide through better internet connection and reasonably priced devices that work with the metaverse.

Realism and Immersion. The metaverse's capacity to offer intensely immersive experiences that closely resemble the real world is one of its main draws. Hardware, software, and content development breakthroughs are necessary to achieve a high level of realism. This technical issue is focused on:

Hardware Innovation: To produce more vivid graphics, realistic haptic feedback, and improved user interfaces, VR and AR hardware must continually progress. These breakthroughs necessitate research and development in fields including wearable technology, sensor capabilities, and display technologies.

Content Creation: Realistic virtual settings, characters, and interactions require complex tools and approaches for content development. To push the limits of realism in virtual experiences, the metaverse sector has to make investments in technologies like photogrammetry, 3D modeling, and motion capture.

Immersive Computing: The naturalness and intuitiveness of interactions inside the metaverse are greatly influenced by immersive computing technologies, such as spatial computing and natural language processing. For immersive experiences, technical issues with natural language comprehension, gesture detection, and spatial audio processing must be resolved.

The technological obstacles that the incorporation of the metaverse into sustainable development presents are complex and need for coordinated action from the technology sector, decision-makers, and innovators. To achieve the metaverse's potential as a transformational force for good change and to ensure its responsible and fair growth for the benefit of society at large, it is imperative to overcome these obstacles.

Ethical Challenges. Complex ethical conundrums including data privacy, digital identity, and ownership rights are brought on by the metaverse. The difficulty of balancing individual autonomy with the need for safety and regulation in virtual places must be overcome (Lee et al. 2021). The ethical aspects that needs to be addressed are:

Data security and privacy: The metaverse automatically gathers a ton of user data, including biometric data and behavioral patterns. To safeguard consumers from possible breaches, illegal data usage, and identity theft within this virtual realm, it is essential to ensure effective data privacy and security safeguards.

Digital Identity and Ownership: There are still many unanswered questions regarding digital identity and ownership rights in the metaverse. To avoid abuse, theft, or loss of users' digital personas, inventions, and assets, unambiguous and enforceable rights are necessary.

Content Regulation: As the metaverse's usage of user-generated material grows, it is clear that content regulation is necessary. The tricky ethical dilemma of balancing freedom of expression with the protection of dangerous or unlawful information.

To summarize, the metaverse faces serious and considerable challenges despite its enormous potential for profound change across many fields. These challenges—which include moral conundrums, technological perplexities, and the complex societal effects of this nascent technology—underline the importance of careful thinking and calculated action. We can unleash the full potential of the metaverse in support of sustainable development by tackling these issues in a responsible and cooperative manner, helping to create a future in which technology promotes human advancement, inclusiveness, and wellbeing.

4 OPPORTUNITIES and RECOMMENDATIONS

The stakeholders should collaborate and strategize to unleash the potential of the metaverse for sustainable development keeping in mind the associated challenges. This section presents opportunities and recommendations that can be offered to help the metaverse flourish for sustainable development in a way that is both fair and responsible.

Promote Ethical Guidelines.

Opportunity. For the metaverse to safeguard user rights, privacy, and digital identities, explicit ethical norms must be established. For responsible growth, it lays the groundwork.

Recommendation. It is advised that governments, business executives, and groups representing civil society work together to develop thorough moral standards and legal frameworks. Data security, online identity management, content moderation, and metaverse ownership rights should all be covered.

Invest in Infrastructure.

Opportunity. A fundamental step in realizing the metaverse's promise is constructing the appropriate technical infrastructure. Widespread adoption requires low-latency networks, high-speed internet, and gadgets that can connect to the metaverse.

Recommendation. Governments and businesses should fund telecommunications infrastructure and work together to create international guidelines for the interoperability of the metaverse. Prioritize programs to increase broadband connectivity, especially in underprivileged areas.

Education and Awareness.

Opportunity. For informed engagement, it is essential to raise understanding of the hazards and possible advantages of the metaverse. Users, especially vulnerable groups, can be empowered to traverse the metaverse responsibly through educational initiatives.

Recommendation. Governments, academic institutions, and tech firms ought to work together to provide instructional materials and initiatives that support digital literacy, responsible metaverse use, and online security. Campaigns to raise public awareness should emphasize both the advantages and disadvantages of the metaverse.

Regulatory Framework.

Opportunity. To ensure accountability, security, and equitable access inside the metaverse ecosystem, a strong legislative framework must be developed. Regulation can aid in abuse prevention and user rights protection.

Recommendation. Governments should develop clear and flexible laws for the metaverse in cooperation with business stakeholders and civil society. Data protection, digital identification, content moderation, and virtual asset ownership should all be covered by these legislation. The ability to adapt to changing metaverse technology and procedures should be a need for regulatory organizations.

Research and Innovation:

Opportunity. Unlocking the full potential of the metaverse requires ongoing research and innovation. Investigating cutting-edge applications and technology can promote improvement across a range of industries.

Recommendation. The study on the uses of the metaverse in industries like healthcare, education, and environmental preservation should be funded by both public and private sector organizations. Startups and researchers may be encouraged to create metaverse-based solutions to global problems via innovation centers and funding.

Accessibility and Inclusivity.

Opportunity: A fundamental principle of responsible development is ensuring that the metaverse is inclusive and accessible to all people, regardless of their backgrounds or skills.

Recommendation. Stakeholders should put inclusion first by keeping accessibility in mind while creating metaverse platforms, content, and user interfaces. By giving underprivileged populations access to the internet and devices compatible with the metaverse, efforts should be made to close the digital divide.

User Empowerment.

Opportunity. To ensure the responsible and fair growth of the metaverse, people must be given the freedom to govern their data, digital identities, and virtual assets.

Recommendation. It is advised that metaverse platforms have tools that are simple to use and enable users to control their data, privacy preferences, and virtual assets. To enable consumers to make wise decisions, education on digital rights and data management should be freely accessible.

Cross-Sector Collaboration.

Opportunity. The potential for sustainable development in the metaverse may be maximized by cooperation between government, technology, academia, and civil society.

Recommendation. It is advised that multi-stakeholder alliances and forums for metaverse development and governance be established in order to promote information exchange and coordinated action. Identifying opportunities across industries and overcoming difficulties should be the main goals of these relationships.

To conclude, the metaverse has enormous potential for sustainable development, but it needs careful thought and responsible action from all the investors. We can overcome the challenges and take advantage of the benefits given by the metaverse by supporting moral standards, making infrastructural investments, developing education and awareness, and putting in place a regulatory framework to enable sustainable development. We can guarantee that the metaverse develops into a potent agent of good change, guiding us toward a more sustainable and just future for all, by working together.

5 CONCLUSION

With cutting-edge answers to some of humanity's most urgent problems, the metaverse is set to become a defining frontier of the digital era. It offers chances to transform socioeconomic inclusion, healthcare, environmental preservation, and education. However, it also raises difficult ethical, technical, and societal issues, as with every significant technology progress.

Stakeholders must navigate the metaverse's evolution with care, foresight, and cooperation. To ensure ethical growth, it is crucial to have a solid technical foundation, thorough regulatory frameworks, and explicit ethical standards. Promoting digital literacy and awareness is equally crucial if users are to be able to traverse the metaverse ethically and safely.

The metaverse has enormous promise for advancing sustainable development goals, providing answers to world problems and opening up fresh possibilities for people and communities. It has the potential to promote a more diverse, connected, and informed society. We must give fairness, accessibility, and user empowerment top priority while developing the metaverse if we are to achieve this ambition.

The proper integration of the metaverse into our lives will require continual discussion, investigation, and invention in the coming years. Together, we can lead the metaverse toward a future in which technology improves human well-being and positively impacts the long-term sustainability of our planet by accepting these opportunities and challenges.

REFERENCES

Allam, Z., Sharifi, A., Bibri, S., Jones, D., & Krogstie, J. (2022). The metaverse as a virtual form of smart cities: Opportunities and challenges for environmental, economic, and social sustainability in urban futures. *Smart Cities*, 5(3), 771–801. DOI: 10.3390/smartcities5030040

Cannavo, A., & Lamberti, F. (2020). "How blockchain, virtual reality, and augmented reality are converging, and why." (2019). *IEEE Consumer Electronics Magazine*, 10(5), 6–13. DOI: 10.1109/MCE.2020.3025753

Dianwei, W.. (2022). *Research on Metaverse: Concept, development and standard system*. IEEE Xplore.

Fowler, C. (2015). Virtual reality and learning: Where is the pedagogy? *British Journal of Educational Technology*, 46(2), 412–422. DOI: 10.1111/bjet.12135

Gartner.com. Elements of a Metaverse. 2022. Available online: https://axveco.com/gartner-metaverse/

Hamacher, Alaric, et al. (2016) "Application of virtual, augmented, and mixed reality to urology." International neurourology journal 20.3: 172.

Han, D.-I. D., Bergs, Y., & Moorhouse, N. (2022). Virtual reality consumer experience escapes: Preparing for the metaverse. *Virtual Reality (Waltham Cross)*, 26(4), 1443–1458. DOI: 10.1007/s10055-022-00641-7

Hwang, G. J., & Chien, S. Y. (2022). Definition, roles, and potential research issues of the metaverse in education: An artificial intelligence perspective. *Computers and Education: Artificial Intelligence*, 3, 100082.

Kemp, J., & Livingstone, D. (2006) "Putting a Second Life "Metaverse" skin on learning management systems". *Proceedings of the Second Life education workshop at the Second Life community convention* (Vol. 20).

Kerdvibulvech, C. (2022) "Exploring the Impacts of COVID-19 on Digital and Metaverse Games." *International Conference on Human-Computer Interaction*. Springer, Cham. DOI: 10.1007/978-3-031-06391-6_69

Kohler, T., Matzler, K., & Füller, J. (2009). Avatar-based innovation: Using virtual worlds for real-world innovation. *Technovation*, 29(6-7), 395–407. DOI: 10.1016/j.technovation.2008.11.004

Lam, T.. (2022). *Metaverse report— Future is here Global XR industry insight*. Deloitte China.

Lea, T. (2022). With good planning, can the metaverse be sustainable. FinTech Weekly, 20.

Lee, H., Woo, D., & Yu, S. (2022). Virtual reality metaverse system supplementing remote education methods: Based on aircraft maintenance simulation. *Applied Sciences (Basel, Switzerland)*, 12(5), 2667.

Lee, L.-H., ... (2021) "All one needs to know about metaverse: A complete survey on technological singularity, virtual ecosystem, and research agenda." arXiv preprint arXiv:2110.05352.

Mystakidis, S. (2022). Metaverse. *Metaverse. Encyclopedia*, 2(1), 486–497. DOI: 10.3390/encyclopedia2010031

Park, S., & Kim, S. (2022). Identifying world types to deliver gameful experiences for sustainable learning in the metaverse. *Sustainability (Basel)*, 14(3), 1361. DOI: 10.3390/su14031361

Sarkady, D., Neuburger, L., & Egger, R. (2021, January). Virtual reality as a travel substitution tool during COVID-19. In Information and communication technologies in tourism 2021: Proceedings of the ENTER 2021 eTourism conference, January 19–22, 2021 (pp. 452-463). Cham: Springer International Publishing.

Singh, R., Akram, S., Gehlot, A., Buddhi, D., Priyadarshi, N., & Twala, B. (2022). Energy System 4.0: Digitalization of the energy sector with inclination towards sustainability. *Sensors (Basel)*, 22(17), 6619. DOI: 10.3390/s22176619 PMID: 36081087

Sourin, A. (2017). *Case study: Shared virtual and augmented environments for creative applications*. SpringerBriefs in Computer Science.

Sykownik, P. (2022) "Something Personal from the Metaverse: Goals, Topics, and Contextual Factors of Self-Disclosure in Commercial Social VR". Conference on Human Factors in Computing Systems – Proceedings. DOI: 10.1145/3491102.3502008

Theodorou, A., Panno, A., Carrus, G., Carbone, G. A., Massullo, C., & Imperatori, C. (2021). Stay home, stay safe, stay green: The role of gardening activities on mental health during the Covid-19 home confinement. *Urban Forestry & Urban Greening*, 61, 127091. DOI: 10.1016/j.ufug.2021.127091 PMID: 35702591

Tozzi, C. (2022). Will the Metaverse Help or Hinder Sustainability. ITPro Today, 10.

United Nations. (2022). *Sustainable Development Goals*. United Nations.

Wiederhold, B. K., & Riva, G. (2022). Metaverse creates new opportunities in healthcare. *Annual Review of Cybertherapy and Telemedicine*, 20, 3–7.

Chapter 13
Penetration Testing:
A Way to Secure IT Industries

Aditya Sharma
Ajeenkya D.Y. Patil University, India

Amna Kausar
https://orcid.org/0009-0000-3940-8845
Ajeenkya D.Y. Patil University, India

Atharva Saraf
https://orcid.org/0009-0006-3842-5508
Ajeenkya D.Y. Patil University, India

Susanta Das
https://orcid.org/0000-0002-9314-3988
Ajeenkya D.Y. Patil University, India

ABSTRACT

To identify system vulnerabilities, pen testing is essential for cybersecurity. Phases of preparation, reconnaissance, scanning, exploitation, and reporting are all involved. Every phase makes use of tools such as Nmap, Nessus, and Metasploit. To guarantee system, network, and data security, businesses should conduct pen tests regularly using skilled testers. This chapter delves into this important domain of cybersecurity as well as information technology industries. The chapter also discusses ethical issues and various challenges associated with it.

DOI: 10.4018/979-8-3693-5728-6.ch013

INTRODUCTION

Testing a system to identify security flaws is known as penetration testing. The goal of the research is to present a thorough understanding of the strategies and tactics applied in this process, emphasizing areas that require improvement and best practices. This covers the stages of a penetration test that involve planning, reconnaissance, scanning, exploiting, and reporting. By offering insightful information about the testing procedure, the research will assist organizations in optimizing their information security posture. Through the examination of current literature on penetration testing techniques, the study will provide a useful tool for enhancing information security assessment procedures and safeguarding confidential data. Improving penetration testing's efficacy and efficiency as a tool for information security assessment is the ultimate objective (Ziro et al., 2023). Several factors can lead to vulnerability, including inadequate programming or an antiquated system (Hasan & Meva, 2018). Information assurance can be impacted by system vulnerabilities and security issues. While achieving a totally secure system may be challenging, lowering the amount of vulnerabilities can greatly improve system security. Nonetheless, penetration testing and vulnerability assessment are frequently undervalued. These actions frequently go unnoticed and are thought of as mere formalities. Organizations can lessen their vulnerability to attacks and have a more secure system by regularly and effectively performing vulnerability assessments (Abu-Dabaseh & Alshammari, 2018). The goal of penetration testing is to identify and fix system vulnerabilities before unauthorized users can take advantage of them. A vulnerability that is left unchecked could be exploited to obtain unauthorized access to business resources and compromise the system. Penetration testing aims to find and address these vulnerabilities to keep the system safe and stop similar compromises (Kesharwani et al., 2018). Access to a wide range of services, such as telehealth, digital transactions, e-commerce, online business, audio/video conferencing, e-commerce, dependable treatment, shipping and aviation services, and mobile payment systems, is becoming more and more dependent on the internet. As a result, a great deal of sensitive and private data is produced online. However, there's also a growing chance of cyberattacks, which can result in lost or corrupted data, compromised user credentials, and interrupted services. Because of this, the business community is starting to take this seriously (Sarker et al., 2023).

Cybercriminals launch attacks by taking advantage of different weaknesses in devices, apps, networks, and user behavior. These vulnerabilities may be the result of human error, complex computing systems, out-of-date hardware and software, poorly configured systems, and design flaws. Governmental bodies, for-profit businesses, and global trade hubs are concentrating on cyber defense in order to combat these threats. An essential component of cyber defense is penetration testing, which looks for vulnerabilities in an organization's policies, procedures, and logical and physical infrastructure. This assists companies in taking corrective action prior to a cyberattack. A routine review procedure called penetration testing assesses the security posture and lowers the possibility of a cyberattack. It can be carried out for recently installed information systems, following a system upgrade, or on a regular basis (Bertoglio & Zorzo, 2017).

A controlled attempt to breach a system or network in order to identify vulnerabilities is known as penetration testing or pentesting. It employs the same strategies as an actual attack, but it does so in a way that is allowed and lawful. Finding weaknesses before unauthorized people can take advantage of them is the aim. Organizations can strengthen their security posture and get rid of vulnerabilities by implementing these steps (Mamilla, 2021). Although there are many advantages to the Internet, there are also new security risks that businesses must deal with. Information security policies are necessary, but they might not be sufficient to stop data breaches. Penetration testing assists companies in identifying and patching vulnerabilities before hackers can take advantage of them. It finds vulnerabilities in an organization's systems and offers a remediation plan to address them by mimicking real-world attacks. Penetration testing is an important tool for safeguarding an organization's information security because of the high cost of data breaches (Shravan et al., 2014). By looking for and locating exploits and vulnerabilities in the IT infrastructure of a company, it reveals the information available to the public or the internet and aids in verifying the efficacy of the security measures in place. The expertise and experience of the testers are critical to the success of security testing, which necessitates a unique level of insight that is difficult to systematize. There is currently no all-encompassing, generic model for security testing, and the majority of security testing is focused on particular domains like web server and firewall testing (Fruhlinger, 2021).

LITERATURE REVIEW

Technical flaws in web applications, such as SQL injection, cross-site scripting, and file inclusion, can compromise the security of organisations. Outdated systems or subpar programming may be the source of these vulnerabilities. Cybercriminals can use these vulnerabilities that can arise from a variety of causes, including shoddy programming or antiquated systems to obtain unauthorized access to private data. Organizations can lessen their vulnerability to attacks and have a more secure system by regularly and effectively performing vulnerability assessments (Hasan & Meva,

2018). They can also lower their risk of cyberattacks by periodically reviewing and upgrading their policies, procedures, and systems with the aid of penetration testing.

Penetration testing is a thorough security assessment technique used to find and fix holes in an organization's IT infrastructure. A range of tasks, such as vulnerability assessment and exploitation, are involved to identify and rank security risks. Finding vulnerabilities in the system, carefully exploiting them, and producing a report with remediation suggestions are all part of the process (Abu-Dabaseh & Alshammari, 2018). It is the process of examining a computer system, network, or web application to identify security flaws that an attacker could exploit. It is also referred to as ethical hacking or pen testing. Software programs can automate this process, or it can be done by hand. It entails learning as much as possible about the target, locating possible points of entry, trying to obtain unauthorized access, and reporting the results (Kesharwani et al., 2018).

Finding weaknesses and strengthening the security system are the objectives in order to stop such attacks in the future. Similar to hacking, penetration testing is carried out with authorization and doesn't endanger the system or its users. Tools for automated penetration testing are also used to learn more about the attacker and the system. There have been cases of malware, SQL injection attacks, and employee permission abuses leading to security breaches in database security. Another issue is denial of service (DoS) attacks, in which hackers take control of systems and data and demand payment until their demands are fulfilled. Nonetheless, a number of approaches have been used to secure databases (Tudosi et al., 2023).

Importance of penetration testing

Pen testing serves to safeguard our data by locating potential points of entry for hackers and bolstering security measures at each turn. Ensuring compliance with industry security policies and establishing a baseline for testing are two benefits it offers. The attacker or ethical hacker's status—internal or external—as well as whether they were employed by the organization or were trying to obtain unauthorized access from the outside influence the test's scope. The company hires an ethical hacker, sometimes referred to as a white hat hacker, to find and fix flaws in the system's security procedures and policies (Shaukat et al., 2016).

Process

There are three primary tasks involved with pen testing, these include determining the scope, carrying out the test, and reporting.

A crucial first step is defining the scope, which establishes the testing range. This could apply to the entire system and network, or just certain hardware like web servers, routers, firewalls, DNS servers, mail servers, and file transfer protocol servers. The test is then carried out, during which all target information is acquired and security flaws are examined. Exploits are used to exploit vulnerabilities once they are found. Should the test be successful, the tester obtains entry to the target and evaluates the consequences of the breach. Reporting is the last step, where all the data collected during the testing process is put together in a report. A prioritized list of risks and vulnerabilities, details on the advantages and disadvantages of the current security setup, risk categorization (high, medium, or low), and vulnerability information for each device are all included in the report. Technical details on how to address vulnerabilities are also included in the report, along with suggestions for doing so. The report might also include useful links to online resources that contain patches for vulnerabilities that have been found. All things considered, penetration testing is an essential procedure that aids businesses in locating and fixing security flaws in their infrastructure (Mamilla, 2021).

Phases

Information gathering: This phase involves collecting information about the target system, such as the web server's IP address, domain, and subdomains, as well as identifying any firewalls in place. Tools like WAFW00F can be used to detect web application firewalls.

Scanning: In this phase, the tester identifies the services running on the web server, their versions, and the ports they are running on. This information is used to determine the operating system of the target system. Tools like NMAP and Metasploit's AUXILIARY/SCANNER facility are commonly used.

Discovering vulnerabilities: The tester uses tools like Nikto, Nessus, or Metasploit's Auxiliary/scanner facility to identify vulnerabilities in the web server or system.

Exploitation: Once vulnerabilities have been discovered, the tester attempts to exploit them in order to gain remote access to the server. Metasploit is commonly used for this purpose.

Report generation: In the final phase, a report is generated detailing the findings of the penetration test, including any vulnerabilities that were discovered and exploited (Kesharwani et al., 2018).

Web Application Pen Testing

Identifying and fixing security flaws, particularly in web applications, is the methodical process of probing a system—which could be any combination of hosts, applications, or networks—through penetration testing.

Known as the discovery phase, vulnerability phase, and attack simulation phase, the test phase comprises three steps: information gathering, vulnerability analysis, and vulnerability exploit. In the information gathering stage, testers gather information such as IP addresses, domain names, network services, and network topology about the target system. When a system is in the vulnerability analysis phase, testers examine its security and look for any possible weaknesses using both automated tools and manual procedures. During the phase of vulnerability exploit, testers try to take advantage of the weaknesses in order to gauge their extent and the possible harm they could cause. The test analysis phase comes after the testing phase, during which testers create a thorough report outlining their findings and providing suggestions for fixing the vulnerabilities. A strong cybersecurity strategy must include pen testing, which gives businesses insight into how to fend off attacks and enhances software security policies (Bacudio et al., 2011).

Standards for pen testing

The goal of penetration testing standards is to give a fundamental description and framework for the testing procedure. A number of standards are available, including PTES, OISSG, OSSTMM, NIST SP 800-115, and ISAAF.

The three types of attacks covered by the OSSTMM v3 standard are communications security, spectrum security, and physical security. It also covers every aspect of penetration testing. It is extremely sophisticated and was first published in 2000. Guidelines for organizing and carrying out information security testing and assessments, interpreting results, and creating mitigation plans are provided by the NIST SP800-115 standard. It offers a summary of the essential components of security testing and assessment along with reports and recommendations for their application, but it does not offer a comprehensive testing program (Abu-Dabaseh & Alshammari, 2018).

Automated pen testing

Automating penetration testing is not a new idea; the earliest approaches used attack graphs. Attack graphs assist in identifying attack paths that result in unauthorized access and model how particular exploits can impact systems. Attackers cannot realistically use this method since it necessitates a thorough understanding of the machine configurations and network topology. It also takes time because the graph needs to be manually configured for every new system that is assessed (Heiding et al., 2023).

Pen testing methodologies

Methodologies for pen testing can be proprietary or open-source, public or private. Two key groups of techniques are enumeration and reconnaissance. While enumeration obtains data directly from target systems, applications, and networks, reconnaissance entails obtaining information for use in penetration testing. Scholars have put forth a number of approaches, such as information (open or closed), roles and responsibilities within the team, toolkit, and justification for the tools. Pre-engagement (deciding on testing type, approach, and focus), engagement (choosing suitable devices and testing techniques), and post-engagement (finishing remediation best practices, retesting found vulnerabilities, and cleaning up) are the standard stages of a penetration test (Tudosi et al., 2023).

Pen testing tools

Pen testers use a variety of tools to look for weaknesses in a system. The following are a few of the most crucial tools:

Metasploit: An instrument for getting past security and into a particular system.

SAINT: A network tool that targets various targets and remotely detects system vulnerabilities.

Nmap: A port-scanning program that determines the computer's operating system.

Core Impact: A tool that lets the tester test more machines in the system while taking advantage of security flaws in the system without crashing it.

Nessus: An extensive library of vulnerabilities and tests to find them are included in this vulnerability scanner.

Codenomicon: The ideal configuration for automated penetration testing and a tool for locating concealed vulnerabilities in a system.

Hydra: A dependable login cracking tool that supports more than 30 protocols for pen-testers.

A network protocol analyzer called Wireshark records and shows comprehensive packet data in real time.

W3af: An instrument for probing and analyzing web applications. Python must be installed in order to use it in a variety of environments.

Pen Testing Techniques

Techniques for penetration testing include White Box, Grey Box, and Black Box testing. When conducting black box testing, the tester is blind to the systems or network architecture of the network being tested. While White Box testing necessitates full knowledge of the network and system configuration, Grey Box testing only requires partial knowledge of the testing network. Better results are obtained from white box testing, which is typically carried out from the internal network and necessitates a thorough understanding of the testing network or system (Abu-Dabaseh & Alshammari, 2018).

White box penetration testing is an elaborate and comprehensive approach to evaluating the security of a system. It calls for in-depth familiarity with the source code, configuration, and architecture of the system. This method provides a thorough assessment of the system's security posture by examining it from both the inside and the outside. White box testing can find even the most subtle flaws because of its in-depth knowledge of the system. White box penetration testing is a comprehensive testing technique that necessitates a deep understanding of the target system. Planning, reconnaissance, vulnerability analysis, exploitation, reporting, remediation, and verification are some of the steps that are involved. This technique helps businesses find and fix security flaws, lowering the chance of data breaches and cyberattacks. It is generally applied during the testing stage of the security operations process. In order to gather information about the target system, analyze vulnerabilities, exploit them, and report findings, the white box testing team collaborates closely with the security team. The security team then makes use of this data to create plans for addressing the vulnerabilities found and enhancing the target system's general security (Ziro et al., 2023).

Black box testing is a kind of penetration testing in which professional hackers pretend to be cyber experts and try to breach a system without knowing anything about its services and infrastructure beforehand. On the other hand, primary data regarding the infrastructure and services is used for white box testing. Limited organization-specific information is used for grey box testing. A pen tester can use online resources and published documents, as well as deceiving stakeholders, customers, suppliers, and staff, to obtain information. In order to investigate vulnerabilities in the physical world, client-side computers, WiFi architecture, devices, web applications, network services, peripheral devices, and media, a variety of tools and techniques are used during the testing process. A report containing vulnerability scores, evidence of vulnerabilities, and recommendations is generated based on the testing results (Sarker et al., 2023).

SQL Injection

One kind of vulnerability that affects dynamic web applications that store data in linked databases is SQL Injection. By inserting malicious code into SQL statements, attackers can take advantage of this vulnerability and retrieve private data from the database. Database versions, usernames, table details, and even private information like passwords can all be included in this information. Apart from pilfering data, hackers have the ability to alter or remove entries from the database, run commands at the system level, and initiate supplementary assaults like denial of service. An SQL Injection attack's effect is contingent upon the role and privilege configuration within the target machine's SQL server (Hasan & Meva, 2018).

Cross Site Scripting

An attack known as cross-site scripting (XSS) occurs when an attacker inserts or runs malicious code on a user's browser with the intention of obtaining their login credentials. This is accomplished by using weak web applications that allow the attacker to run payloads on the client side and steal cookies from other users or send them to phishing websites. XSS attacks come in three flavors: DOM-based, non-persistent, and persistent. These assaults may result in detrimental effects, such as unauthorized access to user accounts and private data (Hasan & Meva, 2018).

Vulnerability Assessment and Penetration Testing

The nine-step Vulnerability Assessment and Penetration Testing (VAPT) procedure assists in locating and resolving security flaws in a network or system. Determining the scope of the assessment is the first step in the process, which is then followed by reconnaissance to learn more about the network, IP address, and operating system. Subsequently, the tester employs diverse vulnerability assessment methodologies to detect vulnerabilities within the system.

The tester then evaluates the vulnerabilities discovered and develops a penetration testing strategy. Next, in an effort to access the system and elevate privileges, the tester tries to take advantage of these vulnerabilities. Testing outcomes are examined, and suggestions are given for resolving any vulnerabilities found. The complete procedure is recorded and submitted to management for necessary action. The system is returned to its initial configuration following the testing process in order to undo any modifications made during the VAPT procedure. VAPT is designed to assist organizations in strengthening their security posture and fortifying themselves against possible intrusions (Abu-Dabaseh & Alshammari, 2018).

The security procedure known as vulnerability assessment and penetration testing, or VAPT, has certain advantages but also some drawbacks. These drawbacks include the testing's time constraint, which may affect its effectiveness, and the fact that the tester's abilities and efforts are what determine VAPT's success. Furthermore, VAPT can be costly because it is frequently carried out by outside parties and is a repetitive procedure that needs to be repeated whenever the system is altered or modified. Additionally, the process may harm the system by, for instance, consuming bandwidth while scanning. There are additional legal restrictions, such as the requirement for authorization to test web applications hosted on shared servers. VAPT is still a useful method for safeguarding web applications in spite of these drawbacks, with a variety of tools and techniques available to assist in the penetration testing process (Hasan & Meva, 2018).

Tools for VAPT

A free SQL Injection exploitation tool with an intuitive interface, Havij 1.16 supports multiple database types and the HTTPS protocol. Post-exploitation activities like hash cracking are also included. A Python command-line tool called Fimap is used to locate and take advantage of LFI and RFI vulnerabilities in web applications. A popular penetration testing tool that can be found on both commercial and open-source platforms, Metasploit offers a sizable database of exploits along with techniques for thorough security auditing. An extensive collection of auditing and exploitation tools for forensics and penetration testing can be found in Kali Linux, an open-source security distribution running on the Linux operating system. Acunetix is a commercial web vulnerability scanner that generates easily readable reports and scans web applications for a variety of vulnerabilities using a black-box methodology. The community edition of Nexpose, a GUI and automated vulnerability assessment tool, has limited features for a year and is available in both commercial and free versions (Hasan & Meva, 2018).

EXPLOITATION OF VULNERABILITIES

Using a variety of tools, penetration testers find and exploit system vulnerabilities during the exploitation phase of the test.

Nessus, a Unix-based network vulnerability scanner, Firewalk, a tool for mapping firewall Access Control Lists and creating network maps.

John the Ripper, a Unix-based password cracker.

Crack/Libcrack an additional Unix-based password cracker.

These resources assist testers in locating system vulnerabilities and assessing the danger they present (Sarker et al., 2023).

Why Vulnerability Assessment

Any organization's primary goal is to generate revenue in order to fund its vision and objectives. Thus, companies have the chance to implement information technology infrastructures. Any organization's primary goal following the implementation of its IT infrastructure is to safeguard its private data from unauthorized access and maintain network security. As a result, vulnerability assessment looks for gaps and weaknesses in a system. System accreditation, risk assessment, network auditing, compliance checking, and continuous monitoring are some possible objectives of vulnerability assessment. Weak passwords, system defects, incorrect and inappropriate configuration, and human error—such as assigning users the wrong permissions, designing networks and devices inappropriately, and so on—are the main causes of vulnerabilities. Certain business standards organizations, such as PCIDSS, mandate that organizations conduct vulnerability assessments (Hussain et al., 2017).

A penetration test differs from a security audit in that its findings should include recommendations for mitigating or eliminating vulnerabilities that are found. Prioritizing the most critical vulnerabilities should be done, and a schedule for confirming their resolution should be established. The vulnerabilities found, the possible harm to the business, and the affordability and efficacy of the available solutions will all influence the solutions that are put into place. Applying the most recent operating system or application patches, centralizing mail delivery, enforcing current policies, and passing a vulnerability test prior to opening a web port at the firewall are a few possible solutions. Confidentiality of the report's findings is necessary to stop unauthorized parties from taking advantage of the vulnerabilities found (Kesharwani et al., 2018).

The process of penetration testing consists of multiple stages, including reconnaissance and footprinting, scanning, enumeration, analysis, and reporting. First, data is gathered from multiple sources regarding the target organization's IT infrastructure. The attacker communicates with the target during the scanning phase in order to find any potential vulnerabilities, and they keep track of the responses for later examination. In the enumeration phase, information from the preceding phase is scanned and attacks are carried out using the strategy and plan created in the first phase. Finding holes in the system and pulling out specific data requires specialized knowledge, expertise, and experience. During the analysis phase, risk analysis is carried out based on the evidence of vulnerability exploitation, and priority is assigned to starting the necessary vulnerability fixes. A formal document summarizing the penetration testing process and offering guidelines for improving organization security is what the pen testers submit to the organization as part of the reporting phase. In order to facilitate additional review, the report should have high-level documentation that is simple to read (Sarker et al., 2023).

Penetration testing is a security procedure that includes examining a system or network to find weaknesses that an attacker could exploit. Penetration testing comes in a variety of forms, including:

Network penetration testing entails assessing a network's hardware, such as switches, routers, firewalls, and servers. Network penetration tests come in three varieties: black-box, white-box, and gray-box. Application penetration testing is the process of looking for security flaws in web-based programs, browsers, and their parts. Periodic Network Vulnerability Assessments: Usually consisting of routinely scanning IP ranges and noting changes, these assessments are meant to supplement a full penetration test. Physical security tests: To find vulnerabilities that could be exploited, these tests mimic an attack on a physical barrier. Client-side Penetration Tests: These tests check for malware infections, clickjacking attacks, cross-site scripting attacks, and other threats that manifest locally on the client side.

Testing all wireless devices linked to the client's Wi-Fi is part of the wireless penetration test process.

Penetration testing using social engineering entails duping a victim into disclosing private information, like passwords. Penetration testing is a crucial security procedure that aids businesses in finding weak points in their security and strengthening it (Mamilla, 2021).

An external third party can verify an organization's IT security by conducting a penetration test. This test offers an overview of the security protocols in effect at a particular moment in time. Frequent testing can help demonstrate the increased security of customer data in web shops and other online applications, even though it cannot guarantee the level of security in the future (Shravan et al., 2014).

Limitation of penetration testing:

Although penetration tests have their limitations, they are a useful tool for enhancing security. Their scope may be project-focused or limited in time, so they might not discover all vulnerabilities. Given more time, attackers might discover weaknesses outside the purview of the test. Since penetration testing is not an exact science, different testers may interpret the same vulnerabilities in different ways. Furthermore, testers might not have the time to create their own custom exploits and instead rely on publicly available exploits. Although successful attacks cannot be completely eliminated, they are considerably less likely when vulnerabilities are fixed thanks to penetration tests. For optimal protection, penetration tests ought to be conducted on a regular basis as an addition to the approved review procedures (Shravan et al., 2014).

Ethical and legal issues

An essential step in protecting IT infrastructure and averting financial losses from data security breaches is penetration testing. Millions of dollars are spent by organizations on notification expenses, corrective actions, and revenue and productivity lost as a result of security breaches. Penetration testing assists in locating and reducing possible threats to the resources and private information of an organization. For computing security systems, adherence to industry standards is required. Failure to do so may result in significant fines or penalties during IT security audit procedures (Hussain et al., 2017).

Penetration testing services and companies

In the tech sector, penetration testing is a specialized field with many companies providing these services. While some businesses provide pen testing as a component of a more comprehensive package, others are experts in ethical hacking. The lack of overlap among the top pen testing companies listed by various research and advisory firms is indicative of the field's diversity. For example, there is very little overlap between Clutch's list of top-rated companies and Cybercrime Magazine's list of companies to watch in 2021 and Explority's list of the top 30 pen testing companies in Hacker Noon (Bacudio et al., 2011).

An essential step in protecting IT infrastructure and averting financial losses from data security breaches is penetration testing. Millions of dollars are spent by organizations on notification expenses, corrective actions, and revenue and productivity lost as a result of security breaches. Penetration testing assists in locating and reducing possible threats to the resources and private information of an organization. For computing security systems, adherence to industry standards is required. Failure to do so may result in significant fines or penalties during IT security audit procedures (Hussain et al., 2017).

CONCLUSION

An organization's information security can be greatly improved by using SecOps methodology, conducting penetration testing, and conducting information security audits. Preparation, reconnaissance, scanning, exploitation, reporting, and other phases are all part of the penetration testing process. Tools and techniques such as Nmap, Ping, Who is, Nessus, OpenVAS, Nexpose, Metasploit, Burp Suite, and Wireshark are required at different stages. White-box penetration testing is a more thorough and focused method of finding weaknesses in the internal operations of an application. Qualitative approaches can provide important insights into the architecture, configuration, and design of a system, assisting in the identification and evaluation of possible security flaws and their possible consequences. Vulnerability identification, potential impact assessment, and effective remediation strategy development can all be greatly aided by integrating qualitative methods into a white-box penetration testing methodology. One essential procedure for determining security is VAPT. Businesses should use VAPT to improve system security as it is an essential procedure for finding security flaws. Penetration testing is a thorough process that finds weaknesses in a system. It has advantages that include preventing financial loss, complying with industry regulations, maintaining customer and shareholder compliance, protecting the company's reputation, and proactively removing risks that are found.

REFERENCES

Abu-Dabaseh, F., & Alshammari, E. (2018). Automated Penetration Testing: An Overview. Dhinaharan Nagamalai et al. (Eds): NATL, CSEA, DMDBS, Fuzzy, ITCON, NSEC, COMIT – 2018. pp. 121–129. DOI: DOI: 10.5121/csit.2018.80610

Bacudio, A. G., Yuan, X., Chu, B. T. B., & Jones, M. (2011). An Overview of Penetration Testing. *International Journal of Network Security & its Applications*, 3(6), 19–38. DOI: 10.5121/ijnsa.2011.3602

Bertoglio, D. D., & Zorzo, A. F. (2017). Overview and open issues on penetration test. *Journal of the Brazilian Computer Society*, 23(1), 2. Advance online publication. DOI: 10.1186/s13173-017-0051-1

Fruhlinger, J. (2021). *Penetration testing explained: How ethical hackers simulate attacks*. CSO Online. https://www.csoonline.com/article/571697/penetration-testing-explained-how-ethical-hackers-simulate-attacks.html

Hasan, A., & Meva, D. (2018). Web Application Safety by Penetration Testing. International Journal of Advanced Studies of Scientific Research (IJASSR), 3(9) (5 pages).

Heiding, F., Süren, E., Olegård, J., & Lagerström, R. (2022). Penetration testing of connected households. *Computers & Security*, 126, 103067. DOI: 10.1016/j.cose.2022.103067

Hussain, M. Z., Hasan, M. Z., & Chughtai, M. T. A. (2017). *Penetration Testing In System Administration*. INTERNATIONAL JOURNAL OF SCIENTIFIC & TECHNOLOGY RESEARCH. https://www.researchgate.net/publication/319876508

Kesharwani, P., Pandey, S., Dixit, V., & Tiwari, L. (2018). A study on Penetration Testing Using Metasploit Framework. *International Research Journal of Engineering and Technology (IRJET)*, 5(12).

Mamilla, S. R. (2021). A Study of Penetration Testing Processes and Tools. Electronic Theses, Projects, and Dissertations. 1220. https://scholarworks.lib.csusb.edu/etd/1220

Sarker, K. U., Yunus, F., & Deraman, A. (2023). Penetration Taxonomy: A Systematic Review on the Penetration Process, Framework, Standards, Tools, and Scoring Methods. *Sustainability (Basel)*, 15(13), 10471. DOI: 10.3390/su151310471

Shaukat, K., Faisal, A., Masood, R., Usman, A., & Shaukat, U. (2016). Security quality assurance through penetration testing. *19th International Multi-Topic Conference (INMIC)*. DOI: DOI: 10.1109/INMIC.2016.7840115

Shravan, K., Neha, B., & Pawan, B. (2014). Penetration Testing: A Review. *International Journal of Advancements in Computing Technology*, 3(4), 752–757.

Tudosi, A., Graur, A., Balan, D. G., & Potorac, A. D. (2023). Research on Security Weakness Using Penetration Testing in a Distributed Firewall. *Sensors (Basel)*, 23(5), 2683. DOI: 10.3390/s23052683 PMID: 36904890

Ziro, A., Gnatyuk, S., & Toibayeva, S. (2023). Improved Method for Penetration Testing of Web Applications. Intel*ITSIS'2023: 4th International Workshop on Intelligent Information Technologies and Systems of Information Security*, March 22–24, Khmelnytskyi, Ukraine.

Chapter 14
The Global Trends and Hotspots of Medical Internet of Things Research:
A Bibliometric Analysis During 2004 to 2023

M. S. Rajeevan
https://orcid.org/0009-0009-3320-3078
Indian Institute of Technology, Hyderabad, India

V. S. Anoop
Kerala University of Digital Sciences, Innovation, and Technology, India

D. Narayana
Central Tribal University of Andhra Pradesh, India

ABSTRACT

The exponential growth of medical data, propelled by technological advancements and societal shifts, underscores the critical role of efficient data management in public health, clinical practice, and medical research. The Medical Internet of Things (MIoT) emerges as a key player, facilitating real-time collection and transmission of patient data through connected devices and sensors. While MIoT revolutionizes healthcare delivery and research, its rapid expansion presents new challenges in data handling. Moreover, the integration of MIoT with 5G technology expands its scope beyond traditional medicine, aiding in diagnosis, treatment, and preventive care. Bibliometrics, an underutilized tool in medical research, offers insights into

DOI: 10.4018/979-8-3693-5728-6.ch014

research trends and collaboration networks, addressing a critical analytical gap in the field. This chapter explores the impact of bibliometric analysis in the medical domain, shedding light on key contributors and research avenues, thus enhancing our understanding of this evolving landscape.

INTRODUCTION

Technological advancements and social developments are responsible for the exponential expansion of medical data. Identifying public health issues, raising the standards for public health care, and researching innovative medical discoveries and medicines all depend on efficient medical data management. In the modern medical business, the Medical Internet of Things is one of the main sources of technological innovation. (Akhtar et al., 2021). Medical devices and sensors can be connected to the internet, enabling the real-time collection, transmission, and recording of medical data. Throughout their treatment, these devices and sensors can keep an eye on their patients' general well-being and physiological parameters. In addition to providing specific therapy and real-time monitoring services, this data is essential for clinical practice and medical research. (Elayan et al., 2017).

However, the Internet of Things' explosive growth poses new challenges for the handling of medical data. (Elhoseny et al., 2021). The intelligent design of medical equipment in the MIoT allows it to connect to the internet to promptly gather, send, and record patient health data, including physiological indications (Dimitrov, 2016). The growing incidence of chronic illnesses and severe aging lead to major problems to the medical system, and the trend is moving away from "passive medical care" and toward "preventive medical care." In the growth of medical rehabilitation, IoT medical has set itself beyond traditional medicine with the broad deployment of 5G technology. The proper diagnosis and treatment of serious crises has been greatly aided by MIoT. However it is also proven useful in assisting with a number of diagnostics (Tao et al., 2024).

Bibliometrics is a multidisciplinary field of research that measures all information carriers using statistical and mathematical techniques. and to analyze the limits and core of the research and the subject matter (van Eck & Waltman, 2017). To showcase the impact of the research field, systematic and metrological features of the literature can be investigated and assessed using VOSviewer, CiteSpace, and the R package "bibliometrix." In the past few years, the medical industry has extensively used bibliometrics. Based on big data analysis, bibliometric analysis assesses important contributors to a field (e.g., authors, nations, institutions, etc.) numerically and visually and ascertains their collaboration networks. These findings aid in analyzing the most popular study avenues in a certain topic. We searched the

Web of Science and found that this field has no bibliometric articles. This research addresses the bibliometric analytical gap in this area.

The developments in Medical Internet of Things (MIoT) is constantly evolving and there are plenty of researches that are being carried out in this area. This has led to several research articles getting published in the literature that discusses the impact, opportunities, and challenges of MIoT. It may be highly difficult for the researchers to analyze large volumes of these literature and then summarize and find the research gap. We may require thematic analysis and summary of what is already been done and what are its implications through a systematic survey which may help the researchers and practitioners in this field to propose new innovative research ideas. The major contributions of this chapter is listed as follows:

- Discusses the impact and opportunities for Medical Internet of Things (MIoT) in developing patient-centric and innovative healthcare applications.
- Conducts and extensive literature analysis using open-source but industry accepted tools for thematic analysis of what is already been published in this innovative area.
- Pose research questions and then try to answer them from the results of the thematic analysis to find out the research trends and future directions.

The remainder of this chapter is organized as follows: Section 2 discusses the impact of MIoT followed by the research gap where the bibliographic analysis would help. The data source and search strategy used are discussed in Section 3, Section 4 discusses the data analysis followed by the results of the analysis in Section 5. The discussion section and conclusions are available in Section 6 and Section 7 respectively.

Impact of MIoT

The technological advancements in the Internet of Things (IoT) is now disrupting every fields such as healthcare, agriculture, manufacturing, to name a few. The Medical Internet of Things (MIoT) is now significantly transforming the healthcare industry by enhancing the patient monitoring both remotely as well as in hospitals (Subhan et al., 2023). This has paved way to the enhanced caregiving process for the patients while keeping a personalized way of treatments. The development and integration of wearable devices, connected implants, and other smart medical equipments lead the healthcare providers and patients to collect real-time health data which facilitates more accurate diagnosis that enables timely interventions. Developments such as improved sensors and sophisticated data analytics are facilitating remote patient monitoring, reducing the need for frequent hospital visits, and supporting proactive management of chronic conditions (Malathi et al., 2024). These innovations promise not only increased efficiency in healthcare delivery but also enhanced quality of life through personalized treatment plans and early disease detection. However, they also introduce challenges such as ensuring data security and addressing ethical concerns related to patient privacy. Balancing these benefits with potential risks will be crucial as MIoT continues to evolve and integrate further into everyday healthcare practices (Al-Suhimat et al., 2024).

Despite the growing body of literature on the Medical Internet of Things (MIoT) and its applications in healthcare, significant research gaps remain, particularly in the integration of patient-centric approaches within innovative healthcare applications. Much of the existing research tends to focus on technological advancements and system architectures, often overlooking the nuanced needs and preferences of patients. Additionally, there is a lack of comprehensive studies that explore the long-term impacts of MIoT on patient outcomes and satisfaction. This gap underscores the need for multidisciplinary research that not only examines technological feasibility but also prioritizes user engagement and addresses ethical considerations. By bridging these gaps, future studies could provide valuable insights into creating more effective and personalized MIoT solutions that truly enhance patient care.

Data Source and Search Strategy

The Web of Science Core Collection (WoSCC) database, which was developed by Clarivate Analytics, was regarded as one of the most comprehensive and authoritative database systems. Hence, in accordance with earlier research, we chose it to collect global academic data for bibliometric analysis. The search was conducted between 2004 and 2023, and all published material was taken from the WOS Science core collection. To prevent the substantial bias resulting from frequent database renewals, all searches were conducted on February 13, 2024, on a single day. We used the following search phrases in order to get the most results: Medical Internet of things (Topic) and 2023 or 2022 or 2021 or 2020 or 2019 or 2018 or 2017 or 2016 or 2015 or 2014 or 2013 or 2012 or 2011 or 2010 or 2009 or 2007 or 2005 or 2004 (Publication Years) and English (Languages) and Article (Document Types). Articles were the only document type, and only English was selected. We were able to acquire 3288 original records by using the criteria. Figure 1 displays a flow chart of the entire procedure.

Figure 1. Flow chart of the identification and screening of data.

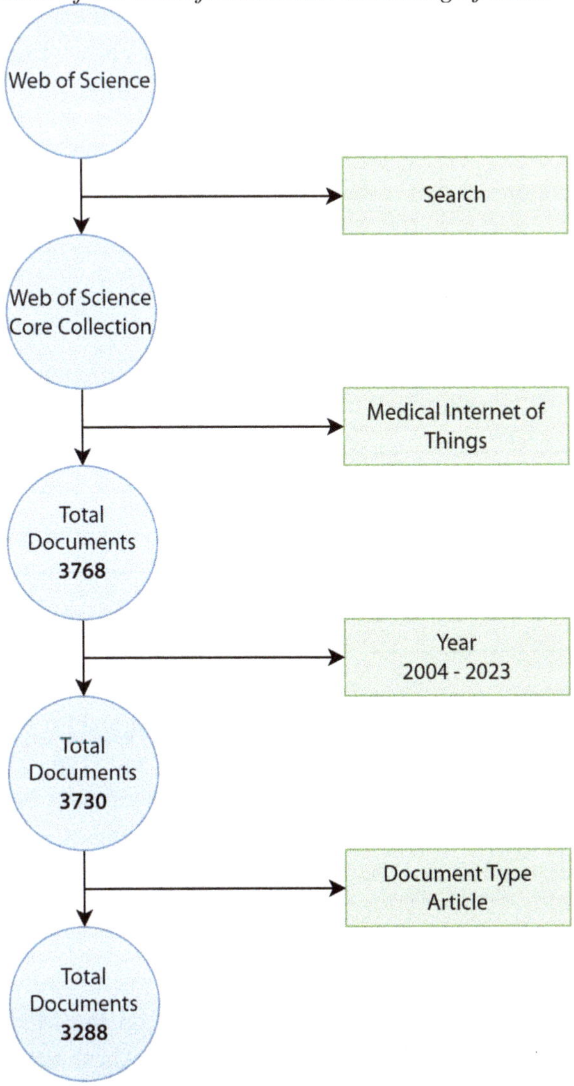

Data Analysis

VOSviewer is a popular software application for creating and viewing bibliometric network maps. It was created in 2009 by Van Eck and Waltman from Leiden University (van Eck & Waltman, 2010)., It is frequently employed to create networks for collaboration, co-occurrence, and co-citation (Pan et al., 2018; Yeung & Mozos, 2020). The software completes the primary analyses in this study. A node on the

VOSviewer map denotes an item, such as an author, nation, journal, or organization. The number and category of these objects are indicated by the node's size and color, accordingly. The line thickness between nodes indicates how closely the items are cited or collaborated (Wu et al., 2021; Zhang et al., 2020).

Table 1. Total publications on the MIoT

Document Types	
Article	3288
Review Article	378
Early Access	123
Editorial Material	42
Data Analysis Proceeding Paper	23
Retracted Publication	22
Meeting Abstract	8
Retraction	4
Data Paper	2
Correction	1

CiteSpace (version 6.1.R1).is an additional tool designed for bibliometric analysis and visualization by Professors Chen C (Synnestvedt et al., 2005). CiteSpace was used in this study to assess the timeline view for references including Citation Bursts. Additionally, a quantitative analysis of publications was carried out using Microsoft Office Excel 365 and the R tool "bibliometrix."

Result

From 2004 to 2023, a total of 3768 publications on the Medical Internet of Things were published on the Web of Science table 1. These titles included 3288 articles, 378 reviews, 123 early access articles, 42 editorial materials, 23 proceeding papers, 22 retracted publications, 8 meeting abstracts, 4 retractions, 2 data papers, and 1 correction. Only articles published between 2004 and 2023 were chosen for this research.

Annual Productivity

This table 2 displays the annual number of publications related to the Medical Internet of Things from 2004 to 2023. It is divided into three periods: the slow growth period (2004–2008), the acceleration period (2009–2016), and the rapid growth period (2017–2023). Before 2008, there was a very modest increase in publications; however, following 2017, there was a quick surge, with over 100 publications annually and another acceleration period in 2021. With 888 publications, the Medical Internet of Things received the most publications in a decade in 2023, and the number of publications kept rising significantly. Figure 2 depicts the graphical representation of annual scientific productivity on Medical IoT

Table 2. Annual publication trends on MIoT.

Sl No	Year	Articles	TC
1	2004	1	0
2	2005	1	0
3	2006	0	2
4	2007	2	4
5	2008	0	10
6	2009	5	9
7	2010	0	8
8	2011	1	14
9	2012	3	9
10	2013	8	24
11	2014	10	57
12	2015	22	129
13	2016	34	376
14	2017	83	817
15	2018	150	1612
16	2019	254	3233
17	2020	383	5885
18	2021	623	11111
19	2022	820	16882
20	2023	888	19202

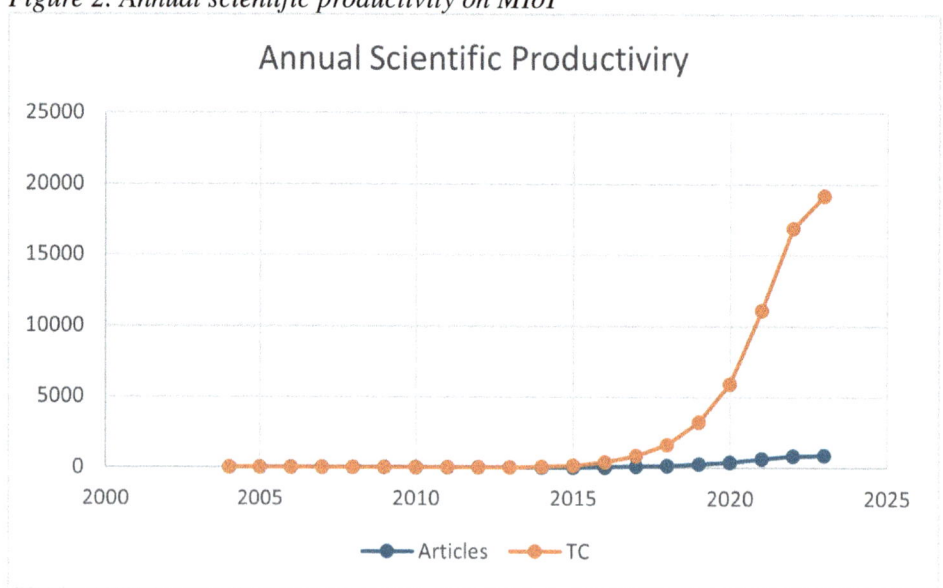

Figure 2. Annual scientific productivity on MIoT

Table 3. Top 10 most productive country

Sl. No.	label	Documents (%)	Citations	Citation Per Document	Total link strength
1	Peoples r China	1009 (30.68)	21021	20.83	5628
2	India	823 (25.03)	14206	17.26	6386
3	Saudi Arabia	485 (14.75)	7439	15.33	4558
4	USA	413 (12.56)	13266	32.12	2878
5	South Korea	294 (8.94)	6086	20.70	2337
6	Pakistan	274 (8.33)	4749	17.33	2725
7	England	219 (6.66)	4379	19.99	1729
8	Taiwan	190 (5.77)	3632	19.11	1449
9	Australia	146 (4.44))	4854	33.24	1516
10	Egypt	135 (4.10)	2847	21.08	1409

Table 3 displayed annual publication trends related to the Medical Internet of Things by country from 2004 to 2023. In this field, China published more papers than any other country (Table 1), with 1009 articles, followed by India (823 publications), Saudi Arabia (485 publications), the USA (413 publications), and South Korea (294 publications.

China was identified first among the top 10 countries/regions in terms of publications, with 21021 citations much more than any other nation, and a high citation/publication ratio of 20.83. India has a lower citation/publication rate (17.26) than other countries while having a large number of citations

(14206). Notably, Australia had the greatest citation/publication ratio (33.24) out of the ten nations, although with comparatively few publications, a sign of the excellent quality of its papers that were published.

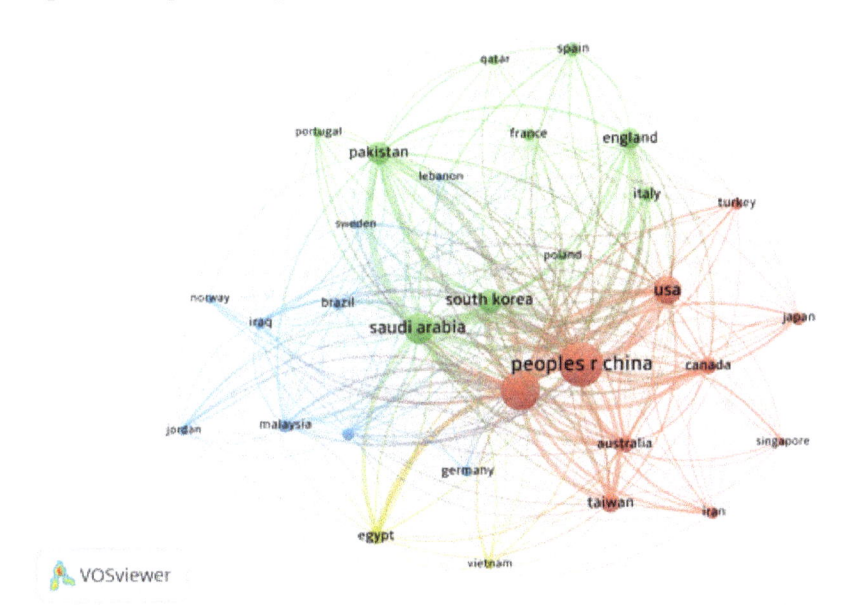

Figure 3. Map showing network visualization of country cooperation.

China has the greatest link network and the most recorded citations, with 5628 total link strengths and 21021 citations., as shown in figure 3 India comes in second with 14206 citations and 6386 overall link strengths, and the USA has 13266 citations. Countries are shown as nodes in the country collaboration network map, and the lines show how nodes are related to one another, representing the cooperative efforts of different countries. Interestingly, the degree of intensity in the collaboration is indicated by the thickness of these lines. In addition, each node's size and font size match the amount of citations in that particular nation. Total link strength (TLS) measures the frequency of international collaboration among experts working on a research project. A higher total link strength (TLS) relative to the number of research papers indicates stronger international co-authorship in scientific publications and better international research collaboration in a given country. (Belli et al., 2020; Sarker & Bartok, 2024).

Top 10 Most Productive Organizations

Table 4 displays the top 10 institutions with the most publications. King Saud University (n = 123) is the university with the highest number of publications, followed by the Vellore Institute of Technology (n = 77) and the Chinese Academy of Sciences (n = 64). The Chinese Academy of Sciences (n = 3678) is the organization with the most cited papers, followed by King Saud University (n = 3252) and the University of Electronic Science and Technology of China (n = 1912).

The Chinese Academy of Sciences (n = 57.46) has the highest average number of citations per document, followed by the University of Electronic Science and Technology of China (n = 34.76) and Sejong University (n = 27.62). Saudi Arabia is domicile to four of the world's ten most productive universities, demonstrating the country's outstanding research capabilities in the area of medical Internet of Things studies. They hold the top spot in this industry.

Table 4. Top 10 institutions with the most publications

Sl No	label	Country	Documents	Citations	Citation Per Document	Total link strength
1	King Saud University	Saudi Arabia	123	3252	26.43	386
2	Vellore Institute of Technology	India	77	1496	19.42	223
3	Chinese Academy of Sciences	China	64	3678	57.46	99
4	Prince Sattam bin Abdulaziz University	Saudi Arabia	63	525	8.33	117
5	Princess Nourah bint Abdulrahman University	Saudi Arabia	62	400	6.45	167
6	king abdulaziz university	Saudi Arabia	60	844	14.06	171
7	Taif University	Saudi Arabia	58	834	14.37	196
8	University of Electronic Science and Technology of China	China	55	1912	34.76	90
9	Sejong University	South Korea	50	1381	27.62	148
10	Xidian University	China	45	1034	22.97	73

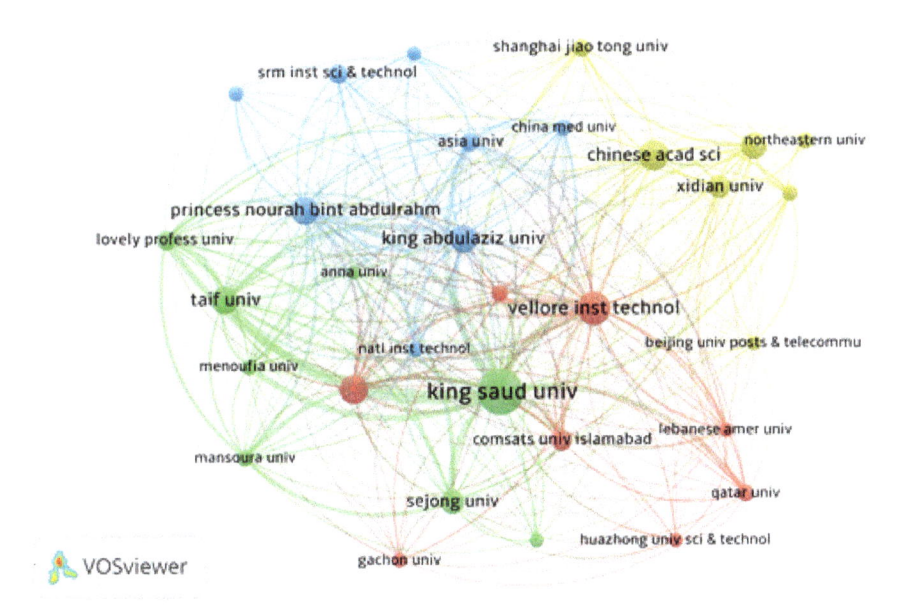

Figure 4. Map showing network visualization institute cooperation

With 3252 citations and 386 total link strengths, King Saud University has the most recorded largest link network, as shown in figure 4 Vellore Institute of Technology is next with 1496 citations and 223 total link strengths, and the Taif University with 834 citations and 196 total link strengths.

Top 10 Journals and Co-Cited Journals

In addition to offering trustworthy references for locating subjects, hot core journals may assist researchers in finding suitable journals for their works easily (Gao et al., 2023). 3 288 articles in this research area were published by 552 journals throughout the study period, according to an analysis of the retrieved publications. Table 5 lists the top 10 journals with the most publications.

Co-cited journals are those that are frequently used as sources of references in research articles. They serve as the main sources of reference for researchers and serve as a side-by-side indicator of the legitimacy and appeal of various journals within the research community. With VOS viewer, the names of the journals involved in the co-citation analysis were identified.

Table 6 displays IEEE Access was the most often co-cited journal with 6258 citations, followed by IEEE Internet Things with 3926 citations and Sensors with 2783 citations. IEEE Communications Surveys & Tutorials had the highest IF of

35.6 and was cited 985 times out of the top 10 co-cited journals. Every journal was in Q1, except Computer Science Lecture Notes. A map of popular journals within the Medical Internet of Things is visualized in the figure 5

Table 5. Top 10 Journals with Most Publications.

Sl No	Journals	Documents	Citations	IF	SJR	Q
1	IEEE access	317	7500	3.9	0.93	Q1
2	IEEE internet of things journal	226	6089	10.6	3.75	Q1
3	Sensors	186	2488	4.1	0.76	Q1
4	Electronics	80	802	2.9	0.63	Q2
5	IEEE transactions on industrial informatics	69	2838	12.3	4	Q1
6	Future generation computer systems	67	3716	7.5	2.04	Q1
7	CMC-computers materials & continua	56	254	3.1	0.53	Q2
8	IEEE journal of biomedical and health informatics	53	1188	7.7	1.67	Q1
9	Applied sciences-basel	52	327	2.9	0.49	Q2
10	Journal of healthcare engineering	47	312	NA	0.4	Q2

Figure 5. The network visualization map of popular journals.

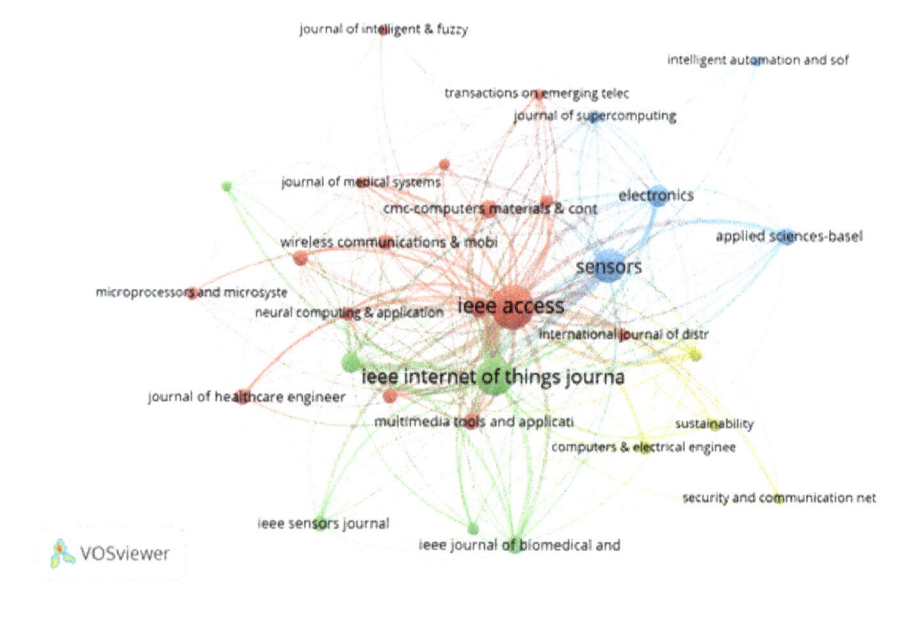

Table 6. Top 10 most co-cited journals

Sl No	Label (Co-Citation)	Citations	IF	SJR	Q
1	IEEE access	6258	3.9	0.93	Q1
2	IEEE internet things	3926	10.6	3.5	Q1
3	Sensors-basel	2783	4.1	0.76	Q1
4	Future generation computer systems	2276	7.5	2.04	Q1
5	Lecture Notes in Computer Science	1895	0.407	0.32	Q3
6	IEEE Transactions on Industrial Informatics	1689	12.3	4	Q1
7	Journal of Medical Systems	1329	5.3	0.91	Q1
8	IEEE journal of biomedical and health informatics	1036	7.7	1.67	Q1
9	IEEE communications surveys & tutorials	985	35.6	14.25	Q1
10	Arxiv	928	NA	NA	NA

Top 10 Authors and Co-Cited Authors.

During the period of our investigation, 10278 authors contributed to the Medical Internet of Things research. Since Web of Science and CiteSpace base their counts on name abbreviations, which are heavily impacted by duplicate names, we proceed with the statistical data from VOSviewer. The table 7 lists the top ten authors with the most publications. Neeraj Kumar (21), Gautam Srivastava (21), and Fadi Al-Turjman (19) are the first three, followed by M. Shamim Hossain (17), Ashok Kumar Das (17), Joel J.P.C. Rodrigues (17), Saraju P. Mohanty (16), Khan Muhammad (15), and Keping Yu (15). The authors with the highest average number of citations per document among the 10 are M. Shamim Hossain (58.88) and Neeraj Kumar (54.67), followed by Khan Muhammad (49.73), Gautam Srivastava (40.29), Keping Yu. (37.73), Ashok Kumar Das (33.71), Fadi Al-Turjman (31.63), Mohsen Guizani (31.33), Saraju P. Mohanty (29.38), and Joel J.P.C. Rodrigues (27.88). Even if these authors don't have many publications, the field may benefit from their research.

The authors in the references, commonly referred to as the "cited authors," were also examined. Of the 66620 co-cited authors, 5 had more than 200 co-citations (table 8). specifically, zhang, y (263), chen, m (261), hossain, ms (244), li, x (214) and he, db (207). In order to investigate the collaboration among authors, we have created a cooperative network graph featuring the authors and co-cited authors on Medical Internet of Things study (Figure 6).

Table 7. Top 10 Authors with Most Publications.

Sl No	Authors	Location	Documents	Citations	Citation Per Document
1	Neeraj Kumar	India	21	1148	54.67
2	Gautam Srivastava	Canada	21	846	40.29
3	Fadi Al-Turjman	Turkey	19	601	31.63
4	Mohsen Guizani	USA	18	564	31.33
5	M. Shamim Hossain	Saudi Arabia	17	1001	58.88
6	Ashok Kumar Das	India	17	573	33.71
7	Joel J.P.C. Rodrigues	Portugal	17	474	27.88
8	Saraju P. Mohanty	USA	16	470	29.38
9	Khan Muhammad	South Korea	15	746	49.73
10	Keping Yu	Japan	15	566	37.73

Table 8. Top 10 most co-cited authors.

Sl No	Authors	Location	Links	Total link strength	Citations
1	zhang, y	1	29	686	263
2	chen, m	3	28	727	261
3	hossain, ms	3	29	797	244
4	li, x	2	29	860	214
5	he, db	2	28	975	207
6	wazid, m	2	29	1228	196
7	yang, y	1	28	390	191
8	liu, y	1	29	387	179
9	islam, smr	1	29	405	178
10	khan, ma	1	29	156	168

Figure 6. The network visualization of authors and co-cited authors on MIoT.

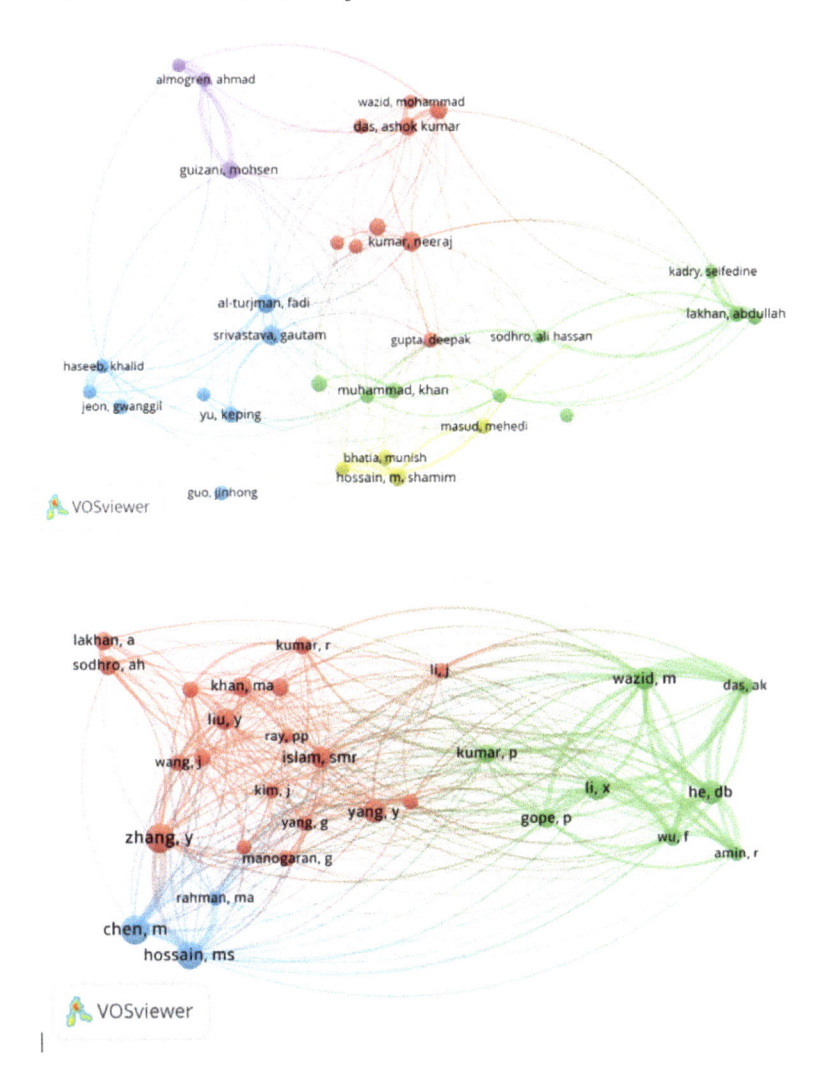

Co-Cited Reference

Table 9 displays the top 10 most highly cited papers in the field of medical internet of things. which received almost 60 citations each. We have used CiteSpace and Vosviewer to do reference co-citation analysis. "The Internet of Things for Health Care: A Comprehensive Survey" (with 171 co-citations) published in IEEE Access (IF=3.9) was the most referenced work by (Islam et al., 2015). Two of the top five co-cited works were also included in the IEEE Access journal. The figure 7 displays the Co-cited reference's network visualization map.

Table 9. Top 10 most highly cited papers in medical internet of things

Sl No.	label	cluster	Links	Total link strength	Citations
1	islam smr, 2015, ieee access, v3, p678, doi 10.1109/access.2015.2437951	2	27	246	171
2	gubbi j, 2013, future gener comp sy, v29, p1645, doi 10.1016/j.future.2013.01.010	2	25	139	90
3	baker sb, 2017, ieee access, v5, p26521, doi 10.1109/access.2017.2775180	1	27	129	79
4	atzori l, 2010, comput netw, v54, p2787, doi 10.1016/j.comnet.2010.05.010	2	19	92	77
5	farahani b, 2018, future gener comp sy, v78, p659, doi 10.1016/j.future.2017.04.036	1	25	145	73
6	dimitrov dv, 2016, healthc inform res, v22, p156, doi 10.4258/hir.2016.22.3.156	1	25	104	70
7	gatouillat a, 2018, ieee internet things, v5, p3810, doi 10.1109/jiot.2018.2849014	3	24	83	68
8	rahmani am, 2018, future gener comp sy, v78, p641, doi 10.1016/j.future.2017.02.014	1	26	111	67
9	dolev d, 1983, ieee t inform theory, v29, p198, doi 10.1109/tit.1983.1056650	4	16	40	65
10	joyia g.j., 2017, j commun, v12, p240, doi 10.12720/jcm.12.4.240-247	3	24	72	62

Figure 7. Network visualization map of co-cited reference

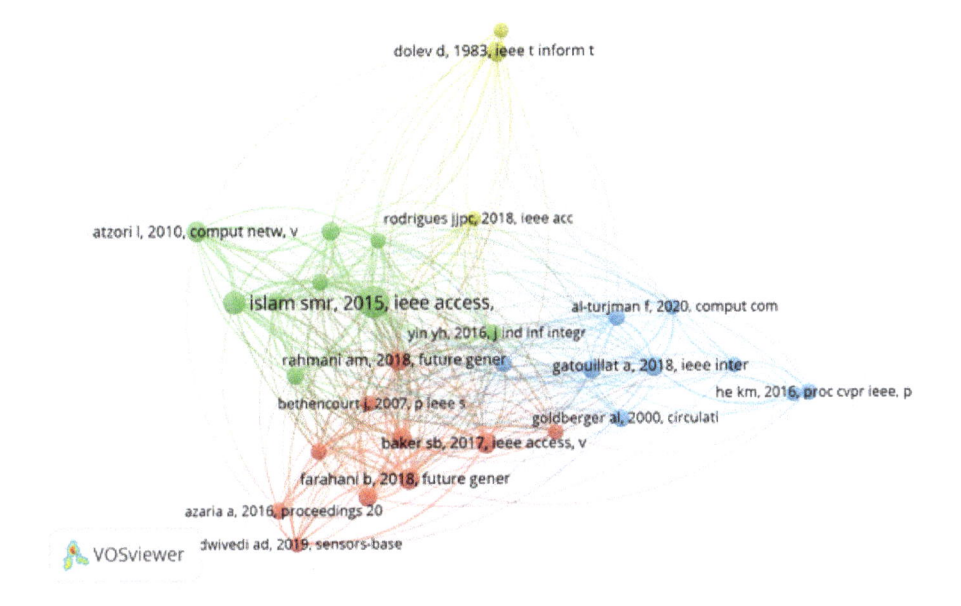

Additionally, We plotted the timeline view of co-cited references after the co-citation analysis. (Figure 8). Most of the studies appeared after 2005, according to the chronological perspective of the co-cited literature. The Medical Internet of Things has consistently focused on Federated Learning (Cluster0), which is the most published literature work and has the darkest color. We found that while "Artificial Intelligence" (Cluster 2) and "Healthcare domain" (Cluster 1) revealed that these clusters are the most recent additions to the field's hotspots for research., "Ubiquitous data" (Cluster 3) and "Smart clothing" (Cluster 5) were comparatively early hotspots.

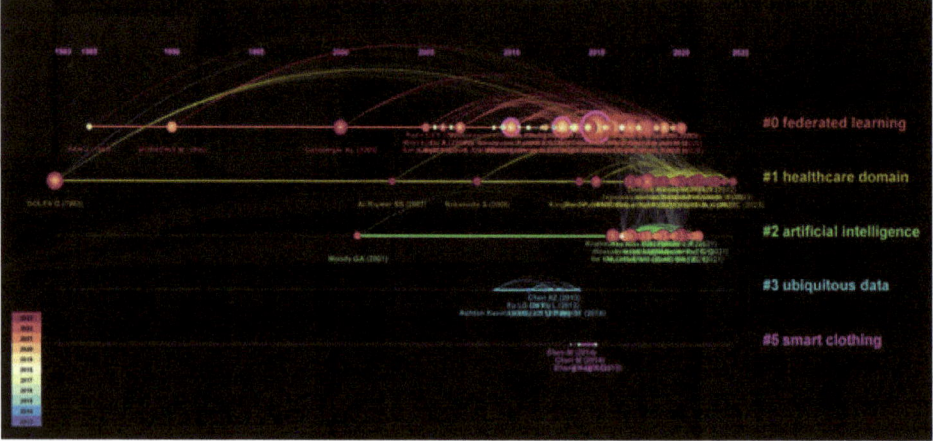

Figure 8. A timeline view for references related to miot that are co-cited.

The reference's initial appearance is shown by the node's location on the horizontal axis, and the node's size has a positive correlation with the reference's citation count. Co-cited relationships are represented by the lines connecting the nodes. A color that is closer to 2013 is indicated by more violet, and a color that is closer to 2023 is indicated by redder. The clusters with larger nodes and redder colors contained more articles indicating that this cluster's topic was prevalent in the field.

Table 10. Top 14 References with the Strongest Citation Bursts

References	Year	Strength	Begin	End	2013 - 2023
Gubbi J, 2013, FUTURE GENER COMP SY, V29, P1645, DOI 10.1016/j.future.2013.01.010, DOI	2013	**17.81**	2016	2020	
Atzori L, 2010, COMPUT NETW, V54, P2787, DOI 10.1016/j.comnet.2010.05.010, DOI	2010	**13.6**	2013	2020	
Islam SMR, 2015, IEEE ACCESS, V3, P678, DOI 10.1109/ACCESS.2015.2437951, DOI	2015	**9.81**	2016	2020	
Dimitrov DV, 2016, HEALTHC INFORM RES, V22, P156, DOI 10.4258/hir.2016.22.3.156, DOI	2016	**8.54**	2018	2020	
Hossain MS, 2016, COMPUT NETW, V101, P192, DOI 10.1016/j.comnet.2016.01.009, DOI	2016	**8.45**	2018	2021	
Gope P, 2016, IEEE SENS J, V16, P1368, DOI 10.1109/JSEN.2015.2502401, DOI	2016	**8.2**	2018	2021	
BURROWS M, 1990, ACM T COMPUT SYST, V8, P18, DOI 10.1145/74851.74852, DOI	1990	**7.76**	2019	2021	

continued on following page

Table 10. Continued

References	Year	Strength	Begin	End	2013 - 2023
Movassaghi S, 2014, IEEE COMMUN SURV TUT, V16, P1658, DOI 10.1109/SURV.2013.121313.00064, DOI	2014	**7.1**	2015	2019	
Xu BY, 2014, IEEE T IND INFORM, V10, P1578, DOI 10.1109/TII.2014.2306382, DOI	2014	**6.81**	2018	2020	
Mutlag AA, 2019, FUTURE GENER COMP SY, V90, P62, DOI 10.1016/j.future.2018.07.049, DOI	2019	**4.84**	2020	2023	
Azaria A, 2016, PROCEEDINGS 20 ...... BIG DATA - OBD 2016, V0, PP25, DOI	2016	**4.42**	2020	2023	
Hassanalieragh M, 2015, 2015 I ...... OMPUTING (SCC 2015), V0, PP285, DOI	2015	**4.38**	2018	2020	
Cavallari R, 2014, IEEE COMMUN SURV TUT, V16, P1635, DOI 10.1109/SURV.2014.012214.00007, DOI	2014	**4.31**	2016	2018	
Esposito C, 2018, IEEE CLOUD COMPUT, V5, P31	2018	**3.58**	2020	2023	

Figure 9 lists the top 14 references with notable bursts in citations., highlighting their importance in the field of Medical Internet of Things research. The red area of the line graph indicates the time period that receives the highest number of citations to the literature. The red section length is correlated with the amount of time that the literature has a high citation rate. The 2013 paper "Internet of Things (IoT): A vision, architectural elements, and future directions" by (Gubbi et al., 2013). in the journal "Future Generation Computer Systems" featured the strongest burst (strength=17.81). The outbreak persisted between 2016 and 2020.

Keyword Co-Occurrence and Burst Analysis

Table 10 provides a summary of the top 15 keywords that were used the most frequently. The top 5 keywords with more than 250 occurrences were internet of things (731 times), security (460 times), medical services (335 times), blockchain (303 times), and system (266 times). Furthermore, figure 10 displayed the network visualization map of the co-occurrence analysis of keywords. Each color represented the merging of clusters corresponding to the closely related keywords.

Additionally, we offered an overlay visualization map (Figure 11) in addition to the network map, where terms were coloured according to their average appearance year, or AAY. From an overall perspective, as Figure 6B illustrates, keywords within the yellow cluster have emerged in recent times, suggesting that scholars have recently given these research directions a great deal of attention.

Furthermore, after 2021, keywords containing AAY include: "Big data (AAY = 2020.68)", "Internet of things (AAY = 2020.96)", "Cloud (AAY = 2021.02)", "Health care (AAY = 2021.33)", "Security (AAY = 2021.47)", "Medical services (AAY = 2021.61)", "Machine learning (AAY = 2021.63)", "Artificial intelligence (AAY = 2021.80)", "Deep learning (AAY = 2021.87)", "Block chain (AAY = 2021.93)", and more.

A keyword burst is one that receives a lot of focus for an extended period of time. We identified the top 25 terms with the strongest citation bursts from CiteSpace by setting the burst duration to at least one year (Figure 7). The red bars represented the starting and ending, and the duration of citation bursts. The term "internet of things" (iot) has the highest burstiness among them, with a burstiness of 11.54. Furthermore, until 2023, the following keywords—temperature sensors, smart contracts, intelligent sensors, smart cities, and health monitoring—were in frequent use. This suggested that these areas of study represent current frontiers and hotspots in the field of medical internet of things research.

Table 11. Top 15 keywords in MIoT research.

Sl No.	label	cluster	Links	Total link strength	Occurrences
1	internet of things	3	30	1472	731
2	security	2	30	1397	460
3	medical services	3	30	1068	335
4	blockchain	2	30	893	303
5	system	1	30	603	266
6	privacy	2	30	834	262
7	internet of medical things	1	29	477	248
8	healthcare	1	30	686	239
9	deep learning	1	29	477	229
10	framework	1	30	631	205
11	machine learning	1	30	458	205
12	scheme	2	30	533	198
13	cloud computing	3	30	598	197
14	authentication	2	30	615	192
15	big data	1	30	486	169

Figure 9. Network visualization map of keywords co-occurrence analysis

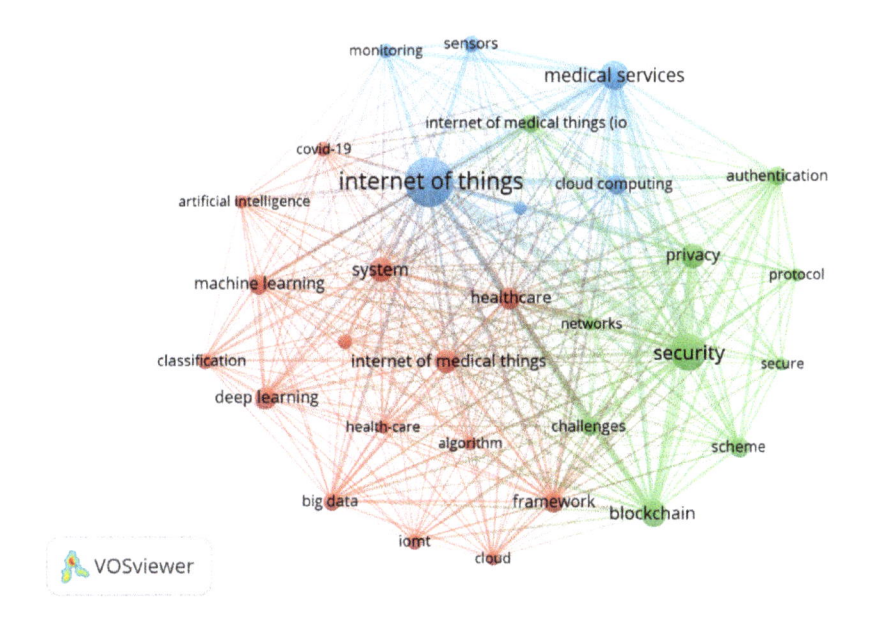

Figure 10. Overlay visualization map of keywords co-occurrence analysis.

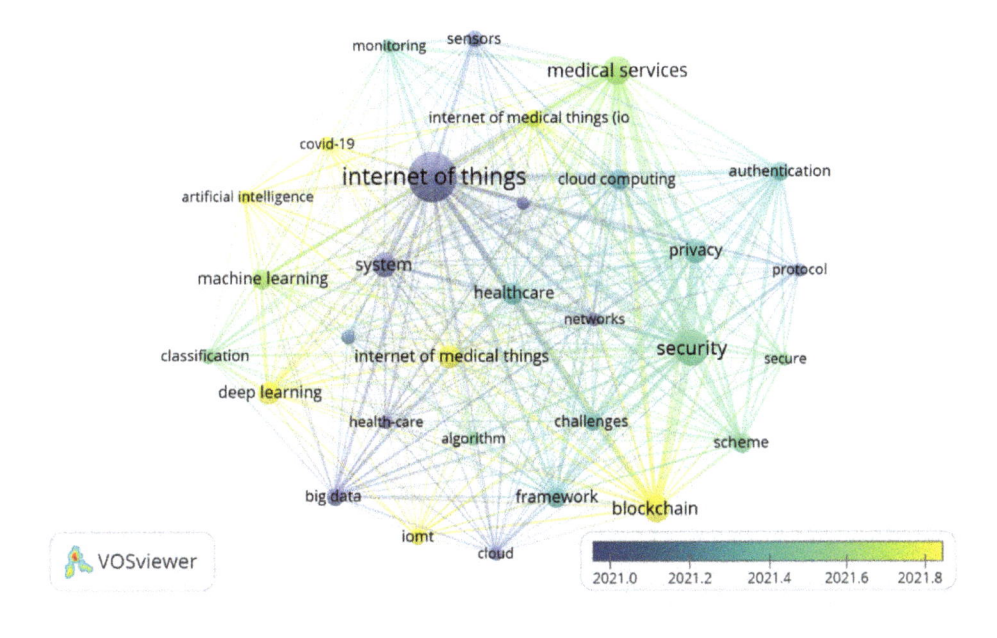

Table 12. Top 25 keywords with the strongest citation bursts

Keywords	Year	Strength	Begin	End	2013 - 2023
internet	2013	6.23	**2013**	2018	
internet of things (iot)	2014	11.54	**2014**	2019	
things	2014	10.49	**2014**	2018	
technology	2014	4.84	**2014**	2020	
design	2014	3.14	**2014**	2017	
internet of things	2013	8.27	**2015**	2017	
privacy	2015	2.61	**2015**	2017	
big data	2016	9.68	**2016**	2019	
care	2016	6.43	**2016**	2019	
health care	2016	5.04	**2016**	2018	
health	2016	2.69	**2016**	2017	
sensor	2017	6.29	**2017**	2020	
wireless	2017	5.09	**2017**	2019	
systems	2017	3.37	**2017**	2018	
sensor networks	2018	6.53	**2018**	2019	
networks	2016	3.48	**2018**	2019	
devices	2018	3.41	**2018**	2020	
patient monitoring	2019	2.91	**2019**	2021	
biomedical monitoring	2020	7.37	**2020**	2021	
data security	2020	3.93	**2020**	2021	
temperature sensors	2021	3.41	**2021**	2023	
smart contracts	2021	3.41	**2021**	2023	
intelligent sensors	2021	3.41	**2021**	2023	
smart city	2021	3.41	**2021**	2023	
health monitoring	2021	2.73	**2021**	2023	

DISCUSSION

The rising popularity of smart medical devices makes data collection easier and opens up new opportunities for medical and healthcare services. The large-scale, diverse, and quick data produced by several devices and sensors included in the medical Internet of Things has made traditional medical data handling more challenging. Various standards and formats applied to data from different sensors and medical equipment may cause medical data to become more diverse and dispersed.

Moreover, maintaining the stringent privacy and security requirements for data from MIoT devices and sensors requires confidentiality and integrity to prevent data leakage and tampering. These new medical technologies and laws have made managing medical data more challenging (Boyi Xu et al., 2014).

We performed a bibliometric analysis on MIoT in this study to compile findings from earlier research and offer suggestions for future avenues. For the study, a total of 3288 MIoT publications from the Web of Science Core Collection database were considered. Over the last 20 years, there has been a major increase in the quantity of publications on MIoT. In terms of publications and citations, China ranked first among the top 10 countries/regions; Australia, on the other hand, had the highest citation/publication ratio, indicating the exceptional caliber of its published articles.

The university with the most publications is King Saud University in Saudi Arabia. The Chinese Academy of Sciences and Vellore Institute of Technology in India are the next two most prolific universities. The Chinese Academy of Sciences (China) is the organization with the highest average number of citations per document and the highest number of cited papers. Saudi Arabia is home to four of the ten most productive universities in the world, providing evidence of the country's outstanding potential for Internet of Things-related medical research. The analysis of the top 10 journals in the Medical Internet of Things field shows that IEEE Access has the most citations and published articles. IEEE Transactions on Industrial Informatics, on the other hand, had the maximum impact factor (12.3). Furthermore, IEEE Access was the MioT field's most co-cited journal.

With the most publications (21) and citations (1148), Neeraj Kumar has made the most notable contributions to the field of medical internet of things. In addition to contributing the same number of papers (21), Gautam Srivastava had fewer citations (846). Neeraj Kumar (Senior Member, IEEE) is currently working as a Full Professor in the Department of Computer Science and Engineering at the Thapar Institute of Engineering and Technology (Deemed to be University), Patiala, India, having a current H-index of 93, he has published over 400 technical research papers in highly referenced journals and conferences, receiving over 29,410 citations from eminent researchers worldwide (*Neeraj Kumar - IEEE Xplore Author Profile*, n.d.). Furthermore, Zhang, y, and Chen, m are the top co-cited authors, and M. Shamim Hossain and Neeraj Kumar have the highest average number of citations per document.

Understanding the present and future directions of medical internet of things research is made easier with the use of the co-citation network and timeline analysis. The key concepts, seminal articles, and developing research trends that define the advancement of knowledge in this particular field are identified. The most frequently referenced article by (Islam et al., 2015) was "The Internet of Things for Health Care: A Comprehensive Survey," which was published in IEEE Access. The IEEE Access magazine also featured two of the top five co-cited articles. According to the timeline analysis, federated learning, artificial intelligence, and healthcare-related topics accounted for the majority of the literature published throughout the research period. The study "Internet of Things (IoT): A vision, architectural elements, and future directions" by (Gubbi et al., 2013). The journal "Future Generation Computer Systems" published in 2013 has the strongest burst of citations, according to the citation burst analysis.

In bibliometrics, keyword co-occurrence analysis is the most powerful way to identify the hot topics of research in a certain field, and keyword-formed clusters could reflect the composition of main research directions and their evolutionary trend. The terms "internet of things", "security," "medical services" and "blockchain" are the most frequently used keywords, and they accurately indicate the primary research direction in this sector. Considering figure 12 the terms "temperature sensors", " smart contracts", "intelligent sensors", "smart city" and "health monitoring" have been popular over the last three years, being the hotspots and possible direction of future research.

CONCLUSION

This research provides an overview of the development of the Medical Internet of Things worldwide and its global trends using bibliometric techniques. Important journals, powerful organizations, and noteworthy and popular articles were found with the aid of bibliometric analytic tools and procedures. The number of publications on the Medical Internet of Things continues to rise quickly. Scholarly interest in this topic has grown rapidly between 2004 and 2023, as evidenced by the publications on this topic during that time. Regarding contribution and influence in this subject, China was the most prominent nation. Three notable institutions that have advanced the field are King Saud University, Vellore Institute of Technology, and Chinese Academy of Sciences. As a result, institutions can collaborate more closely with them to better support the field's future growth. IEEE Access is the most published and cited journal. Moreover, Neeraj Kumar and Zhang, y. were the most productive and co-cited authors, respectively. Through publications in these journals or authors who may become future field collaborators, researchers can thus have a better understanding of the research development in this area. Furthermore, current MIoT research has focused a lot of attention on blockchain, deep learning, and artificial intelligence. Other research areas that need more consideration include temperature sensors, intelligent sensors, smart contracts, smart cities, and health monitoring. These areas might all become potential hotspots for future study. All things taken into account, this bibliometric analysis may offer insightful resources for scholars, particularly those who are just beginning out, as it may aid in their thorough comprehension of the field's knowledge landscape, including major personalities and active journals. Researchers may also draw inspiration from the investigation of future research frontiers and hotspots.

REFERENCES

Akhtar, N., Rahman, S., Sadia, H., & Perwej, Y. (2021). A Holistic Analysis of Medical Internet of Things (MIoT). *Journal of Information and Computational Science*, 11(4), 209–222.

Al-Suhimat, R. I., Ibrahim, A., Maaitah, N. O., & Al-Dmour, N. A. (2024). Review of Security Challenges Encountered in Internet of Things Technology. In 2024 2nd International Conference on Cyber Resilience (ICCR) (pp. 1-6). IEEE. DOI: 10.1109/ICCR61006.2024.10533052

Atzori, L., Iera, A., & Morabito, G. (2010). The Internet of Things: A survey. *Computer Networks*, 54(15), 2787–2805. DOI: 10.1016/j.comnet.2010.05.010

Azaria, A., Ekblaw, A., Vieira, T., & Lippman, A. (2016). MedRec: Using Blockchain for Medical Data Access and Permission Management. *2016 2nd International Conference on Open and Big Data (OBD)*, 25–30. https://doi.org/DOI: 10.1109/OBD.2016.11

Baker, S. B., Xiang, W., & Atkinson, I. (2017). Internet of Things for Smart Healthcare: Technologies, Challenges, and Opportunities. *IEEE Access : Practical Innovations, Open Solutions*, 5, 26521–26544. DOI: 10.1109/ACCESS.2017.2775180

Belli, S., Mugnaini, R., Baltà, J., & Abadal, E. (2020). Coronavirus mapping in scientific publications: When science advances rapidly and collectively, is access to this knowledge open to society? *Scientometrics*, 124(3), 2661–2685. DOI: 10.1007/s11192-020-03590-7 PMID: 32836526

Burrows, M., Abadi, M., & Needham, R. (1989). A logic of authentication. *Operating Systems Review*, 23(5), 1–13. DOI: 10.1145/74851.74852

Cavallari, R., Martelli, F., Rosini, R., Buratti, C., & Verdone, R. (2014). A Survey on Wireless Body Area Networks: Technologies and Design Challenges. *IEEE Communications Surveys and Tutorials*, 16(3), 1635–1657. DOI: 10.1109/SURV.2014.012214.00007

Dimitrov, D. V. (2016a). Medical Internet of Things and Big Data in Healthcare. *Healthcare Informatics Research*, 22(3), 156–163. DOI: 10.4258/hir.2016.22.3.156 PMID: 27525156

Dimitrov, D. V. (2016b). Medical Internet of Things and Big Data in Healthcare. *Healthcare Informatics Research*, 22(3), 156–163. DOI: 10.4258/hir.2016.22.3.156 PMID: 27525156

Dolev, D., & Yao, A. (1983). On the security of public key protocols. *IEEE Transactions on Information Theory*, 29(2), 198–208. DOI: 10.1109/TIT.1983.1056650

Elayan, H., Shubair, R. M., & Kiourti, A. (2017). Wireless sensors for medical applications: Current status and future challenges. *2017 11th European Conference on Antennas and Propagation (EUCAP)*, 2478–2482. https://doi.org/DOI: 10.23919/EuCAP.2017.7928405

Elhoseny, M., Thilakarathne, N. N., Alghamdi, M. I., Mahendran, R. K., Gardezi, A. A., Weerasinghe, H., & Welhenge, A. (2021). Security and Privacy Issues in Medical Internet of Things: Overview, Countermeasures, Challenges and Future Directions. *Sustainability (Basel)*, 13(21), 21. Advance online publication. DOI: 10.3390/su132111645

Esposito, C., De Santis, A., Tortora, G., Chang, H., & Choo, K.-K. R. (2018). Blockchain: A Panacea for Healthcare Cloud-Based Data Security and Privacy? *IEEE Cloud Computing*, 5(1), 31–37. DOI: 10.1109/MCC.2018.011791712

Farahani, B., Firouzi, F., Chang, V., Badaroglu, M., Constant, N., & Mankodiya, K. (2018). Towards fog-driven IoT eHealth: Promises and challenges of IoT in medicine and healthcare. *Future Generation Computer Systems*, 78, 659–676. DOI: 10.1016/j.future.2017.04.036

Gao, Y., Li, Y., Feng, S., & Gu, L. (2023). Bibliometric and visualization analysis of matrix metalloproteinases in ischemic stroke from 1992 to 2022. *Frontiers in Neuroscience*, 17, 1206793. https://www.frontiersin.org/journals/neuroscience/articles/10.3389/fnins.2023.1206793. DOI: 10.3389/fnins.2023.1206793 PMID: 37483355

Gatouillat, A., Badr, Y., Massot, B., & Sejdić, E. (2018). Internet of Medical Things: A Review of Recent Contributions Dealing With Cyber-Physical Systems in Medicine. *IEEE Internet of Things Journal*, 5(5), 3810–3822. DOI: 10.1109/JIOT.2018.2849014

Gope, P., & Hwang, T. (2016). BSN-Care: A Secure IoT-Based Modern Healthcare System Using Body Sensor Network. *IEEE Sensors Journal*, 16(5), 1368–1376. DOI: 10.1109/JSEN.2015.2502401

Gubbi, J., Buyya, R., Marusic, S., & Palaniswami, M. (2013). Internet of Things (IoT): A vision, architectural elements, and future directions. *Future Generation Computer Systems*, 29(7), 1645–1660. DOI: 10.1016/j.future.2013.01.010

Hassanalieragh, M., Page, A., Soyata, T., Sharma, G., Aktas, M., Mateos, G., Kantarci, B., & Andreescu, S. (2015). Health Monitoring and Management Using Internet-of-Things (IoT) Sensing with Cloud-Based Processing: Opportunities and Challenges. *2015 IEEE International Conference on Services Computing*, 285–292. https://doi.org/DOI: 10.1109/SCC.2015.47

Hossain, M. S., & Muhammad, G. (2016). Cloud-assisted Industrial Internet of Things (IIoT) – Enabled framework for health monitoring. *Computer Networks*, 101, 192–202. DOI: 10.1016/j.comnet.2016.01.009

Islam, S. M. R., Kwak, D., Kabir, M. D. H., Hossain, M., & Kwak, K.-S. (2015). The Internet of Things for Health Care: A Comprehensive Survey. *IEEE Access: Practical Innovations, Open Solutions*, 3, 678–708. DOI: 10.1109/ACCESS.2015.2437951

Joyia, G. J., Liaqat, R. M., Farooq, A., & Rehman, S.National University of Sciences and Technology. (2017). Islamabad, Pakistan, Joyia, G. J., Liaqat, R. M., Farooq, A., & Rehman, S. (2017). Internet of Medical Things (IOMT): Applications, Benefits and Future Challenges in Healthcare Domain. *Journal of Communication*. Advance online publication. DOI: 10.12720/jcm.12.4.240-247

Malathi, J., Kusha, K. R., Isaac, S., Ramesh, A., Rajendiran, M., & Boopathi, S. (2024). IoT-Enabled Remote Patient Monitoring for Chronic Disease Management and Cost Savings: Transforming Healthcare. In Advances in Explainable AI Applications for Smart Cities (pp. 371-388). IGI Global.

Movassaghi, S., Abolhasan, M., Lipman, J., Smith, D., & Jamalipour, A. (2014). Wireless Body Area Networks: A Survey. *IEEE Communications Surveys and Tutorials*, 16(3), 1658–1686. DOI: 10.1109/SURV.2013.121313.00064

Mutlag, A. A., Abd Ghani, M. K., Arunkumar, N., Mohammed, M. A., & Mohd, O. (2019). Enabling technologies for fog computing in healthcare IoT systems. *Future Generation Computer Systems*, 90, 62–78. DOI: 10.1016/j.future.2018.07.049

Neeraj Kumar—IEEE Xplore Author Profile. (n.d.). Retrieved 24 February 2024, from https://ieeexplore.ieee.org/author/37395599700

Pan, X., Yan, E., Cui, M., & Hua, W. (2018). Examining the usage, citation, and diffusion patterns of bibliometric mapping software: A comparative study of three tools. *Journal of Informetrics*, 12(2), 481–493. DOI: 10.1016/j.joi.2018.03.005

Rahmani, A. M., Gia, T. N., Negash, B., Anzanpour, A., Azimi, I., Jiang, M., & Liljeberg, P. (2018). Exploiting smart e-Health gateways at the edge of healthcare Internet-of-Things: A fog computing approach. *Future Generation Computer Systems*, 78, 641–658. DOI: 10.1016/j.future.2017.02.014

Sarker, M., & Bartok, I. (2024). Global trends of green manufacturing research in the textile industry using bibliometric analysis. *Case Studies in Chemical and Environmental Engineering*, 9, 100578. DOI: 10.1016/j.cscee.2023.100578

Subhan, F., Mirza, A., Su'ud, M. B. M., Alam, M. M., Nisar, S., Habib, U., & Iqbal, M. Z. (2023). AI-enabled wearable medical internet of things in healthcare system: A survey. *Applied Sciences (Basel, Switzerland)*, 13(3), 1394. DOI: 10.3390/app13031394

Synnestvedt, M. B., Chen, C., & Holmes, J. H. (2005). CiteSpace II: Visualization and Knowledge Discovery in Bibliographic Databases. *AMIA ... Annual Symposium Proceedings - AMIA Symposium. AMIA Symposium*, 2005, 724–728. PMID: 16779135

Tao, P., Liu, N., & Dong, C. (2024). Research progress of MIoT and digital healthcare in the new era. *Clinical eHealth*, 7, 1–4. https://doi.org/DOI: 10.1016/j.ceh.2023.11.004

van Eck, N. J., & Waltman, L. (2010). Software survey: VOSviewer, a computer program for bibliometric mapping. *Scientometrics*, 84(2), 523–538. DOI: 10.1007/s11192-009-0146-3 PMID: 20585380

van Eck, N. J., & Waltman, L. (2017). Citation-based clustering of publications using CitNetExplorer and VOSviewer. *Scientometrics*, 111(2), 1053–1070. DOI: 10.1007/s11192-017-2300-7 PMID: 28490825

Wu, F., Gao, J., Kang, J., Wang, X., Niu, Q., Liu, J., & Zhang, L. (2022). Knowledge Mapping of Exosomes in Autoimmune Diseases: A Bibliometric Analysis (2002–2021). *Frontiers in Immunology*, 13, 939433. https://www.frontiersin.org/journals/immunology/articles/10.3389/fimmu.2022.939433. DOI: 10.3389/fimmu.2022.939433 PMID: 35935932

Wu, H., Cheng, K., Guo, Q., Yang, W., Tong, L., Wang, Y., & Sun, Z. (2021). Mapping Knowledge Structure and Themes Trends of Osteoporosis in Rheumatoid Arthritis: A Bibliometric Analysis. *Frontiers in Medicine*, 8, 787228. DOI: 10.3389/fmed.2021.787228 PMID: 34888333

Wu, Z., Cheng, K., Shen, Z., Lu, Y., Wang, H., Wang, G., Wang, Y., Yang, W., Sun, Z., Guo, Q., & Wu, H. (2023). Mapping knowledge landscapes and emerging trends of sonodynamic therapy: A bibliometric and visualized study. *Frontiers in Pharmacology*, 13, 1048211. https://www.frontiersin.org/journals/pharmacology/articles/10.3389/fphar.2022.1048211. DOI: 10.3389/fphar.2022.1048211 PMID: 36699067

Xu, B., Da Xu, L., Cai, H., Xie, C., Hu, J., & Bu, F. (2014). Ubiquitous Data Accessing Method in IoT-Based Information System for Emergency Medical Services. *IEEE Transactions on Industrial Informatics*, 10(2), 1578–1586. DOI: 10.1109/TII.2014.2306382

Xu, B., Xu, L. D., Cai, H., Xie, C., Hu, J., & Bu, F. (2014). Ubiquitous Data Accessing Method in IoT-Based Information System for Emergency Medical Services. *IEEE Transactions on Industrial Informatics*, 10(2), 1578–1586. DOI: 10.1109/TII.2014.2306382

Yeung, A. W. K., & Mozos, I. (2020). The Innovative and Sustainable Use of Dental Panoramic Radiographs for the Detection of Osteoporosis. *International Journal of Environmental Research and Public Health*, 17(7), 7. Advance online publication. DOI: 10.3390/ijerph17072449 PMID: 32260243

Zhang, X.-L., Zheng, Y., Xia, M.-L., Wu, Y.-N., Liu, X.-J., Xie, S.-K., Wu, Y.-F., & Wang, M. (2020). Knowledge Domain and Emerging Trends in Vinegar Research: A Bibliometric Review of the Literature from WoSCC. *Foods*, 9(2), 2. Advance online publication. DOI: 10.3390/foods9020166 PMID: 32050682

Chapter 15
Real-Time Data Analytics and Decision Making in Cyber-Physical Systems

Vishakha Kuwar
https://orcid.org/0000-0003-1764-2475
Symbiosis International University, India

Vandana Sonwaney
https://orcid.org/0000-0002-2131-2041
Symbiosis International University, India

Shitiz Upreti
Maharishi Markandeshwar, India

Shubham Rajendra Ekatpure
https://orcid.org/0009-0008-2717-885X
Kulicke and Soffa, USA

Prakash Divakaran
Himalayan University, India

Kamal Upreti
https://orcid.org/0000-0003-0665-530X
Christ University, India

Ramesh Chandra Poonia
https://orcid.org/0000-0001-8054-2405
Christ University, India

ABSTRACT

The future of digital innovation lies in Cyber-Physical Systems (CPS), integrating computational capabilities with physical processes. CPS function as interconnected networks, merging physical and digital inputs and outputs. This study defines CPS and highlights real-time data analytics' role in enhancing communication across industries like manufacturing and robotics. CPS depend on data collection via edge computing, IoT, and sensors, involving data cleaning, preparation, and normalization. Real-time analytics, including stream processing, machine learning, and AI, are crucial for CPS. Decision-making systems and algorithms enhance efficiency. Given their sensitivity, security and privacy in real-time analytics are vital. The study addresses threat detection, privacy preservation, and data security, along

DOI: 10.4018/979-8-3693-5728-6.ch015

with challenges like data heterogeneity, latency, and scalability. Future prospects include edge AI, fog computing, and blockchain integration.

1. INTRODUCTION

In the rapidly evolving landscape of modern technology, Cyber-Physical Systems (CPS) represent a groundbreaking paradigm that bridges the gap between the physical and digital worlds. CPS are integrations of computation, networking, and physical processes, where embedded computers and networks monitor and control the physical processes, usually with feedback loops where physical processes affect computations and vice versa. This synergistic interaction enables CPS to perform complex functions, enhancing efficiency, reliability, and automation across a variety of sectors (Yu B. et al.,2020). The concept of CPS has evolved from earlier systems like embedded systems and control systems. The term "Cyber-Physical Systems" was first coined in the early 2000s, driven by advancements in computing, networking, and sensing technologies. The evolution of the Internet of Things (IoT) and advancements in artificial intelligence (AI) have further accelerated the development and deployment of CPS.

Cyber-Physical Systems (CPS) are distinguished from traditional embedded systems by their tight integration and coordination between computational and physical components, enabling real-time processing and response to ensure system stability and performance. CPS involve multiple interconnected subsystems that require seamless interoperability and possess adaptability to changing conditions and requirements in their operational environments (Lesch V. et al.,2023). Cyber-Physical Systems are characterized by their ability to integrate computational algorithms with physical components (Dafflon B. et al.,2021).

Real-time data analytics is pivotal in Cyber-Physical Systems (CPS) as it enables instantaneous decision-making and responsiveness, which are essential for optimizing performance, ensuring safety, and enhancing efficiency across diverse applications. By continuously processing data from sensors and devices, CPS can dynamically adjust operations, predict maintenance needs, and detect security threats, thereby minimizing downtime and reducing operational costs. By using Connected Paper, the Derivative work involved the papers have cited many of the papers in the graph. This typically indicates that they are either surveys of the field or recent relevant works inspired by multiple papers. Selecting a derivative work will highlight all the graph papers it cites, and selecting a graph paper will highlight all the derivative works that cite it. The Base paper considered in the study was "Data-Driven Cyber-Physical Systems via Real-Time Stream Analytics and Machine Learning By Akkaya I. (2016) as shown in Table 1 below.

Figure 1. Connected Paper Analysis Using base paper on Real Time Analysis in Cyber Physical Systems

Title	Last author	Year	Citations	Graph references
Risk-aware motion planning for automated vehicle among human-driven cars	M. Althoff	2019	11	4
On Enabling Technologies for the Internet of Important Things	Edward A. Lee	2019	10	6
Deterministic Timing for the Industrial Internet of Things	Edward A. Lee	2018	6	6
Software-defined control of an emulated hydrogen energy storage for energy internet ecosystems	Josep M. Guerrero	2023	4	5
Service Discovery for the Connected Car with Semantic Accessors	Edward A. Lee	2019	3	3
Randomized Synthesis for Diversity and Cost Constraints with Control Improvisation	Daniel J. Fremont	2022	3	6
Entropy-Guided Control Improvisation	S. Seshia	2021	2	3
Cyber-Physical Systems as Key Element to Industry 4.0: Characteristics, Applications and Related Technologies	André Luis Helleno	2022	1	3
Bachelor's Thesis Computing Science Control Improvisation for infinite regular languages and feature-based parameters	Dr. Jurriaan Rot	2023	0	3
An architectural approach to cyber-physical system design	Tom J. W. Lankhorst	2017	0	3

The applications (Pishdad-Bozorgi P.et al.,2020), of Cyber-Physical Systems span a wide range of industries and domains, including but not limited to: Cyber-Physical Systems (CPS) are transforming multiple sectors by enhancing efficiency, security, and innovation. In cybersecurity, CPS integration has increased vulnerability to cyber threats, prompting advancements in encrypted communication, real-time monitoring, and blockchain technology. Healthcare applications include patient monitoring and medication management, with AI and IoMT driving personalized treatments and surgical planning. In utilities, CPS optimize power and water distribution through real-time data processing, exemplified by smart hydrants preventing water theft. Automotive sectors benefit from CPS in manufacturing, driver assistance systems, and predictive maintenance. Military operations utilize CPS for improved communication and autonomous surveillance. Manufacturing industries leverage IIoT, digital twins, and cobots for enhanced production. Industry 4.0 relies on CPS for smart, connected factories with low-latency connectivity and secure operations. Smart cities use CPS for infrastructure management, traffic control, and public safety. Agriculture employs CPS for precision farming and autonomous operations. In aerospace, CPS improve flight control and maintenance planning. The continued evolution of CPS promises further advancements and efficiencies across these sectors.

Real-time data analytics is the process of extracting insights and making decisions based on data as it is generated. In Cyber-Physical Systems (CPS), this capability is essential for integrating the digital and physical components effectively (Yu T.et al.,2020). The foundation of real-time data analytics in CPS lies in robust data collection and sensing mechanisms. Sensors embedded within the physical components of CPS continuously gather data about various parameters such as temperature, pressure, motion, and more. This raw data serves as the input for subsequent analysis. Efficient data transmission is critical to ensure that the collected data is available for analysis in real-time. This requires high-speed, reliable networks that can handle large volumes of data with minimal latency. Technologies such as 5G, Wi-Fi 6, and advanced networking protocols are often employed to facilitate this (Curry E.et al.,2020).

To process data in real-time, CPS often leverage a combination of edge and cloud computing. Edge computing allows data to be processed close to the source, reducing latency and enabling faster decision-making. Cloud computing, on the other hand, provides scalable resources for more complex data analysis and storage. The hybrid approach ensures both speed and computational power are optimized. Before analysis, data must be integrated from various sources and pre-processed to ensure quality and consistency. This involves data cleaning, normalization, and aggregation. Effective preprocessing is crucial to eliminate noise and handle missing or inconsistent data, ensuring the accuracy of the real-time analytics (Sharma R.et al.,2024).

At the core of real-time data analytics are powerful analytics engines capable of processing and analysing data as it arrives. These engines employ various techniques, including statistical analysis, machine learning algorithms, and artificial intelligence, to extract actionable insights (Khan R., et al.,2024). The choice of algorithms and models is tailored to the specific requirements of the CPS application. Real-time analytics provide the basis for immediate decision-making within CPS. Once insights are derived, the system must act on them promptly. This involves triggering actuators to perform physical actions, adjusting system parameters, or sending alerts. The feedback loop between data analysis and physical actions is what makes CPS responsive and adaptive (Shollo A.et al.,2024). The Vos viewer Cyber Physical System keywords occurrences as shown in the following Figure 2.

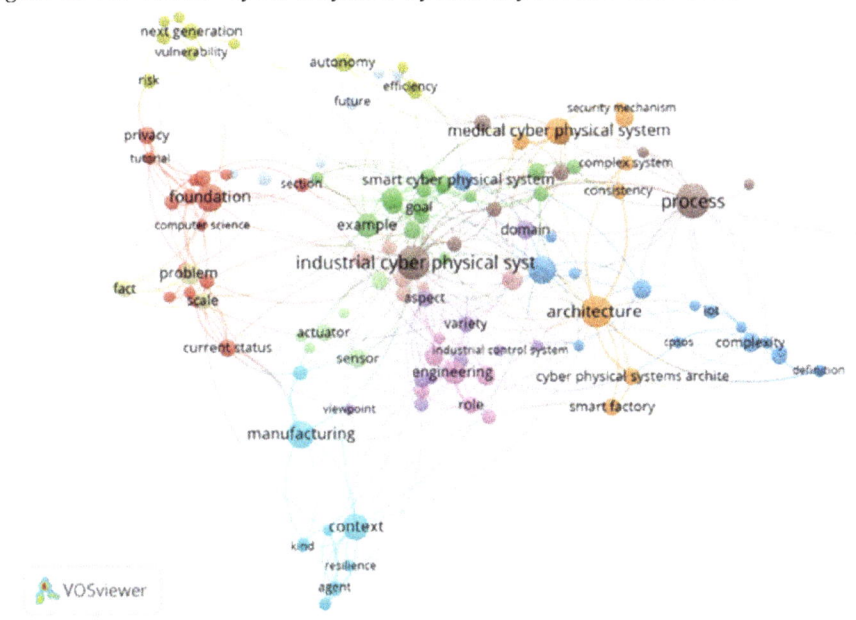

Figure 2. Vos viewer Cyber Physical System keywords occurrences

2. REAL-TIME DATA ANALYTICS

Real-time data analytics refers to the process of collecting, processing, and analysing data as soon as it is created to derive actionable insights immediately. This capability is crucial in various industries, enabling systems and organizations to respond promptly to changing conditions, optimize performance, and make informed decisions. At its core, real-time data analytics relies on the seamless integration of data collection, transmission, processing, and visualization. The process begins with data collection, where sensors and devices continuously gather data from the environment or specific processes (Chen W., et al.,2023). This raw data is then transmitted via high-speed, reliable networks to analytics engines that can handle large volumes of information with minimal delay.

Visualization tools and user interfaces are crucial for presenting real-time data in an accessible and actionable form. They provide stakeholders with intuitive dashboards, real-time alerts, and detailed reports that facilitate informed decision-making. Continuous improvement and learning are vital for maintaining the effectiveness of real-time data analytics, involving regular updates to algorithms and methodologies based on feedback and new technologies (Shete D., & Khobragade P. 2023). The real time Data analytics keywords occurrence as shown Figure 3:

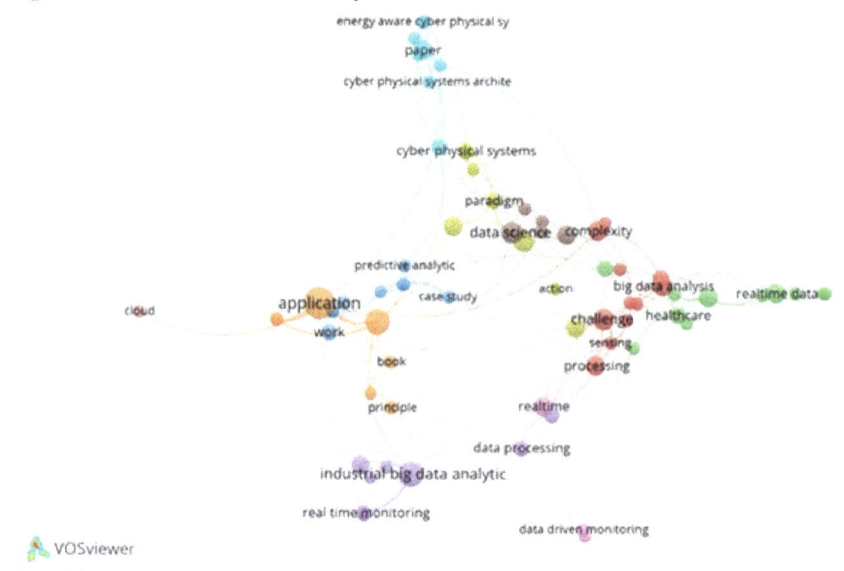

Figure 3. Real Time Data Analytics

3. CHALLENGES OF REAL-TIME PROCESSING

Real-time processing poses several challenges and demands specific requirements to effectively handle the complexities of continuous data streams. One of the primary challenges is latency management, as real-time systems must process data quickly enough to provide timely insights and responses. This necessitates optimized algorithms and infrastructure capable of handling high-speed data transmission and processing. Scalability is another critical requirement, as systems need to expand seamlessly to accommodate growing data volumes without compromising performance. Ensuring data quality and consistency through effective preprocessing is essential to mitigate errors and inaccuracies that can arise from noisy or incomplete data streams. Security and privacy concerns also loom large, demanding robust encryption methods and secure communication protocols to safeguard sensitive information in transit and at rest. Furthermore, maintaining system reliability and availability requires redundancy and fault-tolerant designs to prevent disruptions and ensure continuous operation. Finally, regulatory compliance adds another layer of complexity, requiring adherence to data protection laws and industry standards. Addressing these challenges and meeting these requirements is crucial for deploying robust real-time processing systems across diverse applications, from industrial automation and healthcare to smart cities and financial services. The challenges faced by the system discussed in different.

4. IOT AND EDGE COMPUTING IN CPS

IoT (Internet of Things) and edge computing are integral components of Cyber-Physical Systems (CPS), playing crucial roles in enhancing real-time data processing, decision-making capabilities, and overall system efficiency devices within CPS serve as sensors and actuators, collecting vast amounts of data from physical environments such as industrial machinery, smart infrastructure, healthcare devices, and more. These devices are equipped with sensors that capture data on parameters like temperature, pressure, motion, and environmental conditions. Through wireless communication protocols such as Wi-Fi, Bluetooth, Zigbee, or cellular networks, IoT devices transmit this data to centralized servers or edge computing nodes for analysis and decision-making.

Edge computing complements IoT by decentralizing data processing and analytics closer to where the data is generated, i.e., at the edge of the network. This approach reduces latency by minimizing the distance data needs to travel and allows for real-time analysis and response. Edge computing nodes, located within CPS environments or on IoT devices themselves, host computational resources and applications necessary for data processing. They filter and aggregate incoming data streams, perform initial analysis, and only transmit relevant information to centralized servers or the cloud. This reduces bandwidth usage, optimizes network traffic, and enhances overall system performance. The Vos viewer linkage in the title and abstract of the different studies from the year 2021 to 2024 as shown in the figure 4:

Figure 4. IoT and Edge Computing in Cyber Physical Systems

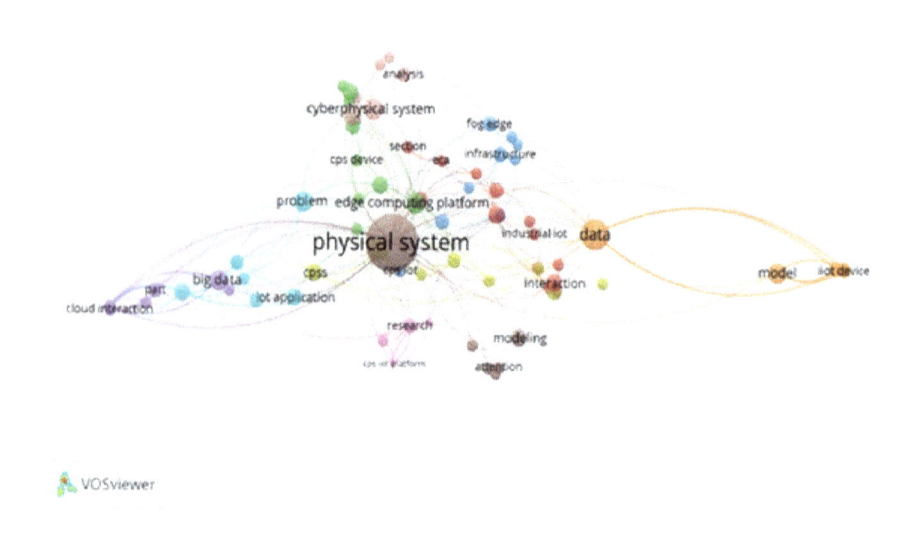

The synergy between IoT and edge computing enhances CPS in several ways. First, it enables rapid decision-making by processing data closer to the source, reducing response times for critical applications such as industrial automation or healthcare monitoring. Second, it improves data privacy and security by limiting the amount of sensitive data transmitted over networks. Edge computing nodes can preprocess and anonymize data locally before sending it to centralize servers, minimizing exposure to potential cyber threats. Third, it supports scalability by distributing computational load across edge devices, ensuring CPS can handle increasing data volumes without overwhelming centralized infrastructure. Moreover, IoT and edge computing facilitate autonomous operation within CPS. Edge devices can execute local control algorithms based on real-time data inputs, enabling autonomous decision-making and action without constant reliance on centralized systems. This autonomy is particularly valuable in applications like autonomous vehicles, where split-second decisions are critical for safety and efficiency.

5. DATA INTEGRATION AND MANAGEMENT STRATEGIES

Effective data integration and management are pivotal for optimizing the performance and reliability of data within Cyber-Physical Systems (CPS). These systems rely on seamless integration and efficient management of diverse data sources to support real-time decision-making and operational efficiency across various applications. Data integration strategies in CPS focus on creating unified architectures that consolidate data from multiple sources. This involves designing cohesive frameworks that ensure consistency in data formats, standards, and protocols. Event-Driven Architecture (EDA) is particularly effective for CPS, enabling real-time processing of data streams and immediate responses to events. Middleware solutions play a crucial role by acting as intermediaries between CPS components, facilitating smooth communication and data exchange (Aldoseri A. et al.,2023). They standardize interfaces and protocols, simplifying integration efforts and ensuring interoperability across heterogeneous systems.

Additionally, data federation techniques allow CPS to integrate data from distributed sources without physically consolidating it, providing a virtual view of data across different systems. In terms of data management, ensuring data quality is paramount. This involves rigorous processes such as data cleaning, validation, and normalization to eliminate errors and inconsistencies. Scalable storage solutions, including cloud storage and distributed databases, accommodate the vast volumes of data generated by CPS while ensuring high availability and efficient retrieval.

Data security and privacy are critical considerations, with CPS implementing robust measures such as encryption, authentication, and access control to protect sensitive information. Privacy-preserving techniques anonymize personal data, ensuring compliance with data protection regulations. Metadata management is essential for providing descriptive information about CPS data, including its origin, structure, and usage. Effective metadata management enhances data discovery, understanding, and traceability, supporting governance and usability. Finally, integrating real-time analytics capabilities within CPS enables continuous monitoring, analysis, and visualization of data. Advanced analytics, such as machine learning and predictive modelling, uncover insights and patterns that facilitate proactive decision-making and operational optimization (Scquizzato S.2024). By implementing these integrated approaches to data integration and management, CPS can leverage data effectively to enhance operational efficiency, enable real-time decision-making, and drive innovation across diverse applications and industries.

6. REAL-TIME DATA ANALYTICS TECHNIQUES

Real-time data analytics techniques are crucial in Cyber-Physical Systems (CPS) for processing and analysing data streams promptly (Raji A.et al.,2024) Key techniques include stream processing with frameworks like Apache Kafka and Apache Flink, complex event processing (CEP) for identifying patterns and triggering actions, and real-time machine learning models for continuous updates and predictive insights (Ziehn A.et al.,2024). Edge analytics reduces latency by processing data locally, while visualization tools and anomaly detection algorithms provide real-time insights for quick decision-making and system optimization. These techniques collectively enhance CPS operations by enabling rapid responses, proactive maintenance, and efficient resource management.

Stream processing frameworks are integral to Cyber-Physical Systems (CPS), enabling efficient handling of continuous data streams in real-time. Apache Kafka, known for its scalability and fault tolerance, facilitates high-throughput data ingestion and processing through its distributed messaging system (Garcia M.et al.,2023). Apache Flink offers low-latency processing and stateful computations, ideal for complex analytics and event-driven applications. Apache Spark Streaming extends Spark's capabilities to process live data streams in micro-batches, integrating seamlessly with batch processing jobs. Amazon Kinesis provides a managed service for scalable ingestion, processing, and analysis of streaming data, supporting real-time insights at scale. Apache Storm offers fault-tolerant, distributed real-time computation for processing unbounded data streams with guaranteed message processing (Fragkoulis M.et al.,2024). Streamlio (formerly Heron) excels in low-latency processing and exactly once semantics, making it suitable for critical CPS applications. These frameworks empower CPS to perform real-time analytics, complex event processing, and anomaly detection, enhancing operational efficiency and decision-making capabilities across various domains.

Machine learning (ML) and artificial intelligence (AI) are essential components of real-time analytics in Cyber-Physical Systems (CPS), enabling these systems to process and analyse data swiftly to derive actionable insights and make informed decisions. ML algorithms within CPS continuously learn from incoming data streams to predict outcomes, classify events, and detect anomalies in real-time. This predictive capability is crucial for proactive maintenance, optimizing resource allocation, and enhancing operational efficiency. AI techniques like natural language processing (NLP) facilitate real-time interaction with human-generated data, enabling applications such as chatbots and sentiment analysis for immediate decision support. Moreover, AI-driven reinforcement learning empowers CPS to autonomously adapt and optimize operations based on real-time feedback, enhancing system responsiveness and reliability (Bharadiya P.2023). By leveraging ML and AI for real-time analytics,

CPS can achieve greater efficiency, reliability, and innovation across various domains including manufacturing, healthcare, transportation, and smart cities.

6. APPLICATION OF CPS WITH BLOCKCHAIN INTEGRATION

Healthcare: Cyber-Physical Systems (CPS) are revolutionizing healthcare through real-time data from smart devices and advanced robotics, improving surgical precision and recovery times. Blockchain integration can enhance data security and traceability, ensuring accurate medical records and tamper-proof information sharing (Ali R.et al.,2022). This fusion promises faster patient care, reduced errors, and a growing market but also presents challenges such as job displacement, environmental impact, and the need for updated privacy regulations.

Agriculture and Food Supply: CPS is transforming agriculture with sensors and autonomous machines for precision farming (Yadav, V.et al.,2022). Blockchain can track food safety from farm to table, ensuring transparency and reducing fraud. This technology promises environmental benefits, efficient resource use, and job creation in tech roles, though it may lead to job losses in traditional farming and raise privacy concerns with smart packaging systems.

Manufacturing: In manufacturing, CPS and blockchain enable smart factories with customized products and efficient production. Blockchain can ensure data integrity across manufacturing networks and support secure end-to-end engineering. This advancement is expected to boost economy and increase high-skilled jobs but may also increase power consumption and disrupt existing business models.

Energy and Critical Infrastructure: Blockchain-enhanced CPS will create smart grids and decentralized energy systems, allowing for efficient grid management and secure energy transactions. This integration supports environmental benefits and self-sustaining communities but may widen the digital divide and raise privacy concerns (Rahman Z.et al.,2021). Blockchain will enable remote energy management and support shifts from urban to rural living.

Logistics and Transport: CPS, combined with blockchain, will advance logistics and transport through automation and enhanced tracking systems. Blockchain can provide secure data on vehicle operations and logistics chains, optimizing efficiency and safety (Chen Y.et al.,2022). This integration promises reduced emissions and improved accessibility, though it may disrupt current business models and insurance systems.

7. ETHICAL CONSIDERATIONS OF CYBER-PHYSICAL SYSTEMS (CPS)

The integration of Cyber-Physical Systems (CPS) into society offers substantial benefits, including improved efficiency, sustainability, and the creation of new markets. Expected advantages include automated vehicles that enhance traffic flow and reduce pollution, mass-customized products that align with consumer preferences and reduce waste, and telecare systems that support independent living for the elderly and sick (Sobb T.et al.,2023). CPS also promises smart aids for disabled individuals, optimized agricultural practices, and enhanced safety through drones and search-and-rescue robots. However, potential unintended consequences, such as increased waste from the widespread use of 3D printing, must be anticipated to ensure these technologies provide the intended benefits (Grady C.et al.,2021).

As CPS systems become more widespread, they will reshape the job market by eliminating some roles and creating new ones, such as in robot maintenance and human-robot interaction. This shift raises concerns about whether routine tasks should be delegated to robots, potentially diminishing the personal satisfaction derived from meaningful work. Additionally, safety, liability, and privacy issues must be addressed, particularly as robots operate closely with humans and collect extensive data (Khargonekar P. and Sampath M. 2020). Determining responsibility for system failures and protecting personal information are critical challenges that need to be managed to fully realize the benefits of CPS while mitigating negative impacts.

8. IMPACT OF FOG COMPUTING AND 5G ON CYBER-PHYSICAL SYSTEMS

The adoption of fog computing and 5G will profoundly impact Cyber-Physical Systems (CPS) by enhancing their efficiency and responsiveness. Fog computing will bring data processing and analysis closer to the source, reducing latency and enabling real-time decision-making crucial for CPS applications such as smart grids and autonomous systems (Salim N.et al.,2023). 5G will provide ultra-fast, low-latency connectivity, supporting the high-bandwidth needs of CPS for seamless data transmission and coordination in dynamic environments. Concurrently, Wi-Fi 6 will offer improved network capacity and efficiency, accommodating a higher density of connected devices and ensuring robust performance in densely populated areas (Segura D.et al.,2024). Together, these technologies will enable more agile, reliable, and scalable CPS solutions, driving innovations across various domains while also introducing new challenges in terms of security, data management, and infrastructure requirements.

9. FUTURE TRENDS AND INNOVATIONS

The future of Cyber-Physical Systems (CPS) is poised for significant advancements, driven by cutting-edge technologies and methodologies. Future progress in machine learning (ML) will greatly enhance real-time analytics within CPS. Techniques like reinforcement learning and online learning will enable these systems to continuously adapt to dynamic environments. Sophisticated ML models will support complex decision-making processes and predictive analytics with increased accuracy and efficiency, making CPS more responsive and intelligent. Edge AI and fog computing are set to play pivotal roles in advancing CPS capabilities. By performing advanced analytics locally, edge devices will reduce latency and bandwidth usage, enabling real-time decision-making at the network's edge. Fog computing will integrate these edge devices into a cohesive framework, supporting distributed data processing and analytics closer to the data source, thereby enhancing system efficiency and responsiveness (Iftikhar S.et al.,2023).

Blockchain technology will be increasingly integrated into CPS to bolster security, transparency, and trust in data transactions. Smart contracts will automate secure transactions and agreements, while blockchain's immutable ledger will ensure data integrity and traceability across CPS networks. This will be particularly beneficial in sectors like supply chain management, healthcare data sharing, and decentralized energy grids. New standards and protocols will streamline interoperability and data exchange among diverse CPS components. Standards such as OPC UA and emerging communication protocols like 5G and Wi-Fi 6 will enable seamless integration of devices and systems. These protocols will support efficient data transmission, real-time analytics, and interoperability across heterogeneous CPS environments. These trends and innovations in CPS are set to transform industries by creating more intelligent, autonomous, and efficient systems. By leveraging advancements in ML, edge AI, blockchain integration, and standardized protocols, CPS will evolve into more resilient, secure, and interconnected frameworks capable of addressing complex challenges and driving innovation across various domains.

8. CONCLUSION

Cyber-Physical Systems (CPS) integrate physical processes with computational capabilities, enabling real-time monitoring, analysis, and control. This introduction delves into CPS, defining its components and emphasizing the importance of real-time data analytics. It explores diverse applications across healthcare, transportation, manufacturing, and smart cities, showcasing how CPS optimize operations and enhance efficiency. Fundamental to CPS is real-time data analytics, which involves immediate processing of data streams from sensors and IoT devices. Challenges such as scalability, latency, and data heterogeneity are addressed through advanced techniques like edge computing and standardized protocols. Looking forward, CPS will evolve with advancements in machine learning, edge AI, and blockchain integration, fostering smarter, more resilient systems across various sectors.

REFERENCES

Akkaya, I. (2016). Data-driven cyber-physical systems via real-time stream analytics and machine learning (Doctoral dissertation, UC Berkeley)

Aldoseri, A., Al-Khalifa, K. N., & Hamouda, A. M. (2023). Re-thinking data strategy and integration for artificial intelligence: Concepts, opportunities, and challenges. *Applied Sciences (Basel, Switzerland)*, 13(12), 7082. DOI: 10.3390/app13127082

Ali, R. A., Ali, E. S., Mokhtar, R. A., & Saeed, R. A. (2022). Blockchain for IoT-based cyber-physical systems (CPS): Applications and challenges. In Blockchain based internet of things (pp. 81-111)

Bharadiya, J. P. (2023). Machine learning and AI in business intelligence: Trends and opportunities. [IJC]. *International Journal of Computer*, 48(1), 123–134.

Chen, W., Milosevic, Z., Rabhi, F. A., & Berry, A. (2023). Real-Time Analytics: Concepts, Architectures and ML/AI Considerations. *IEEE Access : Practical Innovations, Open Solutions*, 11, 71634–71657. DOI: 10.1109/ACCESS.2023.3295694

Chen, Y., Lu, Y., Bulysheva, L., & Kataev, M. Y. (2022). Applications of blockchain in industry 4.0: A review. *Information Systems Frontiers*, •••, 1–15. DOI: 10.1007/s10796-022-10248-7

Curry, E., & Curry, E. (2020). Fundamentals of real-time linked dataspaces. In Real-time linked dataspaces: Enabling data ecosystems for intelligent systems (pp. 63-80) DOI: 10.1007/978-3-030-29665-0_4

Dafflon, B., Moalla, N., & Ouzrout, Y. (2021). The challenges, approaches, and used techniques of CPS for manufacturing in Industry 4.0: A literature review. *International Journal of Advanced Manufacturing Technology*, 113(7-8), 2395–2412. DOI: 10.1007/s00170-020-06572-4

Fragkoulis, M., Carbone, P., Kalavri, V., & Katsifodimos, A. (2024). A survey on the evolution of stream processing systems. *The VLDB Journal*, 33(2), 507–541. DOI: 10.1007/s00778-023-00819-8

Garcia, A. M., Griebler, D., Schepke, C., & Fernandes, L. G. (2023). SPBench: A framework for creating benchmarks of stream processing applications. *Computing*, 105(5), 1077–1099. DOI: 10.1007/s00607-021-01025-6

Grady, C., Rajtmajer, S., & Dennis, L. (2021). When smart systems fail: The ethics of cyber–physical critical infrastructure risk. *IEEE Transactions on Technology and Society*, 2(1), 6–14. DOI: 10.1109/TTS.2021.3058605

Huang, L., Qin, J., Zhou, Y., Zhu, F., Liu, L., & Shao, L. (2023). Normalization techniques in training DNNs: Methodology, analysis and application. *IEEE Transactions on Pattern Analysis and Machine Intelligence*, 45(8), 10173–10196. DOI: 10.1109/TPAMI.2023.3250241 PMID: 37027763

Hung, C. C., Tu, M. Y., Chien, T. W., Lin, C. Y., Chow, J. C., & Chou, W. (2023). The model of descriptive, diagnostic, predictive, and prescriptive analytics on 100 top-cited articles of nasopharyngeal carcinoma from 2013 to 2022: Bibliometric analysis. *Medicine*, 102(6), e32824. DOI: 10.1097/MD.0000000000032824 PMID: 36820592

Iftikhar, S., Gill, S. S., Song, C., Xu, M., Aslanpour, M. S., Toosi, A. N., Du, J., Wu, H., Ghosh, S., Chowdhury, D., Golec, M., Kumar, M., Abdelmoniem, A. M., Cuadrado, F., Varghese, B., Rana, O., Dustdar, S., & Uhlig, S. (2023). AI-based fog and edge computing: A systematic review, taxonomy and future directions. *Internet of Things : Engineering Cyber Physical Human Systems*, 21, 100674. DOI: 10.1016/j.iot.2022.100674

Khan, R., Usman, M., & Moinuddin, M. (2024). From raw data to actionable insights: Navigating the world of data analytics. *International Journal of Advanced Engineering Technologies and Innovations*, 1(4), 142–166.

Khargonekar, P. P., & Sampath, M. (2020). A framework for ethics in cyber-physical-human systems. *IFAC-PapersOnLine*, 53(2), 17008–17015. DOI: 10.1016/j.ifacol.2020.12.1251

Lesch, V., Züfle, M., Bauer, A., Iffländer, L., Krupitzer, C., & Kounev, S. (2023). A literature review of IoT and CPS—What they are, and what they are not. *Journal of Systems and Software*, 200, 111631. DOI: 10.1016/j.jss.2023.111631

Li, Y., & Mardani, A. (2023). Digital twins and blockchain technology in the industrial Internet of Things (IIoT) using an extended decision support system model: Industry 4.0 barriers perspective. *Technological Forecasting and Social Change*, 195, 122794. DOI: 10.1016/j.techfore.2023.122794

Pishdad-Bozorgi, P., Gao, X., & Shelden, D. R. (2020). Introduction to cyber-physical systems in the built environment. In Construction 4.0 (pp. 23-41) DOI: 10.1201/9780429398100-2

Rahman, Z., Khalil, I., Yi, X., & Atiquzzaman, M. (2021). Blockchain-based security framework for a critical industry 4.0 cyber-physical system. *IEEE Communications Magazine*, 59(5), 128–134. DOI: 10.1109/MCOM.001.2000679

Raji, M. A., Olodo, H. B., Oke, T. T., Addy, W. A., Ofodile, O. C., & Oyewole, A. T. (2024). Real-time data analytics in retail: A review of USA and global practices. GSC Advanced Research and Reviews, 18(3), 059-065

Salim, N., Singh, M. S., Abed, A. T., & Islam, M. T. (2023). 4X4 MIMO slot antenna spanner shaped low mutual coupling for Wi-Fi 6 and 5G communications. *Alexandria Engineering Journal*, 78, 141–148. DOI: 10.1016/j.aej.2023.07.042

Scquizzato, S. (2024). Data integration and data-driven evaluation for the non-profit sector: A study case: Response innovation lab

Segura, D., Damsgaard, S. B., Kabaci, A., Mogensen, P., Khatib, E. J., & Barco, R. (2024). An empirical study of 5G, Wi-Fi 6 and multi-connectivity scalability in an indoor industrial scenario. *IEEE Access : Practical Innovations, Open Solutions*, 12, 74406–74416. DOI: 10.1109/ACCESS.2024.3404870

Sharma, R., & Garg, P. (2024). Fundamentals of data analytics and lifecycle. In *Data Analytics and Machine Learning: Navigating the Big Data Landscape* (pp. 19–37). Springer Nature Singapore. DOI: 10.1007/978-981-97-0448-4_2

Shete, D., & Khobragade, P. (2023, September). An empirical analysis of different data visualization techniques from statistical perspective. In AIP Conference Proceedings (Vol. 2839, No. 1). AIP Publishing DOI: 10.1063/5.0167670

Shollo, A., & Galliers, R. D. (2024). Constructing actionable insights: The missing link between data, artificial intelligence, and organizational decision-making. In *Research Handbook on Artificial Intelligence and Decision Making in Organizations* (pp. 195–213). Edward Elgar Publishing. DOI: 10.4337/9781803926216.00020

Sobb, T., Turnbull, B., & Moustafa, N. (2023). A holistic review of cyber–physical–social systems: New directions and opportunities. *Sensors (Basel)*, 23(17), 7391. DOI: 10.3390/s23177391 PMID: 37687846

Strengholt, P. (2023). *Data management at scale*. O'Reilly Media, Inc.

Yadav, V. S., Singh, A. R., Raut, R. D., Mangla, S. K., Luthra, S., & Kumar, A. (2022). Exploring the application of Industry 4.0 technologies in the agricultural food supply chain: A systematic literature review. *Computers & Industrial Engineering*, 169, 108304. DOI: 10.1016/j.cie.2022.108304

Yu, B., Zhou, J., & Hu, S. (2020). Cyber-physical systems: An overview. In Big data analytics for cyber-physical systems (pp. 1-11)

Yu, T., & Wang, X. (2022). Real-time data analytics in Internet of Things systems. In *Handbook of real-time computing* (pp. 541–568). Springer Nature Singapore. DOI: 10.1007/978-981-287-251-7_38

Ziehn, A., Grulich, P. M., Zeuch, S., & Markl, V. (2024). Bridging the gap: Complex event processing on stream processing systems. In EDBT (pp. 447-460)

Compilation of References

Abawajy, J. H., & Kim, T. H. (2019). Artificial intelligence and cybersecurity: Trends, challenges, and future directions. *Journal of Cybersecurity*, 5(1), 1–14.

Aboualola, M., Abualsaud, K., Khattab, T., Zorba, N., & Hassanein, H. S. (2023, January 1). Edge Technologies for Disaster Management: A Survey of Social Media and Artificial Intelligence Integration. *IEEE Access : Practical Innovations, Open Solutions*, 11, 73782–73802. DOI: 10.1109/ACCESS.2023.3293035

Abu-Dabaseh, F., & Alshammari, E. (2018). Automated Penetration Testing: An Overview. Dhinaharan Nagamalai et al. (Eds): NATL, CSEA, DMDBS, Fuzzy, ITCON, NSEC, COMIT – 2018. pp. 121–129. DOI: DOI: 10.5121/csit.2018.80610

Adedeji, K. B., & Hamam, Y. (2020, November 17). Cyber-Physical Systems for Water Supply Network Management: Basics, Challenges, and Roadmap. *Sustainability (Basel)*, 12(22), 9555–9555. DOI: 10.3390/su12229555

Ahmed, A A., Nazzal, M A., & Darras, B M. (2021, October 29). Cyber-Physical Systems as an Enabler of Circular Economy to Achieve Sustainable Development Goals: A Comprehensive Review. Springer Science+Business Media, 9(3), 955-975. DOI: 10.1007/s40684-021-00398-5

Akhtar, N., Rahman, S., Sadia, H., & Perwej, Y. (2021). A Holistic Analysis of Medical Internet of Things (MIoT). *Journal of Information and Computational Science*, 11(4), 209–222.

Akhuseyinoglu, N B., & Joshi, J. (2020, September 1). A constraint and risk-aware approach to attribute-based access control for cyber-physical systems. Elsevier BV, 96, 101802-101802. DOI: 10.1016/j.cose.2020.101802

Akkaya, I. (2016). Data-driven cyber-physical systems via real-time stream analytics and machine learning (Doctoral dissertation, UC Berkeley)

Aldoseri, A., Al-Khalifa, K. N., & Hamouda, A. M. (2023). Re-thinking data strategy and integration for artificial intelligence: Concepts, opportunities, and challenges. *Applied Sciences (Basel, Switzerland)*, 13(12), 7082. DOI: 10.3390/app13127082

Alguliyev, R., İmamverdiyev, Y., & Sukhostat, L. (2018). Cyber-physical systems and their security issues. *Computers in Industry*, 100, 212–223. DOI: 10.1016/j.compind.2018.04.017

Ali, N., Hussain, M., Kim, Y., & Hong, J. (2020, January 1). A Generic Framework For Capturing Reliability in Cyber Physical Systems. Cornell University. https://doi.org//arxiv.2010.05490DOI: 10.48550

Ali, R. A., Ali, E. S., Mokhtar, R. A., & Saeed, R. A. (2022). Blockchain for IoT-based cyber-physical systems (CPS): Applications and challenges. In Blockchain based internet of things (pp. 81-111)

Ali, M., Naeem, F., Adam, N., Kaddoum, G., Huda, N. U., Adnan, M., & Tariq, M. U. (2023, January 1). Integration of Data Driven Technologies in Smart Grids for Resilient and Sustainable Smart Cities: A Comprehensive Review. Cornell University. https://doi.org/DOI: 10.48550/arXiv.2301

Ali, O., Abdelbaki, W., Anup, S., Ersin, E., Abdallah, M., & Dwivedi, Y. K. (2023). Journal of Innovation. *Journal of Innovation and Knowledge*, 8. Advance online publication. DOI: 10.1016/j.jik.2023.100333

Allahrakha, N. (2023). Balancing Cyber-security and Privacy: Legal and Ethical Considerations in the Digital Age. *Legal Issues in the Digital Age*, 4(2), 78–121. DOI: 10.17323/10.17323/2713-2749.2023.2.78.121

Allam, Z., Sharifi, A., Bibri, S., Jones, D., & Krogstie, J. (2022). The metaverse as a virtual form of smart cities: Opportunities and challenges for environmental, economic, and social sustainability in urban futures. *Smart Cities*, 5(3), 771–801. DOI: 10.3390/smartcities5030040

Alshammari, K., Beach, T., & Rezgui, Y. (2021, August 11). Cybersecurity for Digital Twins in the Built Environment: Research Landscape, Industry Attitudes and Future Direction. Hilaris, 15(8), 382-387. https://publications.waset.org/10012172/pdf

Al-Suhimat, R. I., Ibrahim, A., Maaitah, N. O., & Al-Dmour, N. A. (2024). Review of Security Challenges Encountered in Internet of Things Technology. In 2024 2nd International Conference on Cyber Resilience (ICCR) (pp. 1-6). IEEE. DOI: 10.1109/ICCR61006.2024.10533052

Alzubi, K. M., Alaloul, W. S., & Qureshi, A. H. (2022). Applications of cyber-physical systems in construction projects. In *Cyber-Physical Systems in the Construction Sector* (pp. 153–171). CRC Press. DOI: 10.1201/9781003190134-9

Amel-Zadeh, A., & Serafeim, G. (2018). Why and how investors use ESG information: Evidence from a global survey. *Financial Analysts Journal*, 74(3), 87–103. DOI: 10.2469/faj.v74.n3.2

Angelopoulos, A., Michailidis, E. T., Νομικός, N., Trakadas, P., Hatziefremidis, A., Voliotis, S., & Zahariadis, T. (2019, December 23). Tackling Faults in the Industry 4.0 Era—A Survey of Machine-Learning Solutions and Key Aspects. *Sensors (Basel)*, 20(1), 109–109. DOI: 10.3390/s20010109 PMID: 31878065

Anthi, E., Williams, L., Rhode, M., Burnap, P., & Wedgbury, A. (2021). Adversarial attacks on machine learning cybersecurity defences in industrial control systems. *Journal of Information Security and Applications*, 58, 102717. DOI: 10.1016/j.jisa.2020.102717

Arulkumaran, K., Deisenroth, M. P., & Brundage, M. (2017). *Deep reinforcement learning: a brief survey*. IEEE., DOI: 10.1109/MSP.2017.2743240

Aslan, Ö., Aktuğ, S. S., Ozkan-Okay, M., Yilmaz, A. A., & Akin, E. (2023). A comprehensive review of cyber security vulnerabilities, threats, attacks, and solutions. *Electronics (Basel)*, 12(6), 1333. DOI: 10.3390/electronics12061333

Atzori, L., Iera, A., & Morabito, G. (2010). The Internet of Things: A survey. *Computer Networks*, 54(15), 2787–2805. DOI: 10.1016/j.comnet.2010.05.010

Ayerdi, J., Valle, P., Segura, S., Arrieta, A., Sagardui, G., & Arratibel, M. (2023). Performance-Driven metamorphic testing of Cyber-Physical systems. *IEEE Transactions on Reliability*, 72(2), 827–845. DOI: 10.1109/TR.2022.3193070

Azaria, A., Ekblaw, A., Vieira, T., & Lippman, A. (2016). MedRec: Using Blockchain for Medical Data Access and Permission Management. *2016 2nd International Conference on Open and Big Data (OBD)*, 25–30. https://doi.org/DOI: 10.1109/OBD.2016.11

Bacudio, A. G., Yuan, X., Chu, B. T. B., & Jones, M. (2011). An Overview of Penetration Testing. *International Journal of Network Security & its Applications*, 3(6), 19–38. DOI: 10.5121/ijnsa.2011.3602

Baheti, R. and Gill, H. (2011). Cyber-physical systems. The Impact of Control Technology, 161-166.

Baker, S. B., Xiang, W., & Atkinson, I. (2017). Internet of Things for Smart Healthcare: Technologies, Challenges, and Opportunities. *IEEE Access : Practical Innovations, Open Solutions*, 5, 26521–26544. DOI: 10.1109/ACCESS.2017.2775180

Baliyan, A., Kaswan, K. S., Kumar, N., Upreti, K., & Kannan, R. (Eds.). (2023). Cyber Physical Systems: Concepts and Applications.

Banik, S., Ramachandran, T., Bhattacharya, A., & Bopardikar, S. D. (2023). Automated Adversary-in-the-Loop Cyber-Physical Defense Planning. *ACM Transactions on Cyber-Physical Systems*, 7(3), 1–25. DOI: 10.1145/3596222

Bashendy, M., Tantawy, A., & Erradi, A. (2023). Intrusion response systems for cyber-physical systems: A comprehensive survey. *Computers & Security*, 124, 102984. DOI: 10.1016/j.cose.2022.102984

Bebbington, J., & Unerman, J. (2020). Advancing research into accounting and the UN Sustainable Development Goals. *Accounting, Auditing & Accountability Journal*, 33(7), 1657–1670. DOI: 10.1108/AAAJ-05-2020-4556

Belli, S., Mugnaini, R., Baltà, J., & Abadal, E. (2020). Coronavirus mapping in scientific publications: When science advances rapidly and collectively, is access to this knowledge open to society? *Scientometrics*, 124(3), 2661–2685. DOI: 10.1007/s11192-020-03590-7 PMID: 32836526

Berger, S., Häckel, B., & Häfner, L. (2021). Organizing self-organizing systems: A terminology, taxonomy, and reference model for enti- ties in cyber-physical production systems. *Information Systems Frontiers*, 23(2), 391–414. DOI: 10.1007/s10796-019-09952-8

Bertoglio, D. D., & Zorzo, A. F. (2017). Overview and open issues on penetration test. *Journal of the Brazilian Computer Society*, 23(1), 2. Advance online publication. DOI: 10.1186/s13173-017-0051-1

Bhangale, U., Patil, S., Vishwanath, V., Thakker, P., Bansode, A., & Navandhar, D. (2020). Near real-time crowd counting using deep learning approach. *Procedia Computer Science*, 171, 770–779. DOI: 10.1016/j.procs.2020.04.084

Bharadiya, J. P. (2023). Machine learning and AI in business intelligence: Trends and opportunities. [IJC]. *International Journal of Computer*, 48(1), 123–134.

Bhuiyan, M. R., Abdullah, J., Hashim, N., & Farid, F. A. (2022). Video analytics using deep learning for crowd analysis: A review. *Multimedia Tools and Applications*, 81(19), 27895–27922. DOI: 10.1007/s11042-022-12833-z

Bittl, S. (2014). Attack potential and efficient security enhancement of automotive bus networks using short MACs with rapid key change. In Communication Technologies for Vehicles: 6th International Workshop, Nets4Cars/Nets4Trains/Nets4Aircraft 2014, Offenburg, Germany, May 6-7, 2014. [Springer International Publishing.]. *Proceedings*, 6, 113–125.

Bodkhe, U., Mehta, D., Tanwar, S., Bhattacharya, P., Singh, P. K., & Hong, W. (2020). A survey on Decentralized Consensus Mechanisms for Cyber Physical Systems. *IEEE Access : Practical Innovations, Open Solutions*, 8, 54371–54401. DOI: 10.1109/ACCESS.2020.2981415

Broo, D. G., Boman, U., & Törngren, M. (2021). Cyber-physical systems research and education in 2030: Scenarios and strategies. *Journal of Industrial Information Integration*, 21, 100192. DOI: 10.1016/j.jii.2020.100192

Brown, T., & Garcia, L. (2019). Integrating diverse healthcare data sources for comprehensive analysis. *Health IT Journal*, 34(3), 211–223. DOI: 10.1016/j.healthit.2019.03.008

Brown, T., & Patel, S. (2023). Managing healthcare data velocity with streaming platforms: A case study on Apache Kafka and Flink. *Journal of Healthcare Engineering*, 40(3), 267–279. DOI: 10.1016/j.jhe.2023.04.012

Bujorianu, M., & Barringer, H. (2009). An integrated specification logic for cyber-physical systems. In *Proceedings of 14th IEEE International Conference on Engineering of Complex Computer Systems*, 291-300. DOI: 10.1109/ICECCS.2009.36

Burg, A., Chattopadhyay, A., & Lam, K. (2018, January 1). Wireless Communication and Security Issues for Cyber–Physical Systems and the Internet-of-Things. *Proceedings of the IEEE*, 106(1), 38–60. DOI: 10.1109/JPROC.2017.2780172

Burrows, M., Abadi, M., & Needham, R. (1989). A logic of authentication. *Operating Systems Review*, 23(5), 1–13. DOI: 10.1145/74851.74852

Cabello, J. C., Karimipour, H., Jahromi, A. N., Dehghantanha, A., & Parizi, R. M. (2020). Big-data and cyber-physical systems in healthcare: Challenges and opportunities. Handbook of Big Data Privacy, 255-283.

Cachada, A., Barbosa, J., Leitao, P., Deusdado, L., Costa, J., Teixeira, J., . . . Moreira, P. M. (2019). Development of ergonomic user interfaces for the human integration in cyber-physical systems. In 2019 IEEE 28th International Symposium on Industrial Electronics (ISIE) (pp. 1632-1637). IEEE DOI: 10.1109/ISIE.2019.8781101

Cai, Y., & Qi, D. (2017, July 1). Physical control framework and protocol design for cyber-physical control system. *International Journal of Distributed Sensor Networks*, 13(7), 155014771772269–155014771772269. DOI: 10.1177/1550147717722692

California Attorney General. (n.d.). *California Consumer Privacy Act (CCPA)*. https://oag.ca.gov/privacy/ccpa

Calinescu, R., Camara, J., & Paterson, C. (2019). Socio-cyber-phys- ical systems: Models, opportunities, open challenges. In *5th IEEE/ACM International Workshop on Software Engineering for Smart Cyber-Physical Systems (SEsCPS)* (pp. 2–6). IEEE. https://doi.org/DOI: 10.1109/SEsCPS.2019.00008

Cannavo, A., & Lamberti, F. (2020). "How blockchain, virtual reality, and augmented reality are converging, and why." (2019). *IEEE Consumer Electronics Magazine*, 10(5), 6–13. DOI: 10.1109/MCE.2020.3025753

Castaño, F., Strzelczak, S., Villalonga, A., Haber, R. E., & Kossakowska, J. (2019). Sensor reliability in cyber-physical systems using internet-of-things data: A review and case study. *Remote Sensing (Basel)*, 11(19), 2252. DOI: 10.3390/rs11192252

Cavallari, R., Martelli, F., Rosini, R., Buratti, C., & Verdone, R. (2014). A Survey on Wireless Body Area Networks: Technologies and Design Challenges. *IEEE Communications Surveys and Tutorials*, 16(3), 1635–1657. DOI: 10.1109/SURV.2014.012214.00007

Cavelty, D. M. (2018), *The Routledge Handbook of New Security Studies*, 2.

Cederbladh, J., Eramo, R., Muttillo, V., & Strandberg, P. E. (2024). Experiences and challenges from developing cyber-physical systems in industry-academia collaboration. *Software, Practice & Experience*, 54(6), 1193–1212. DOI: 10.1002/spe.3312

Chakraborty, T., Reddy K S, U., Naik, S. M., Panja, M., & Manvitha, B. (2024). Ten years of generative adversarial nets (GANs): A survey of the state-of-the-art. *Machine Learning: Science and Technology*, 5(1), 011001. DOI: 10.1088/2632-2153/ad1f77

Chamberlain, L B., Davis, L E., Stanley, M., & Gattoni, B. (2020, May 1). Automated Decision Systems for Cybersecurity and Infrastructure Security. DOI: 10.1109/SPW50608.2020.00048

Chaudhuri, A., & Kahyaoğlu, S B. (2023, March 8). CYBERSECURITY ASSURANCE IN SMART CITIES: A RISK MANAGEMENT PERSPECTIVE. Taylor & Francis, 67(4), 1-22. DOI: 10.1080/07366981.2023.2165293

Chen, J., de Mendonça, J. L. V., Ayele, B. S., Bekele, B. N., Jalili, S., Sharma, P., . . . Jeannin, J. B. (2024). Synchronous Programming with Refinement Types. Proceedings of the ACM on Programming Languages, 8(ICFP), 938-972 DOI: 10.1145/3674657

Chen, C., Hasan, M., & Mohan, S. (2018, December 10). Securing Real-Time Internet-of-Things. *Sensors (Basel)*, 18(12), 4356–4356. DOI: 10.3390/s18124356 PMID: 30544673

Cheng, B., Ioannou, I., & Serafeim, G. (2014). Corporate social responsibility and access to finance. *Strategic Management Journal*, 35(1), 1–23. DOI: 10.1002/smj.2131

Chen, K., Loy, C. C., Gong, S., & Xiang, T. (2012). *Feature mining for localised crowd counting*. BMVC., DOI: 10.5244/C.26.21

Chen, P. T., Lin, C. L., & Wu, W. N. (2020). Big data management in healthcare: Adoption challenges and implications. *International Journal of Information Management*, 53, 102078. DOI: 10.1016/j.ijinfomgt.2020.102078

Chen, W., Milošević, Z., Rabhi, F., & Berry, A. J. (2023, January 1). Real-Time Analytics: Concepts, Architectures, and ML/AI Considerations. *IEEE Access : Practical Innovations, Open Solutions*, 11, 71634–71657. DOI: 10.1109/ACCESS.2023.3295694

Chen, Y., Lu, Y., Bulysheva, L., & Kataev, M. Y. (2022). Applications of blockchain in industry 4.0: A review. *Information Systems Frontiers*, •••, 1–15. DOI: 10.1007/s10796-022-10248-7

Cheong, K. H., Poeschmann, S., Lai, J. W., Koh, J., Acharya, U. R., Yu, S. C. M., & Tang, K. J. W. (2019). Practical automated video analytics for crowd monitoring and counting. *IEEE Access : Practical Innovations, Open Solutions*, 7, 183252–183261. DOI: 10.1109/ACCESS.2019.2958255

Chikurtev, D., Ivanov, V., Yosifova, V., & Dimitrov, D. (2022). Cyber-physical system for intelligent control of infrared heating. *IFAC-PapersOnLine*, 55(11), 37–41. DOI: 10.1016/j.ifacol.2022.08.045

Chohlas-Wood, A., Coots, M., Goel, S., & Nyarko, J. (2023, July 24). Designing equitable algorithms. *Nature Computational Science*, 3(7), 601–610. DOI: 10.1038/s43588-023-00485-4 PMID: 38177749

Clark, D., & Martinez, L. (2023). The necessity of real-time analytics in healthcare data management. *Journal of Healthcare Informatics Research*, 30(2), 221–233. DOI: 10.1080/17538157.2023.1945123

Clark, G. L., Feiner, A., & Viehs, M. (2015). *From the stockholder to the stakeholder: How sustainability can drive financial outperformance*. Oxford University Press.

Claure, R. E. M., Heynssens, J. B., Burke, I., Alam, M., & Cambou, B. (2024). Enhancing cyber-physical systems (CPS) robustness through sensor-pair health indicator. In Autonomous Systems: Sensors, Processing, and Security for Ground, Air, Sea, and Space Vehicles and Infrastructure 2024 (Vol. 13052, pp. 21-28). SPIE

Conversation on Intellectual Property and Artificial Intelligence. (2019), *WIPO*, 2, https://www.wipo.int/edocs/mdocs/mdocs/en/wipo_ip_ai_ge_19/wipo_ip_ai_ge_19_inf_4.pdf

Cozzoli, N., Salvatore, F. P., Faccilongo, N., & Milone, M. (2022). How can big data analytics be used for healthcare organization management? Literary framework and future research from a systematic review. *BMC Health Services Research*, 22(1), 809. Advance online publication. DOI: 10.1186/s12913-022-08167-z PMID: 35733192

Curry, E., & Curry, E. (2020). Fundamentals of real-time linked dataspaces. In Real-time linked dataspaces: Enabling data ecosystems for intelligent systems (pp. 63-80) DOI: 10.1007/978-3-030-29665-0_4

Cyberspace Administration of China. (2021, August 25). *Regulation on the management of network data security*. https://www.cac.gov.cn/2021-08/25/c_1631480920680924.htm

Dabla, P. K., Sharma, S., Dabas, A., Tyagi, V., Agrawal, S., Jhamb, U., Begos, D., Upreti, K., & Mir, R. (2022). Ionized Blood Magnesium in Sick Children: An Overlooked Electrolyte. *Journal of Tropical Pediatrics*, 68(2), fmac022. Advance online publication. DOI: 10.1093/tropej/fmac022 PMID: 35265997

Dabla, P. K., Upreti, K., Singh, D., Singh, A., Sharma, J., Dabas, A., Gruson, D., Gouget, B., Bernardini, S., Homsak, E., & Stankovic, S. (2022). Target association rule mining to explore novel paediatric illness patterns in emergency settings. *Scandinavian Journal of Clinical and Laboratory Investigation*, 82(7–8), 595–600. DOI: 10.1080/00365513.2022.2148121 PMID: 36399102

Dafflon, B., Moalla, N., & Ouzrout, Y. (2021). The challenges, approaches, and used techniques of CPS for manufacturing in Industry 4.0: A literature review. *International Journal of Advanced Manufacturing Technology*, 113(7-8), 2395–2412. DOI: 10.1007/s00170-020-06572-4

Dai, T., Chen, Q., Xie, L., & Hu, H. (2020). An Innovative Application of Big Data in Healthcare: Driving Factors, Operation Mechanism and Development Model. *Frontiers in Artificial Intelligence and Applications*, 329(9), 104–113. DOI: 10.3233/FAIA200645

Damaševičius, R., Bačanin, N., & Misra, S. (2023, May 16). From Sensors to Safety: Internet of Emergency Services (IoES) for Emergency Response and Disaster Management. *Multidisciplinary Digital Publishing Institute*, 12(3), 41–41. DOI: 10.3390/jsan12030041

Das, S. K., & Bhattacharjee, S. (2024). Science of Cyber Physical Security in Smart Living CPS Applications. In 2024 IEEE International Conference on Smart Computing (SMARTCOMP) (pp. 5-5). IEEE de Niz, D., Andersson, B., Klein, M., Lehoczky, J., Vasudevan, A., Kim, H., & Moreno, G. Mixed-Trust Computing: Safe and Secure Real-Time Systems. ACM Transactions on Cyber-Physical Systems DOI: 10.1109/SMARTCOMP61445.2024.00022

Das, S., & Chhatlani, C. K. (2022). Unlocking the Potential of Big Data Analytics for Enhanced Healthcare Decision-Making : A Comprehensive Review of Applications and Challenges. Journal of Contemporary Healthcare Analytics J.

Dave, R., Seliya, N., & Siddiqui, N. (2021, October 29). The Benefits of Edge Computing in Healthcare, Smart Cities, and IoT. *The Benefits of Edge Computing in Healthcare, Smart Cities, and IoT.*, 9(1), 23–34. DOI: 10.12691/jcsa-9-1-3

De-Arteaga, M., Fogliato, R., & Chouldechova, A. (2020, April 21). A Case for Humans-in-the-Loop: Decisions in the Presence of Erroneous Algorithmic Scores. DOI: 10.1145/3313831.3376638

Debala Chanu, A., & Sharma, B. (2019). TCP Connection Monitoring System. *International Journal of Computational Intelligence & IoT*, 2(1).

Dianwei, W.. (2022). *Research on Metaverse: Concept, development and standard system*. IEEE Xplore.

Dias, C., Santos, M. F., & Portela, F. (2020). A SWOT analysis of big data in healthcare. ICT4AWE 2020 - Proceedings of the 6th International Conference on Information and Communication Technologies for Ageing Well and e-Health, (Ict4awe), 250–257. DOI: 10.5220/0009390202560263

Dieterle, E., Dede, C., & Walker, M E. (2022, September 27). The cyclical ethical effects of using artificial intelligence in education. Springer Nature. DOI: 10.1007/s00146-022-01497-w

Dimitrov, D. V. (2016a). Medical Internet of Things and Big Data in Healthcare. *Healthcare Informatics Research*, 22(3), 156–163. DOI: 10.4258/hir.2016.22.3.156 PMID: 27525156

Dolev, D., & Yao, A. (1983). On the security of public key protocols. *IEEE Transactions on Information Theory*, 29(2), 198–208. DOI: 10.1109/TIT.1983.1056650

Dsouza, J., Elezabeth, L., Mishra, V. P., & Jain, R. (2019). Security in Cyber-Physical Systems. *2019 Amity International Conference on Artificial Intelligence (AICAI)*. DOI: 10.1109/AICAI.2019.8701411

Dunkels, A. (2001). Design and Implementation of the lwIP TCP/IP Stack. Swedish Institute of Computer Science, 2(77).

Dunmore, A., Jang-Jaccard, J., Sabrina, F., & Kwak, J. (2023). A comprehensive survey of generative adversarial networks (GANs) in cybersecurity intrusion detection. *IEEE Access : Practical Innovations, Open Solutions*, 11, 76071–76094. DOI: 10.1109/ACCESS.2023.3296707

Eccles, R. G., Ioannou, I., & Serafeim, G. (2014). The impact of corporate sustainability on organizational processes and performance. *Management Science*, 60(11), 2835–2857. DOI: 10.1287/mnsc.2014.1984

Elattar, M. (2019, August 7). Reliable Communications within Cyber-Physical Systems Using the Internet (RC4CPS). Springer Nature. DOI: 10.1007/978-3-662-59793-4

Elayan, H., Shubair, R. M., & Kiourti, A. (2017). Wireless sensors for medical applications: Current status and future challenges. *2017 11th European Conference on Antennas and Propagation (EUCAP)*, 2478–2482. https://doi.org/DOI: 10.23919/EuCAP.2017.7928405

Elhoseny, M., Thilakarathne, N. N., Alghamdi, M. I., Mahendran, R. K., Gardezi, A. A., Weerasinghe, H., & Welhenge, A. (2021). Security and Privacy Issues in Medical Internet of Things: Overview, Countermeasures, Challenges and Future Directions. *Sustainability (Basel)*, 13(21), 21. Advance online publication. DOI: 10.3390/su132111645

Eperjesi, A. (2022), Automated Incident Response: Everything You Need to Know, https://securityboulevard.com/2022/10/automated-incident-response-everything-you-need-to-know

Esposito, C., De Santis, A., Tortora, G., Chang, H., & Choo, K.-K. R. (2018). Blockchain: A Panacea for Healthcare Cloud-Based Data Security and Privacy? *IEEE Cloud Computing*, 5(1), 31–37. DOI: 10.1109/MCC.2018.011791712

Ethernet_Frame. Available online: https://en.wikipedia.org/wiki/Ethernet_frame

European Union General Data Protection Regulation (GDPR). (n.d.). *What is GDPR, the EU's new data protection law?*https://gdpr.eu/what-is-gdpr/

Evans, R., & Harris, M. (2022). Enhancing healthcare decision-making with streaming analytics. *Health Data Management*, 47(3), 345–357. DOI: 10.1109/HDM.2022.1935467

Fadlullah, Z. M., Nishiyama, H., Kato, N., & Fouda, M. M. (2013). Intrusion Detection System (IDS) for /Combating Attacks Against Cognitive Radio Networks. *IEEE Network*, 31(3), 51–56. DOI: 10.1109/MNET.2013.6523809

Farahani, B., Firouzi, F., Chang, V., Badaroglu, M., Constant, N., & Mankodiya, K. (2018). Towards fog-driven IoT eHealth: Promises and challenges of IoT in medicine and healthcare. *Future Generation Computer Systems*, 78, 659–676. DOI: 10.1016/j.future.2017.04.036

Federal Trade Commission (FTC). (n.d.). *Federal Trade Commission (FTC)*.https://www.ftc.gov/

Feng, F., Sun, J., Zhang, L., Cao, C., & Yang, Q. (2016), A support vector machine based naive Bayes algorithm for spam filtering, *IEEE*, https://www.computer.org/csdl/proceedings-article/ipccc/2016/07820655/12OmNzWx0bb

Feng, W., Sun, J., Zhang, L., Cao, C., & Yang, Q. A., (2016). support vector machine based naive Bayes algorithm for spam filtering, *IEEE*, https://www.computer.org/csdl/proceedings-article/ipccc/2016/07820655/12OmNzWx0bb

Feng, F., Zhou, Q., Shen, Z., Yang, X., Han, L., & Wang, J. (2018). *The application of a novel neural network in the detection of phishing websites*. Springer.

Feng, F., Zhou, Q., Shen, Z., Yang, X., Han, L., & Wang, J. (2024). The application of a novel neural network in the detection of phishing websites. *Journal of Ambient Intelligence and Humanized Computing*, 15(3), 1–15. DOI: 10.1007/s12652-018-0786-3

Fernando, J. (2023, November 7). Discounted cash flow (DCF) explained with formula and examples. Investopedia. https://www.investopedia.com/terms/d/dcf.asp

Filios, G., Katsidimas, I., Nikoletseas, S., Panagiotou, S H., & Raptis, T P. (2022, December 9). Agnostic learning for packing machine stoppage prediction in smart factories., 3(3), 793-807. DOI: 10.52953/LEDZ3942

FlexRay Consortium. FlexRay Communications System—Protocol Specification—Version 2.1 Revision A. Available online: https://www.google.com.hk/url?sa=t&rct=j&q=&esrc=s&source=web&cd=1&ved=2ahUKEwi--6TIo5nnAhVjL6YKHSzcDb8QFjAAegQIBBAB&url=https%3A%2F%2Fsvn.ipd.kit.edu%2Fnlrp%2Fpublic%2FFlexRay%2FFlexRay%25E2%2584%25A2%2520Protocol%2520Specification%2520V2.1%2520Rev.A.pdf&usg=AOvVaw0snIyyfkFMHWc7KHRLdS82

Fowler, C. (2015). Virtual reality and learning: Where is the pedagogy? *British Journal of Educational Technology*, 46(2), 412–422. DOI: 10.1111/bjet.12135

Fragkoulis, M., Carbone, P., Kalavri, V., & Katsifodimos, A. (2024). A survey on the evolution of stream processing systems. *The VLDB Journal*, 33(2), 507–541. DOI: 10.1007/s00778-023-00819-8

Franki, V., Majnarić, D., & Višković, A. (2023, January 18). A Comprehensive Review of Artificial Intelligence (AI) Companies in the Power Sector. *Energies*, 16(3), 1077–1077. DOI: 10.3390/en16031077

Friede, G., Busch, T., & Bassen, A. (2015). ESG and financial performance: Aggregated evidence from more than 2000 empirical studies. *Journal of Sustainable Finance & Investment*, 5(4), 210–233. DOI: 10.1080/20430795.2015.1118917

Fruhlinger, J. (2021). *Penetration testing explained: How ethical hackers simulate attacks*. CSO Online. https://www.csoonline.com/article/571697/penetration-testing-explained-how-ethical-hackers-simulate-attacks.html

Frysak, J. (2020, January 1). S-BPM Diagrams as Decision Aids in a Decision Based Framework for CPS Development. Springer Science+Business Media, 23-32. DOI: 10.1007/978-3-030-64351-5_2

Fu-xing, L., Li, L., & Peng, Y. (2021, December 3). Research on Digital Twin and Collaborative Cloud and Edge Computing Applied in Operations and Maintenance in Wind Turbines of Wind Power Farm. IOS Press. DOI: 10.3233/ATDE210263

Gaggatur, J. S. (2023). *Designing and Implementing a Cyber-Physical Systems (CPS) Education Program for Pre-University Students. In 2023 IEEE Technology & Engineering Management Conference-Asia Pacific (TEMSCON-ASPAC)*. IEEE.

Gao, Y., Li, Y., Feng, S., & Gu, L. (2023). Bibliometric and visualization analysis of matrix metalloproteinases in ischemic stroke from 1992 to 2022. *Frontiers in Neuroscience*, 17, 1206793. https://www.frontiersin.org/journals/neuroscience/articles/10.3389/fnins.2023.1206793. DOI: 10.3389/fnins.2023.1206793 PMID: 37483355

Garcia, A. M., Griebler, D., Schepke, C., & Fernandes, L. G. (2023). SPBench: A framework for creating benchmarks of stream processing applications. *Computing*, 105(5), 1077–1099. DOI: 10.1007/s00607-021-01025-6

Garraghan, P., McKee, D., Ouyang, X., Webster, D. E., & Xu, J. (2016). SEED: A Scalable Approach for Cyber-Physical System Simulation. *IEEE Transactions on Services Computing*, 9(2), 199–212. DOI: 10.1109/TSC.2015.2491287

Gartner.com. Elements of a Metaverse. 2022. Available online: https://axveco.com/gartner-metaverse/

Gatouillat, A., Badr, Y., Massot, B., & Sejdić, E. (2018). Internet of Medical Things: A Review of Recent Contributions Dealing With Cyber-Physical Systems in Medicine. *IEEE Internet of Things Journal*, 5(5), 3810–3822. DOI: 10.1109/JIOT.2018.2849014

Ghahramani, M., Zhou, M., Mölter, A., & Pilla, F. (2022, July 15). IoT-Based Route Recommendation for an Intelligent Waste Management System. *IEEE Internet of Things Journal*, 9(14), 11883–11892. DOI: 10.1109/JIOT.2021.3132126

Gilad, M., Fishbein, D., Nave, G., & Packin, N. G. (2023, March 31). Science for policy to protect children in cyberspace. *Science*, 379(6639), 1294–1297. DOI: 10.1126/science.ade9447 PMID: 36996216

Gope, P., & Hwang, T. (2016). BSN-Care: A Secure IoT-Based Modern Healthcare System Using Body Sensor Network. *IEEE Sensors Journal*, 16(5), 1368–1376. DOI: 10.1109/JSEN.2015.2502401

Goss, A., & Roberts, G. S. (2011). The impact of corporate social responsibility on the cost of bank loans. *Journal of Banking & Finance*, 35(7), 1794–1810. DOI: 10.1016/j.jbankfin.2010.12.002

Grace, V., Mofokeng, J T., Olutola, A A., & Morero, M. (2022, March 22). An evaluation of the challenges encountered by the South African police service with regard to the fourth industrial revolution. Ümit Hacıoğlu, 11(2), 447-453. DOI: 10.20525/ijrbs.v11i2.1606

Grady, C., Rajtmajer, S., & Dennis, L. (2021). When smart systems fail: The ethics of cyber–physical critical infrastructure risk. *IEEE Transactions on Technology and Society*, 2(1), 6–14. DOI: 10.1109/TTS.2021.3058605

Gubbi, J., Buyya, R., Marusic, S., & Palaniswami, M. (2013). Internet of Things (IoT): A vision, architectural elements, and future directions. *Future Generation Computer Systems*, 29(7), 1645–1660. DOI: 10.1016/j.future.2013.01.010

Gunawan, G., Nasution, B B., Zarlis, M., Sari, M., Lubis, A R., & Solikhun. (2021, April 1). Design of Earthquake Early Warning System Based on Internet of Thing. IOP Publishing, 1830(1), 012010-012010. DOI: 10.1088/1742-6596/1830/1/012010

Gupta, M., Akiri, C., Aryal, K., Parker, E., & Praharaj, L. (2023). From chatgpt to threatgpt: Impact of generative ai in cybersecurity and privacy. *IEEE Access : Practical Innovations, Open Solutions*, 11, 80218–80245. DOI: 10.1109/ACCESS.2023.3300381

Ha, K. M. (2023). Reviewing the Itaewon Halloween crowd crush, Korea 2022: Qualitative content analysis. *F1000 Research*, 12, 829. DOI: 10.12688/f1000research.135265.1 PMID: 38037564

Halder, S., Afsari, K., & Akanmu, A. (2024, February 10). A Robotic Cyber-Physical System for Automated Reality Capture and Visualization in Construction Progress Monitoring. Cornell University. https://doi.org/DOI: 10.48550/arXiv.2402

Haleem, A., Javaid, M., Singh, R. P., Suman, R., & Rab, S. (2021). Blockchain technology applications in healthcare: An overview. *International Journal of Intelligent Networks*, 2, 130–139. DOI: 10.1016/j.ijin.2021.09.005

Hamacher, Alaric, et al. (2016) "Application of virtual, augmented, and mixed reality to urology." International neurourology journal 20.3: 172.

Han, D.-I. D., Bergs, Y., & Moorhouse, N. (2022). Virtual reality consumer experience escapes: Preparing for the metaverse. *Virtual Reality (Waltham Cross)*, 26(4), 1443–1458. DOI: 10.1007/s10055-022-00641-7

Han, Y., Chen, J., Dou, M., Wang, J., & Feng, K. (2023, May 20). The Impact of Artificial Intelligence on the Financial Services Industry. *The Impact of Artificial Intelligence on the Financial Services Industry.*, 2(3), 83–85. DOI: 10.54097/ajmss.v2i3.8741

Haque, M., Kumar, V. V., Singh, P., Goyal, A. A., Upreti, K., & Verma, A. (2023). A systematic meta-analysis of blockchain technology for educational sector and its advancements towards education 4.0. *Education and Information Technologies*, 28(10), 13841–13867. DOI: 10.1007/s10639-023-11744-2

Harkat, H., Camarinha-Matos, L. M., Goes, J., & Ahmed, H. F. T. (2024). Cyber-physical systems security: A systematic review. *Computers & Industrial Engineering*, 188, 109891. DOI: 10.1016/j.cie.2024.109891

Hasan, A., & Meva, D. (2018). Web Application Safety by Penetration Testing. International Journal of Advanced Studies of Scientific Research (IJASSR), 3(9) (5 pages).

Hasan, M. (2024). Regulating Artificial Intelligence: A Study in the Comparison between South Asia and Other Countries. *Legal Issues in the digital. The Age (Melbourne, Vic.)*, (1), 122–149.

Hassanalieragh, M., Page, A., Soyata, T., Sharma, G., Aktas, M., Mateos, G., Kantarci, B., & Andreescu, S. (2015). Health Monitoring and Management Using Internet-of-Things (IoT) Sensing with Cloud-Based Processing: Opportunities and Challenges. *2015 IEEE International Conference on Services Computing*, 285–292. https://doi.org/DOI: 10.1109/SCC.2015.47

Hayes, A. (2023, December 14). Scenario analysis: How it works and examples. Investopedia. https://www.investopedia.com/terms/s/scenario_analysis.asp

Heiding, F., Süren, E., Olegård, J., & Lagerström, R. (2022). Penetration testing of connected households. *Computers & Security*, 126, 103067. DOI: 10.1016/j.cose.2022.103067

Hemani, N., Singh, D., & Dwivedi, R. K. (2024). Designing blockchain based secure autonomous vehicular internet of things (IoT) architecture with efficient smart contracts. *International Journal of Information Technology : an Official Journal of Bharati Vidyapeeth's Institute of Computer Applications and Management*. Advance online publication. DOI: 10.1007/s41870-023-01712-x

Herber, C., Richter, A., Wild, T., & Herkersdorf, A. (2015, March). Real-time capable can to avb ethernet gateway using frame aggregation and scheduling. In 2015 Design, Automation & Test in Europe Conference & Exhibition (DATE) (pp. 61-66). IEEE.

Hercegová, K., Baranovskaya, T. P., & Efanova, N. (2021, January 1). Smart technologies for energy consumption management. *EDP Sciences*, 128, 02005–02005. DOI: 10.1051/shsconf/202112802005

Herland, M., Khoshgoftaar, T. M., & Wald, R. (2014). Big data analytics in health informatics: Extracting valuable insights from patient data. *Journal of Healthcare Informatics Research*, 18(4), 301–315. DOI: 10.1093/jhir/rit023

He, Z. W. (2022, August 21). Research on the Civic Policy Model and Reform Innovation of Intelligent Sensor Technology Course. *Journal of Sensors*, 2022, 1–8. DOI: 10.1155/2022/2499421

Hirsch, M., Krahmer, D., & Stroh, P. (2020). Scenario-based investment under uncertainty: A method to include demand-side and sustainability risks. *Journal of Business Research*, 120(5), 36–47.

Hoffmann, M. W., Malakuti, S., Grüner, S., Finster, S., Gebhardt, J., Tan, R., Schindler, T., & Gamer, T. (2021). Developing industrial cps: A multi-disciplinary challenge. *Sensors (Basel)*, 21(6), 1991. DOI: 10.3390/s21061991 PMID: 33799891

Hossain, M. S., & Muhammad, G. (2016). Cloud-assisted Industrial Internet of Things (IIoT) – Enabled framework for health monitoring. *Computer Networks*, 101, 192–202. DOI: 10.1016/j.comnet.2016.01.009

Houmb, S. H., Iversen, F., Ewald, R., & Færaas, E. (2023, February). Intelligent risk-based cybersecurity protection for industrial systems control-a feasibility study. In International Petroleum Technology Conference (p. D021S014R001). IPTC DOI: 10.2523/IPTC-22795-MS

Huang, J., & Zhang, L. (2017, January 1). Research and challenges of CPS. American Institute of Physics. DOI: 10.1063/1.4992851

Huang, L., Qin, J., Zhou, Y., Zhu, F., Liu, L., & Shao, L. (2023). Normalization techniques in training DNNs: Methodology, analysis and application. *IEEE Transactions on Pattern Analysis and Machine Intelligence*, 45(8), 10173–10196. DOI: 10.1109/TPAMI.2023.3250241 PMID: 37027763

Humayun, M., Niazi, M., Jhanjhi, N. Z., Alshayeb, M., & Mahmood, S. (2020). Cyber Security Threats and Vulnerabilities: A Systematic Mapping study. Arabian Journal for Science and Engineering (2011. Online), 45(4), 3171–3189. DOI: 10.1007/s13369-019-04319-2

Hung, C. C., Tu, M. Y., Chien, T. W., Lin, C. Y., Chow, J. C., & Chou, W. (2023). The model of descriptive, diagnostic, predictive, and prescriptive analytics on 100 top-cited articles of nasopharyngeal carcinoma from 2013 to 2022: Bibliometric analysis. *Medicine*, 102(6), e32824. DOI: 10.1097/MD.0000000000032824 PMID: 36820592

Hung, C. W., & Hsu, W. T. (2018). Power Consumption and Calculation Requirement Analysis of AES for WSN IoT. *Sensors (Basel)*, 18(6), 1675. DOI: 10.3390/s18061675 PMID: 29882865

Hu, S., Yu, S., Li, H., & Piuri, V. (2022). Guest Editorial Special Issue on Security, Privacy, and Trustworthiness in Intelligent Cyber–Physical Systems and Internet of Things. *IEEE Internet of Things Journal*, 9(22), 22044–22047. DOI: 10.1109/JIOT.2022.3207335

Hussain, M. Z., Hasan, M. Z., & Chughtai, M. T. A. (2017). *Penetration Testing In System Administration*. INTERNATIONAL JOURNAL OF SCIENTIFIC & TECHNOLOGY RESEARCH. https://www.researchgate.net/publication/319876508

Hwang, G. J., & Chien, S. Y. (2022). Definition, roles, and potential research issues of the metaverse in education: An artificial intelligence perspective. *Computers and Education: Artificial Intelligence*, 3, 100082.

IBM. (n.d.). *What is cybersecurity?* https://www.ibm.com/topics/cybersecurity

Iftikhar, S., Gill, S. S., Song, C., Xu, M., Aslanpour, M. S., Toosi, A. N., Du, J., Wu, H., Ghosh, S., Chowdhury, D., Golec, M., Kumar, M., Abdelmoniem, A. M., Cuadrado, F., Varghese, B., Rana, O., Dustdar, S., & Uhlig, S. (2023). AI-based fog and edge computing: A systematic review, taxonomy and future directions. *Internet of Things : Engineering Cyber Physical Human Systems*, 21, 100674. DOI: 10.1016/j.iot.2022.100674

Imen, S., & Chang, N. (2016, April 1). Developing a cyber-physical system for smart and sustainable drinking water infrastructure management. DOI: 10.1109/ICNSC.2016.7478983

Indian Computer Emergency Response Team (CERT-In). (n.d.). *CERT-In – National nodal agency for cyber security.* https://www.cert-in.org.in/

Indian Penal Code. 1860, No. 45, The Acts of Parliament, 1862 (India).

Information Technology Act, 2000 S. 62, No. 21, Acts of Parliament, 2000 (India).

Information Technology Act, 2000, S 72A 43A, No. 21, Acts of Parliament, 2000 (India).

Insurance Regulatory and Development Authority of India. https://irdai.gov.in/

Islam, S. M. R., Kwak, D., Kabir, M. D. H., Hossain, M., & Kwak, K.-S. (2015). The Internet of Things for Health Care: A Comprehensive Survey. *IEEE Access : Practical Innovations, Open Solutions*, 3, 678–708. DOI: 10.1109/ACCESS.2015.2437951

IT Governance UK. (n.d.). *What is cybersecurity?* https://www.itgovernance.co.uk/what-is-cybersecurity

Jadoon, A. K., Wang, L., Li, T., & Zia, M. A. (2018). Lightweight Cryptographic Techniques for Automotive Cybersecurity. *Wireless Communications and Mobile Computing*, 2018(1), 1640167. DOI: 10.1155/2018/1640167

Jalali, M. S., Siegel, M., & Madnick, S. (2019). Decision-making and biases in cybersecurity capability development: Evidence from a simulation game experiment. *The Journal of Strategic Information Systems*, 28(1), 66–82. DOI: 10.1016/j.jsis.2018.09.003

Jardim-Gonçalves, R., Romero, D., Gonçalves, D. M., & Mendonça, J. P. (2020). Interoperability enablers for cyber-physical enterprise systems. *Enterprise Information Systems*, 14(8), 1061–1070. DOI: 10.1080/17517575.2020.1815084

Javaid, M., Haleem, A., Singh, R. P., & Suman, R. (2023). An integrated outlook of Cyber–Physical Systems for Industry 4.0: Topical practices, architecture, and applications. *Green Technologies and Sustainability*, 1(1), 100001.

Jawhar, I., Al-Jaroodi, J., Noura, H., & Mohamed, N. (2017, June 1). Networking and Communication in Cyber Physical Systems. DOI: 10.1109/ICDCSW.2017.31

Jeffrey, N., Tan, Q., & Villar, J. R. (2023). A review of anomaly detection strategies to detect threats to cyber-physical systems. *Electronics (Basel)*, 12(15), 3283. DOI: 10.3390/electronics12153283

Jena, A. K., & Dash, S. P. (2021). Blockchain Technology: introduction, applications, challenges. In Intelligent systems reference library (pp. 1–11). DOI: 10.1007/978-3-030-69395-4_1

Jin, X. (2023, September 1). Art Interactive Design of Public Leisure Space Environment Based on PSO Algorithm. Taylor & Francis, 43-58. DOI: 10.14733/cadaps.2024.S7.43-58

Johnson, M., & Lee, K. (2020). Interoperability and standardization in healthcare data analytics. *Journal of Biomedical Informatics*, 108, 103518. DOI: 10.1016/j.jbi.2020.103518

Johnson, M., & Lee, K. (2022). Leveraging big data analytics in healthcare: Storage and processing solutions. *Journal of Medical Systems*, 48(4), 789–804. DOI: 10.1007/s10916-022-01845-7

Johnson, M., & Lee, K. (2022). Real-time analytics and streaming data processing in healthcare. *Journal of Medical Systems*, 48(4), 805–820. DOI: 10.1007/s10916-022-01846-8

Jo, S., Han, H., Leem, Y., & Lee, S. (2021, September 3). Sustainable Smart Cities and Industrial Ecosystem: Structural and Relational Changes of the Smart City Industries in Korea. *Sustainability (Basel)*, 13(17), 9917–9917. DOI: 10.3390/su13179917

Joyia, G. J., Liaqat, R. M., Farooq, A., & Rehman, S.National University of Sciences and Technology. (2017). Islamabad, Pakistan, Joyia, G. J., Liaqat, R. M., Farooq, A., & Rehman, S. (2017). Internet of Medical Things (IOMT): Applications, Benefits and Future Challenges in Healthcare Domain. *Journal of Communication*. Advance online publication. DOI: 10.12720/jcm.12.4.240-247

Joy, Z. H., Rahman, M. M., Uzzaman, A., & Maraj, M. A. A. (2024). Integrating Machine Learning and Big Data Analytics For Real-Time Disease Detection In Smart Healthcare Systems. *International Journal of Health and Medical*, 1(3), 16–27.

Jozinović, D., Lomax, A., Štajduhar, I., & Michelini, A. (2020, May 31). Rapid prediction of earthquake ground shaking intensity using raw waveform data and a convolutional neural network. Oxford University Press, 222(2), 1379-1389. DOI: 10.1093/gji/ggaa233

Kankanamge, M. W., Hasan, S. M., Shahid, A. R., & Yang, N. (2024, July 2)... *Large Language Model Integrated Healthcare Cyber-Physical Systems Architecture.*, 1540-1541, 1540–1541. Advance online publication. DOI: 10.1109/COMPSAC61105.2024.00228

Kannan, N., Upreti, K., Pradhan, R., Dhingra, M., Kalimuthukumar, S., Mahaveerakannan, R., & Gayathri, R. (2023). Future perspectives on new innovative technologies comparison against hybrid renewable energy systems. *Computers & Electrical Engineering*, 111, 108910. DOI: 10.1016/j.compeleceng.2023.108910

Kanwade, A. B., Sardey, M. P., Panwar, S. A., Gajare, M. P., Chaudhari, M. N., & Upreti, K. (2023). Combined weighted feature extraction and deep learning approach for chronic obstructive pulmonary disease classification using electromyography. *International Journal of Information Technology : an Official Journal of Bharati Vidyapeeth's Institute of Computer Applications and Management*, 16(3), 1485–1494. DOI: 10.1007/s41870-023-01498-y

Kapucu, N., Ge, Y., Martín, Y., & Williamson, Z. (2021, October 12). Urban resilience for building a sustainable and safe environment. Elsevier BV, 1(1), 10-16. DOI: 10.1016/j.ugj.2021.09.001

Karaaslan, E., Bağcı, U., & Çatbaş, F. N. (2021, June 25). A Novel Decision Support System for Long-Term Management of Bridge Networks. *Applied Sciences (Basel, Switzerland)*, 11(13), 5928–5928. DOI: 10.3390/app11135928

Kaspersky. (n.d.). *What is cybersecurity?* https://www.kaspersky.co.in/resource-center/definitions/what-is-cyber-security

Kathiravelu, P., Van Roy, P., & Veiga, L. (2019). SD-CPS: Software-defined cyber-physical systems. Taming the challenges of CPS with workflows at the edge. *Cluster Computing*, 22(3), 661–677. DOI: 10.1007/s10586-018-2874-8

Kee, K., Lau, S., Lim, Y. S., Ting, Y. P., & Rashidi, R. B. (2022, February 1). Universal cyber physical system, a prototype for predictive maintenance. [IAES]. *Institute of Advanced Engineering and Science*, 11(1), 42–49. DOI: 10.11591/eei.v11i1.3216

Keerthi, C., Jabbar, M. A., & Seetharamulu, B. (2017). Cyber Physical Systems(CPS):Security Issues, Challenges and Solutions. *2017 IEEE International Conference on Computational Intelligence and Computing Research (ICCIC)*. DOI: 10.1109/ICCIC.2017.8524312

Kempf, A., & Osthoff, P. (2007). The effect of socially responsible investing on portfolio performance. *European Financial Management*, 13(5), 908–922. DOI: 10.1111/j.1468-036X.2007.00402.x

Kemp, J., & Livingstone, D. (2006) "Putting a Second Life "Metaverse" skin on learning management systems". *Proceedings of the Second Life education workshop at the Second Life community convention* (Vol. 20).

Kenton, W. (2024, July 1). Capital asset pricing model (CAPM): Definition, formula, and assumptions. Investopedia. https://www.investopedia.com/terms/c/capm.asp

Kerdvibulvech, C. (2022) "Exploring the Impacts of COVID-19 on Digital and Metaverse Games." *International Conference on Human-Computer Interaction*. Springer, Cham. DOI: 10.1007/978-3-031-06391-6_69

Kesharwani, P., Pandey, S., Dixit, V., & Tiwari, L. (2018). A study on Penetration Testing Using Metasploit Framework. *International Research Journal of Engineering and Technology (IRJET)*, 5(12).

Keshk, M., Turnbull, B., Sitnikova, E., Vatsalan, D., & Moustafa, N. (2021, January 1). Privacy-Preserving Schemes for Safeguarding Heterogeneous Data Sources in Cyber-Physical Systems. *IEEE Access : Practical Innovations, Open Solutions*, 9, 55077–55097. DOI: 10.1109/ACCESS.2021.3069737

Khan, F., Ramasamy, L. K., Kadry, S., Nam, Y., & Meqdad, M. N. (2021, August 1). Cyber physical systems: A smart city perspective. [IAES]. *Institute of Advanced Engineering and Science*, 11(4), 3609–3609. DOI: 10.11591/ijece.v11i4.pp3609-3616

Khan, M. A., Menouar, H., & Hamila, R. (2023). LCDnet: A lightweight crowd density estimation model for real-time video surveillance. *Journal of Real-Time Image Processing*, 20(2), 29. DOI: 10.1007/s11554-023-01286-8

Khan, R., Usman, M., & Moinuddin, M. (2024). From raw data to actionable insights: Navigating the world of data analytics. *International Journal of Advanced Engineering Technologies and Innovations*, 1(4), 142–166.

Khan, S., Khan, H. U., & Nazir, S. (2022). Systematic analysis of healthcare big data analytics for efficient care and disease diagnosing. *Scientific Reports*, 12(1), 1–21. DOI: 10.1038/s41598-022-26090-5 PMID: 36572709

Khan, W., Abidin, S., Arif, M., Ishrat, M., Haleem, M., Shaikh, A. A., Farooqui, N. A., & Faisal, S. M. (2024). Anomalous node detection in attributed social networks using dual variational autoencoder with generative adversarial networks. *Data Science and Management*, 7(2), 89–98. DOI: 10.1016/j.dsm.2023.10.005

Khargonekar, P. P., & Sampath, M. (2020). A framework for ethics in cyber-physical-human systems. *IFAC-PapersOnLine*, 53(2), 17008–17015. DOI: 10.1016/j.ifacol.2020.12.1251

Kim, J. H., Seo, S. H., Nguyen, T., Cheon, B. M., Lee, Y. S., & Jeon, J. W. (2015). Gateway Framework for In-Vehicle Networks based on CAN, FlexRay and Ethernet. *IEEE Transactions on Vehicular Technology*, 64(10), 4472–4486. DOI: 10.1109/TVT.2014.2371470

Kohler, T., Matzler, K., & Füller, J. (2009). Avatar-based innovation: Using virtual worlds for real-world innovation. *Technovation*, 29(6-7), 395–407. DOI: 10.1016/j.technovation.2008.11.004

Konstantopoulos, G. C., Alexandridis, A. T., & Papageorgiou, P. (2020, May 1). Towards the Integration of Modern Power Systems into a Cyber–Physical Framework. *Energies*, 13(9), 2169–2169. DOI: 10.3390/en13092169

Koutsoukos, X., Karsai, G., Laszka, A., Neema, H., Potteiger, B., Volgyesi, P., Vorobeychik, Y., & Sztipanovits, J. (2017). SURE: A modeling and simulation integration platform for evaluation of secure and resilient cyber–physical systems. *Proceedings of the IEEE*, 106(1), 93–112. DOI: 10.1109/JPROC.2017.2731741

Kula, S., Choraś, M., & Kozik, R. (2021). Application of the bert-based architecture in fake news detection. In 13th International Conference on Computational Intelligence in Security for Information Systems (CISIS 2020) 12 (pp. 239-249). Springer International Publishing. DOI: 10.1007/978-3-030-57805-3_23

Kumar, N., & Raubal, M. (2021, November 1). Applications of deep learning in congestion detection, prediction and alleviation: A survey. Elsevier BV, 133, 103432-103432. DOI: 10.1016/j.trc.2021.103432

Lam, T.. (2022). *Metaverse report— Future is here Global XR industry insight*. Deloitte China.

LaSorda, M M., Borky, J M., & Sega, R M. (2020, March 1). Model-Based Systems Architecting with Decision Quantification for Cybersecurity, Cost, and Performance. DOI: 10.1109/AERO47225.2020.9172283

Lea, T. (2022). With good planning, can the metaverse be sustainable. FinTech Weekly, 20.

Lee, L.-H., ... (2021) "All one needs to know about metaverse: A complete survey on technological singularity, virtual ecosystem, and research agenda." arXiv preprint arXiv:2110.05352.

Lee, H., Woo, D., & Yu, S. (2022). Virtual reality metaverse system supplementing remote education methods: Based on aircraft maintenance simulation. *Applied Sciences (Basel, Switzerland)*, 12(5), 2667.

Lee, T. Y., Lin, I. A., & Liao, R. H. (2020). Design of a FlexRay/Ethernet gateway and security mechanism for in-vehicle networks. *Sensors (Basel)*, 20(3), 641.

Lee, Y. S., Kim, J. H., & Jeon, J. W. (2017). FlexRay and Ethernet AVB Synchronization for High QoS Automotive Gateway. *IEEE Transactions on Vehicular Technology*, 66(7), 5737–5751. DOI: 10.1109/TVT.2016.2636867

Lenka, R. K., Rath, A. K., Tan, Z., Sharma, S., Puthal, D., Simha, N. V. R., Prasad, M., Raja, R., & Tripathi, S. S. (2018). Building scalable Cyber-Physical-Social networking infrastructure using IoT and low power sensors. *IEEE Access : Practical Innovations, Open Solutions*, 6, 30162–30173. DOI: 10.1109/ACCESS.2018.2842760

Lesch, V., Züfle, M., Bauer, A., Iffländer, L., Krupitzer, C., & Kounev, S. (2023). A literature review of IoT and CPS—What they are, and what they are not. *Journal of Systems and Software*, 200, 111631. DOI: 10.1016/j.jss.2023.111631

Li, H. (2016). Introduction to cyber physical systems. Communications for Control in Cyber Physical Systems, 1–8. DOI: 10.1016/B978-0-12-801950-4.00001-9

Li, N. (2023). Ethical considerations in artificial intelligence: A comprehensive discussion from the perspective of computer vision. SHS Web of Conferences, 179, 04024. DOI: 10.1051/shsconf/202317904024

Li, D., Li, D., Liu, J., Song, Y., & Ji, Y. (2022). *Backstepping sliding mode control for cyberphysical systems under false data injection attack*. IEEE International.

Lim, W., Chek, K. Y. S., Theng, L. B., & Lin, C. T. C. (2024). Future of generative adversarial networks (GAN) for anomaly detection in network security: A review. *Computers & Security*, 139, 103733. DOI: 10.1016/j.cose.2024.103733

Lin, J., Sedigh, S., and et.al. (2011). A semantic agent framework for cyber-physical systems.

Liu, J. J., Zhang, S. B., Sun, W., & Shi, Y. P. (2017). In-Vehicle Network Attacks and Countermeasures: Challenges and Future Directions. *IEEE Network*, 31(5), 55–58. DOI: 10.1109/MNET.2017.1600257

Liu, L., Lu, S., Ren, Z., Wu, B., Yao, Y., Zhang, Q., & Shi, W. (2020, December 10). Computing Systems for Autonomous Driving: State of the Art and Challenges. *IEEE Internet of Things Journal*, 8(8), 6469–6486. DOI: 10.1109/JIOT.2020.3043716

Liu, S., Trivedi, A., Yin, X., & Zamani, M. (2022, January 1). Secure-by-Construction Synthesis of Cyber-Physical Systems. Cornell University. https://doi.org/DOI: 10.48550/arXiv.2202

Liu, Y., Peng, Y., Wang, B., Yao, S., & Liu, Z. (2017, January 1). Review on cyber-physical systems. *Institute of Electrical and Electronics Engineers*, 4(1), 27–40. DOI: 10.1109/JAS.2017.7510349

Liu, Y., & Yang, K. (2022, September 12). Communication, sensing, computing and energy harvesting in smart cities. *IET Smart Cities*, 4(4), 265–274. DOI: 10.1049/smc2.12041

Li, Y., & Mardani, A. (2023). Digital twins and blockchain technology in the industrial Internet of Things (IIoT) using an extended decision support system model: Industry 4.0 barriers perspective. *Technological Forecasting and Social Change*, 195, 122794. DOI: 10.1016/j.techfore.2023.122794

Li, Z., Wang, S., Ding, W., Chen, Y., Chen, M., Zhang, S., Liu, Z., Yang, W., & Li, Y. (2023). Mechanically robust, flexible hybrid tactile sensor with microstructured sensitive composites for human-cyber-physical systems. *Composites Science and Technology*, 244, 110303. DOI: 10.1016/j.compscitech.2023.110303

Luo, Y., Xiao, Y., Cheng, L., Peng, G., & Yao, D. (2021). Deep learning-based anomaly detection in cyber-physical systems: Progress and opportunities. *ACM Computing Surveys*, 54(5), 1–36. DOI: 10.1145/3453155

Malathi, J., Kusha, K. R., Isaac, S., Ramesh, A., Rajendiran, M., & Boopathi, S. (2024). IoT-Enabled Remote Patient Monitoring for Chronic Disease Management and Cost Savings: Transforming Healthcare. In Advances in Explainable AI Applications for Smart Cities (pp. 371-388). IGI Global.

Maleh, Y., Lakkineni, S., Tawalbeh, L., & El-Latif, A. A. (2022). Blockchain for Cyber-Physical Systems: Challenges and applications. In Internet of things (pp. 11–59). DOI: 10.1007/978-3-030-93646-4_2

Mamilla, S. R. (2021). A Study of Penetration Testing Processes and Tools. Electronic Theses, Projects, and Dissertations. 1220. https://scholarworks.lib.csusb.edu/etd/1220

Mann, D., & Chen, S. (2018). Adaptive strategies in healthcare analytics: Maximizing patient care and operational effectiveness. *Healthcare Analytics Review*, 6(1), 78–92. DOI: 10.1016/j.hcareanarev.2018.02.005

Marda, V. (2018). Artificial intelligence policy in India: A framework for engaging the limits of data-driven decision-making. *Philosophical Transactions. Series A, Mathematical, Physical, and Engineering Sciences*, 376(2133), 20180087. DOI: 10.1098/rsta.2018.0087 PMID: 30323001

Markowitz, H. (1952). Portfolio selection. *The Journal of Finance*, 7(1), 77–91.

Marsden, M., McGuinness, K., Little, S., & O'Connor, N. E. (2016). Holistic features for real-time crowd behaviour anomaly detection. arXiv. /arxiv.1606.05310DOI: 10.1109/ICIP.2016.7532491

Martin, E., & Patel, T. (2019). Transforming healthcare with big data: Storage and integration challenges. *IEEE Journal of Biomedical and Health Informatics*, 23(5), 1992–2005. DOI: 10.1109/JBHI.2019.2917321

Martinez Spessot, C. (2023). Cyber-Physical Automation. In *Springer Handbook of Automation* (pp. 379–404). Springer International Publishing. DOI: 10.1007/978-3-030-96729-1_17

Mashayekhi, A N., & Heravi, G. (2020, August 1). A decision-making framework opted for smart building's equipment based on energy consumption and cost trade-off using BIM and MIS. Elsevier BV, 32, 101653-101653. DOI: 10.1016/j.jobe.2020.101653

Mateus, K., & Königseder, T. (2014). *Automotive Ethernet* (1st ed.). Cambridge University Press. DOI: 10.1017/CBO9781107414884

Maurya, J., Pant, H., Dwivedi, S., & Jaiswal, M. (2021, May 1). FLOOD AVOIDANCE USING IOT. *IJEAST*, 6(1). Advance online publication. DOI: 10.33564/IJEAST.2021.v06i01.021

Mehndiratta, M. (2022) Information Technology Act 2000, Ipleaders, https://blog.ipleaders.in/information-technology-act-2000

Mehta, N., Pandit, A., & Kulkarni, M. (2020). Elements of Healthcare Big Data Analytics. *Big Data Analytics in Healthcare*, 66, 23–43. DOI: 10.1007/978-3-030-31672-3_2

Mesa, D. (2023, February 21). Digital divide, e-government and trust in public service: The key role of education. *Frontiers of Medicine*, 8, 1140416. Advance online publication. DOI: 10.3389/fsoc.2023.1140416 PMID: 36895333

Ministry of Electronics and Information Technology, Government of India. (2013). *National Cyber Security Policy - 2013*.https://www.meity.gov.in/writereaddata/files/downloads/National_cyber_security_policy-2013(1).pdf

Mitake. (n.d.). Phishing URLs and benign URLs [Dataset]. Hugging Face. https://huggingface.co/datasets/Mitake/PhishingURLsANDBenignURLs. Accessed August 7, 2024.

Mohammad, S. (2019). *Jalali & Michael Siegel, Stuart Madnick, Decision- making and biases in cybersecurity capability development: Evidence from a simulation game experiment.* Elsevier.

Moik, B., Bobek, V., & Horvat, T. (2021, June 30). India's National Smart City Mission. *Analysis of Project Dimensions Including Sources of Funding.*, 13(1), 50–59. DOI: 10.32015/JIBM/2021.13.1.50-59

Morel, B. (2011), Anomaly Based Intrusion Detection and Artificial Intelligence, Intechopen, https://www.intechopen.com/chapters/14355

Morovat, K., & Panda, B. (2020). A survey of artificial intelligence in cybersecurity. In 2020 *International conference on computational science and computational intelligence* (CSCI) (pp. 109-115). IEEE. DOI: 10.1109/CSCI51800.2020.00026

Movassaghi, S., Abolhasan, M., Lipman, J., Smith, D., & Jamalipour, A. (2014). Wireless Body Area Networks: A Survey. *IEEE Communications Surveys and Tutorials*, 16(3), 1658–1686. DOI: 10.1109/SURV.2013.121313.00064

Mushtaq, A., Haq, I. U., Sarwar, M. A., Khan, A., & Shafiq, O. (2022, January 1). Traffic Management of Autonomous Vehicles using Policy Based Deep Reinforcement Learning and Intelligent Routing. Cornell University. https://doi.org/DOI: 10.48550/arXiv.2206

Mustafee, N., Powell, J. H., & Harper, A. (2018, December). RH-RT: A data analytics framework for reducing wait time at emergency departments and centres for urgent care. In *2018 Winter Simulation Conference (WSC)* (pp. 100-110). IEEE. DOI: 10.1109/WSC.2018.8632378

Müter, M., Groll, A., & Freiling, F. C. (2010, August). A structured approach to anomaly detection for in-vehicle networks. In *2010 Sixth International Conference on Information Assurance and Security* (pp. 92-98). IEEE.

Mutlag, A. A., Abd Ghani, M. K., Arunkumar, N., Mohammed, M. A., & Mohd, O. (2019). Enabling technologies for fog computing in healthcare IoT systems. *Future Generation Computer Systems*, 90, 62–78. DOI: 10.1016/j.future.2018.07.049

Mystakidis, S. (2022). Metaverse. *Metaverse. Encyclopedia*, 2(1), 486–497. DOI: 10.3390/encyclopedia2010031

Nardelli, P. H. (2022). *Cyber-physical systems: Theory, methodology, and applications*. John Wiley & Sons. DOI: 10.1002/9781119785194

National Critical Infrastructure Protection Centre. https://nciipc.gov.in/

National Cyber Security Policy. (2013), 1, 2-4, https://www.meity.gov.in/writereaddata/files/downloads/National_cyber_security_policy-2013%281%29.pdf

National Institute of Standards and Technology (NIST). (n.d.). *NIST – Cybersecurity*. https://www.nist.gov/

National Institute of Standards and Technology. Recommendation for Block Cipher Mode of Operation: The CCM Mode for Authentication and Confidentiality. Available online: https://www.nist.gov/publications/recommendation-block-cipher-modes-operation-ccm-mode-authentication-and-confidentiality (accessed on 23 January 2020).

Nayak, M. M., & Dash, S. K. (2020). A review paper on crowd estimation. International Journal of Advanced Research in Engineering and Technology (IJARET), 2020. https://ssrn.com/abstract=3878696

Nazari-Heris, M., Esfehankalateh, A. T., & Ifaei, P. (2023, June 15). Hybrid Energy Systems for Buildings: A Techno-Economic-Enviro Systematic Review. *Energies*, 16(12), 4725–4725. DOI: 10.3390/en16124725

Neeraj Kumar—IEEE Xplore Author Profile. (n.d.). Retrieved 24 February 2024, from https://ieeexplore.ieee.org/author/37395599700

Nejković, V., Visa, A., Tošić, M., Petrović, N., Valkama, M., Koivisto, M., Talvitie, J., Rančić, S., Grzonka, D., Tchórzewski, J., Kuonen, P., & Gortázar, F. (2019, January 1). Big Data in 5G Distributed Applications. Springer Science+Business Media, 138-162. DOI: 10.1007/978-3-030-16272-6_5

Nguyen, D. C., Cheng, P., Ding, M., López-Pérez, D., Pathirana, P. N., Li, J., Seneviratne, A., Li, Y., & Poor, H. V. (2021, January 1). Enabling AI in Future Wireless Networks: A Data Life Cycle Perspective. *IEEE Communications Surveys and Tutorials*, 23(1), 553–595. DOI: 10.1109/COMST.2020.3024783

Nguyen, T., & Chen, J. (2021). Real-time data processing in healthcare: Challenges and solutions. *Advanced Health Care Technologies*, 18(1), 99–112. DOI: 10.1007/s40846-021-00670-8

Noureldin, H. F., & Fadel, M. (2021, June 1). Rationalizing Resource Utilization in Cloud Computing Using Coalition Formation Strategy. *Science Publications*, 17(6), 539–555. DOI: 10.3844/jcssp.2021.539.555

O'Shaughnessy, M. (2023, January 1). Five policy uses of algorithmic transparency and explainability. Cornell University. https://doi.org/DOI: 10.48550/arXiv.2302

Odyurt, U., Pimentel, A. D., & Alonso, I. G. (2022). Improving the robustness of industrial Cyber–Physical Systems through machine learning-based performance anomaly identification. *Journal of Systems Architecture*, 131, 102716. DOI: 10.1016/j.sysarc.2022.102716

Ostad-Ali-Askari, K., Gholami, H., Dehghan, S., & Ghane, M. (2021, February 1). The Role of Public Participation in Promoting Urban Planning. *American Journal of Engineering and Applied Sciences*, 14(2), 177–184. DOI: 10.3844/ajeassp.2021.177.184

Palermo, S. A., Maiolo, M., Brusco, A. C., Turco, M., Pirouz, B., Greco, E., Spezzano, G., & Piro, P. (2022, August 19). Smart Technologies for Water Resource Management: An Overview. *Sensors (Basel)*, 22(16), 6225–6225. DOI: 10.3390/s22166225 PMID: 36015982

Pan, X., Yan, E., Cui, M., & Hua, W. (2018). Examining the usage, citation, and diffusion patterns of bibliometric mapping software: A comparative study of three tools. *Journal of Informetrics*, 12(2), 481–493. DOI: 10.1016/j.joi.2018.03.005

Papageorgiou, A., Cheng, B., & Kovács, E. (2015, November 1). Real-time data reduction at the network edge of Internet-of-Things systems. DOI: 10.1109/CNSM.2015.7367373

Pardeshi, T., Vekariya, D., & Gandhi, A. (2023, December). Exploring the Synergy of GAN and CNN Models for Robust Intrusion Detection in Cyber Security. In 2023 3rd International Conference on Innovative Mechanisms for Industry Applications (ICIMIA) (pp. 1470-1475). IEEE. DOI: 10.1109/ICIMIA60377.2023.10426063

Paredes, C. M., Martínez Castro, D., González Potes, A., Rey Piedrahita, A., & Ibarra Junquera, V. (2024). Design Procedure for Real-Time Cyber–Physical Systems Tolerant to Cyberattacks. *Symmetry*, 16(6), 684. DOI: 10.3390/sym16060684

Park, H., Oh, H., & Choi, J. K. (2023, January 1). A Consent-Based Privacy-Compliant Personal Data-Sharing System. *IEEE Access : Practical Innovations, Open Solutions*, 11, 95912–95927. DOI: 10.1109/ACCESS.2023.3311823

Park, S., & Kim, S. (2022). Identifying world types to deliver gameful experiences for sustainable learning in the metaverse. *Sustainability (Basel)*, 14(3), 1361. DOI: 10.3390/su14031361

PATAK Engineering FlexRay Controller Documentation. Available online: http://patakengineering.eu/download/FlexRayController.pdf

Patel, S B., & Lam, K. (2023, March 1). ChatGPT: the future of discharge summaries?. Elsevier BV, 5(3), e107-e108. DOI: 10.1016/S2589-7500(23)00021-3

Pathak, S., Srivastava, K B L., & Dewangan, R L. (2023, January 14). Decision styles and their association with heuristic cue and decision-making rules. Taylor & Francis, 10(1). DOI: 10.1080/23311908.2023.2166307

Patwal, A., Diwakar, M., Tripathi, V., & Singh, P. (2023). Crowd counting analysis using deep learning: A critical review. *Procedia Computer Science*, 218, 2448–2458. DOI: 10.1016/j.procs.2023.01.220

Pavlovic, M. (2020). Designing for ambient UX: design framework for managing user experience within cyber-physical systems. L. RAMPINO, I. MARIANI, 39

Perera, D., Seidou, O., Agnihotri, J., Mehmood, H., & Rasmy, M. (2020, December 16). Challenges and Technical Advances in Flood Early Warning Systems (FEWSs). IntechOpen. DOI: 10.5772/intechopen.93069

Peros, S., Delbruel, S., Michiels, S., Joosen, W., & Hughes, D. (2020, June 1). Simplifying CPS Application Development through Fine-grained, Automatic Timeout Predictions. *Association for Computing Machinery*, 1(3), 1–30. DOI: 10.1145/3385960

Pietrosémoli, E., Rainone, M., Zennaro, M., & Mikeka, C. (2022, January 1). Massive RF Simulation Applied to School Connectivity in Malawi. Cornell University. https://doi.org/DOI: 10.48550/arXiv.2207

Pishdad-Bozorgi, P., Gao, X., & Shelden, D. R. (2020). Introduction to cyber-physical systems in the built environment. In Construction 4.0 (pp. 23-41) DOI: 10.1201/9780429398100-2

PitchBook Profile – Microsoft. (n.d.). PitchBook. https://my.pitchbook.com/profile/11026-45/company/profile

PitchBook Profile – Nvidia. (n.d.). PitchBook. https://my.pitchbook.com/profile/41161-24/company/profile

Pop, E., Iliuta, M. C., Constantin, N., Gîfu, D., & Dumitraşcu, A. (2023). Interoperability Framework for Cyber-Physical Systems Based Capabilities. 2023 24th International Conference on Control Systems and Computer Science (CSCS). DOI: 10.1109/CSCS59211.2023.00062

Pradhan, N. R., Singh, A. P., Verma, S., Kavita, N., Wozniak, M., Shafi, J., & Ijaz, M. F. (2022). A blockchain based lightweight peer-to-peer energy trading framework for secured high throughput micro-transactions. *Scientific Reports*, 12(1), 14523. Advance online publication. DOI: 10.1038/s41598-022-18603-z PMID: 36008545

Prenzel, L., & Steinhorst, S. (2021). *Decentralized Autonomous Architecture for Resilient Cyber-Physical Production Systems. 2021 Design, Automation & Test in Europe Conference & Exhibition*. DATE., DOI: 10.23919/DATE51398.2021.9473954

Proactive Measures to Avoid Failure. (2021, April 21)., 309-320. DOI: 10.1002/9781119615606.ch17

Puhm, A., Rössler, P., Wimmer, M., Swierczek, R., & Balog, P. (2008, September). Development of a flexible gateway platform for automotive networks. In *2008 IEEE International Conference on Emerging Technologies and Factory Automation* (pp. 456-459). IEEE.

Pullen, D., Anagnostopoulus, N. A., Arul, T., & Katzenbeisser, S. (2020). *"Securing FlexRay-based in-vehicle networks", Microprocessors and Microsystems*. Elsevier.

Pupillo, L., Fantin, S., Ferreira, A., & Polito, C. (2021), Artificial Intelligence and Cybersecurity, CEPS 2, https://www.ceps.eu/wp-content/uploads/2021/05/CEPS-TFR-Artificial-Intelligence-and-Cybersecurity.pdf

Quincozes, S. E., Mossé, D., Passos, D., Albuquerque, C., Ochi, L. S., & Santos, V F D. (2022, March 1). On the Performance of GRASP-Based Feature Selection for CPS Intrusion Detection. *IEEE Transactions on Network and Service Management*, 19(1), 614–626. DOI: 10.1109/TNSM.2021.3088763

Quintana, I., Tsiopoulos, A., Lema, M. A., Sardis, F., Sequeira, L., Arias, J., Raman, A., Azam, A., & Döhler, M. (2018, December 1). The Making of 5G: Building an End-to-End 5G-Enabled System. *Institute of Electrical and Electronics Engineers*, 2(4), 88–96. DOI: 10.1109/MCOMSTD.2018.1800024

Ragunthar, T., Kaliappan, S., & Ali, H. M. (2024). Detection of Feedback Control Through Optimization in the Cyber Physical System Through Big Data Analysis and Fuzzy Logic System. In *AI Approaches to Smart and Sustainable Power Systems* (pp. 299–313). IGI Global. DOI: 10.4018/979-8-3693-1586-6.ch016

Rahmani, A. M., Gia, T. N., Negash, B., Anzanpour, A., Azimi, I., Jiang, M., & Liljeberg, P. (2018). Exploiting smart e-Health gateways at the edge of healthcare Internet-of-Things: A fog computing approach. *Future Generation Computer Systems*, 78, 641–658. DOI: 10.1016/j.future.2017.02.014

Rahman, Z., Khalil, I., Yi, X., & Atiquzzaman, M. (2021). Blockchain-based security framework for a critical industry 4.0 cyber-physical system. *IEEE Communications Magazine*, 59(5), 128–134. DOI: 10.1109/MCOM.001.2000679

Raisin, S. N., Jamaludin, J., Rahalim, F. M., Mohamad, F. A. J., & Naeem, B. (2020). Cyber-physical system (CPS) application-a review. REKA ELKOMIKA: Jurnal Pengabdian kepada Masyarakat, 1(2), 52-65

Rajhans, A., Bhave, A., Ruchkin, I., Krogh, B. H., Garlan, D., Platzer, A., & Schmerl, B. (2014). Supporting heterogeneity in Cyber-Physical systems architectures. *IEEE Transactions on Automatic Control*, 59(12), 3178–3193. DOI: 10.1109/TAC.2014.2351672

Raji, M. A., Olodo, H. B., Oke, T. T., Addy, W. A., Ofodile, O. C., & Oyewole, A. T. (2024). Real-time data analytics in retail: A review of USA and global practices. GSC Advanced Research and Reviews, 18(3), 059-065

Rajkumar, R. (2023, September 24). Cyber-physical systems. https://dl.acm.org/doi/10.1145/1837274.1837461

Ranade, P., Piplai, A., Joshi, A., & Finin, T. (2021, December). Cybert: Contextualized embeddings for the cybersecurity domain. In 2021 IEEE International Conference on Big Data (Big Data) (pp. 3334-3342). IEEE.

Rathore, H., Mohamed, A., & Guizani, M. (2020). A survey of blockchain enabled Cyber-Physical systems. *Sensors (Basel)*, 20(1), 282. DOI: 10.3390/s20010282 PMID: 31947860

Raut, R., Narkhede, B., & Gardas, B. B. (2017). To identify the critical success factors of sustainable supply chain management practices in the context of oil and gas industries: ISM approach. *Renewable & Sustainable Energy Reviews*, 68, 33–47. DOI: 10.1016/j.rser.2016.09.067

Ravichandiran, S. (2021). *Getting Started with Google BERT: Build and train state-of-the-art natural language processing models using BERT*. Packt Publishing Ltd.

Rawat, A. (2024). Enhancing Abstractive and Extractive Reviews Text Summarization using NLP and Neural Networks (Doctoral dissertation, Dublin Business School).

Ray, A. (2013). Autonomous perception and decision-making in cyber-physical systems. 2013 8th International Conference on Computer Science & Education. https://doi.org/DOI: 10.1109/ICCSE.2013.6554173

Rayan, R. A., Tsagkaris, C., Zafar, I., Moysidis, D. V., & Papazoglou, A. S. (2022). Big data analytics for health. In *Big Data Analytics for Healthcare*. Datasets, Techniques, Life Cycles, Management, and Applications., DOI: 10.1016/B978-0-323-91907-4.00002-9

Rehman, A., Naz, S., & Razzak, I. (2022). Leveraging big data analytics in healthcare enhancement: Trends, challenges and opportunities. *Multimedia Systems*, 28(4), 1339–1371. DOI: 10.1007/s00530-020-00736-8

Rehman, S. F. (2022). Practical Implementation of Artificial Intelligence in Cybersecurity–A Study. *International Journal of Advanced Research in Computer and Communication Engineering*, 11(11). Advance online publication. DOI: 10.17148/IJARCCE.2022.111103

Rein, A., Rieke, R., Jäger, M., Kuntze, N., & Coppolino, L. (2016). Trust establishment in cooperating Cyber-Physical systems. In Lecture Notes in Computer Science (pp. 31–47). DOI: 10.1007/978-3-319-40385-4_3

Rejeb, A., Rejeb, K., Appolloni, A., Jagtap, S., Iranmanesh, M., Alghamdi, S., Alhasawi, Y., & Kayıkçı, Y. (2024, January 1). Unleashing the power of internet of things and blockchain: A comprehensive analysis and future directions. Elsevier BV, 4, 1-18. DOI: 10.1016/j.iotcps.2023.06.003

Rifat, N., Ahsan, M., Chowdhury, M., & Gomes, R. (2022, May). Bert against social engineering attack: Phishing text detection. In 2022 IEEE International Conference on Electro Information Technology (eIT) (pp. 1-6). IEEE.

Riggs, H., Tufail, S., Parvez, I., Tariq, M., Khan, M. A., Amir, A., Vuda, K. V., & Sarwat, A. I. (2023, April 17). Impact, Vulnerabilities, and Mitigation Strategies for Cyber-Secure Critical Infrastructure. *Sensors (Basel)*, 23(8), 4060–4060. DOI: 10.3390/s23084060 PMID: 37112400

Rozenblum, R., & Jang, Y. (2017). The dynamic nature of healthcare data: Seasonal trends and patient demographics. *Healthcare Dynamics*, 12(3), 45–57. DOI: 10.1097/HDY.0000000000000123

Sakiz, F., & Sen, S. (2017). A survey of attacks and detection mechanisms on intelligent transportation systems: VANETs and IoV. *Ad Hoc Networks*, 61, 33–50. DOI: 10.1016/j.adhoc.2017.03.006

Saleh, S. A. M., Suandi, S. A., & Ibrahim, H. (2015). Recent survey on crowd density estimation and counting for visual surveillance. *Engineering Applications of Artificial Intelligence*, 41, 103–114. DOI: 10.1016/j.engappai.2015.01.007

Salim, N., Singh, M. S., Abed, A. T., & Islam, M. T. (2023). 4X4 MIMO slot antenna spanner shaped low mutual coupling for Wi-Fi 6 and 5G communications. *Alexandria Engineering Journal*, 78, 141–148. DOI: 10.1016/j.aej.2023.07.042

Samalna, D A., Ngossaha, J M., Ari, A A A., & Kolyang. (2022, December 22). Cyber-Physical Urban Mobility Systems. IGI Global, 11(1), 1-21. DOI: 10.4018/IJSI.315662

Sander, O., Hubner, M., Becker, J., & Traub, M. (2008, December). Reducing latency times by accelerated routing mechanisms for an FPGA gateway in the automotive domain. In *2008 International Conference on Field-Programmable Technology* (pp. 97-104). IEEE.

Šarak, E., Dobrojević, M., & Sedmak, S. (2020, January 1). IoT based early warning system for torrential floods. *Faculty of Agronomy in Čačak*, 48(3), 511–515. DOI: 10.5937/fme2003511S

Sarkady, D., Neuburger, L., & Egger, R. (2021, January). Virtual reality as a travel substitution tool during COVID-19. In Information and communication technologies in tourism 2021: Proceedings of the ENTER 2021 eTourism conference, January 19–22, 2021 (pp. 452-463). Cham: Springer International Publishing.

Sarker, I. H. (2024). Generative AI and Large Language Modeling in Cybersecurity. In *AI-Driven Cybersecurity and Threat Intelligence: Cyber Automation, Intelligent Decision-Making and Explainability* (pp. 79–99). Springer Nature Switzerland. DOI: 10.1007/978-3-031-54497-2_5

Sarker, K. U., Yunus, F., & Deraman, A. (2023). Penetration Taxonomy: A Systematic Review on the Penetration Process, Framework, Standards, Tools, and Scoring Methods. *Sustainability (Basel)*, 15(13), 10471. DOI: 10.3390/su151310471

Sarker, M., & Bartok, I. (2024). Global trends of green manufacturing research in the textile industry using bibliometric analysis. *Case Studies in Chemical and Environmental Engineering*, 9, 100578. DOI: 10.1016/j.cscee.2023.100578

Schilberg, D., Hoffmann, M., Schmitz, S., & Meisen, T. (2016). Interoperability in smart automation of cyber physical systems. In Springer series in wireless technology (pp. 261–286). DOI: 10.1007/978-3-319-42559-7_10

Scquizzato, S. (2024). Data integration and data-driven evaluation for the non-profit sector: A study case: Response innovation lab

Securities and Exchange Board of India (SEBI). (n.d.). *SEBI – Protecting the interests of investors in securities.*https://www.sebi.gov.in

Segura, D., Damsgaard, S. B., Kabaci, A., Mogensen, P., Khatib, E. J., & Barco, R. (2024). An empirical study of 5G, Wi-Fi 6 and multi-connectivity scalability in an indoor industrial scenario. *IEEE Access : Practical Innovations, Open Solutions*, 12, 74406–74416. DOI: 10.1109/ACCESS.2024.3404870

Sengupta, S., & Ayan, R. (2019). Ethical Considerations in Artificial Intelligence: A Perspective from India. *Computer Science and Information Technology*, 9(1), 21–29.

Shaik, T., Tao, X., Higgins, N., Li, L., Gururajan, R., Zhou, X., & Acharya, U R. (2023, January 5). Remote patient monitoring using artificial intelligence: Current state, applications, and challenges. Wiley, 13(2). DOI: 10.1002/widm.1485

Shane, T., Martin, G., Ciaran, H., Edward, J., Mohan, T., & Liam, K. (2015). Intra-Vehicle Networks: A Review. *IEEE Transactions on Intelligent Transportation Systems*, 16(2), 534–545. DOI: 10.1109/TITS.2014.2320605

Sharma, M. (2024). Enhancing Security and Privacy in Cyber-Physical Systems: Challenges and Solutions. 2024 IEEE 14th Annual Computing and Communication Workshop and Conference (CCWC). DOI: 10.1109/CCWC60891.2024.10427691

Sharma, A., Rani, S., Shah, S. H., Sharma, R., Yu, F., & Hassan, M. M. (2023). An efficient hybrid deep learning model for denial of service detection in cyber physical systems. *IEEE Transactions on Network Science and Engineering*, 10(5), 2419–2428. DOI: 10.1109/TNSE.2023.3273301

Sharma, N., & Garg, R. D. (2022). Real-time computer vision for transportation safety using deep learning and IoT. *2022 International Conference on Engineering and Emerging Technologies (ICEET)*, Kuala Lumpur, Malaysia, 1-5. DOI: 10.1109/ICEET56468.2022.10007226

Sharma, R., & Garg, P. (2024). Fundamentals of data analytics and lifecycle. In *Data Analytics and Machine Learning: Navigating the Big Data Landscape* (pp. 19–37). Springer Nature Singapore. DOI: 10.1007/978-981-97-0448-4_2

Sharpe, W. F. (1964). Capital asset prices: A theory of market equilibrium under conditions of risk. *The Journal of Finance*, 19(3), 425–442.

Shaukat, K., Faisal, A., Masood, R., Usman, A., & Shaukat, U. (2016). Security quality assurance through penetration testing. *19th International Multi-Topic Conference (INMIC)*. DOI: DOI: 10.1109/INMIC.2016.7840115

Shermon, G. (2011). *Competency based HRM: A strategic resource for competency mapping, assessment, and development centers.* Tata McGraw-Hill Education.

Shete, D., & Khobragade, P. (2023, September). An empirical analysis of different data visualization techniques from statistical perspective. In AIP Conference Proceedings (Vol. 2839, No. 1). AIP Publishing DOI: 10.1063/5.0167670

Shih, C., Chou, J., Reijers, N., & Kuo, T. (2016, December 1). Designing CPS/IoT applications for smart buildings and cities. *IET Cyber-Physical Systems*, 1(1), 3–12. DOI: 10.1049/iet-cps.2016.0025

Shivanshu, (2024), What is Cybersecurity? Definition and Types Explained, Intellippat, https://intellipaat.com/blog/what-is-cyber-security/#no1

Shollo, A., & Galliers, R. D. (2024). Constructing actionable insights: The missing link between data, artificial intelligence, and organizational decision-making. In *Research Handbook on Artificial Intelligence and Decision Making in Organizations* (pp. 195–213). Edward Elgar Publishing. DOI: 10.4337/9781803926216.00020

Shravan, K., Neha, B., & Pawan, B. (2014). Penetration Testing: A Review. *International Journal of Advancements in Computing Technology*, 3(4), 752–757.

Shreejith, S., Mundhenk, P., Ettner, A., Fahmy, A., Steinhorst, S., Lukasiewycz, M., & Chakraborty, S. (2017). VEGa: A High Performance Vehicular Ethernet Gateway on Hybrid FPGA. *IEEE Transactions on Computers*, 66(10), 1790–1803. DOI: 10.1109/TC.2017.2700277

Silva, E. M., & Jardim-Goncalves, R. (2021). Cyber-Physical Systems: A multi-criteria assessment for Internet-of-Things (IoT) systems. *Enterprise Information Systems*, 15(3), 332–351. DOI: 10.1080/17517575.2019.1698060

Singh, A., & Jain, A. (2018, April). Study of cyber-attacks on cyber-physical system. In *Proceedings of 3rd International Conference on Internet of Things and Connected Technologies (ICIoTCT)* (pp. 26-27)

Singh, R., Akram, S., Gehlot, A., Buddhi, D., Priyadarshi, N., & Twala, B. (2022). Energy System 4.0: Digitalization of the energy sector with inclination towards sustainability. *Sensors (Basel)*, 22(17), 6619. DOI: 10.3390/s22176619 PMID: 36081087

Singh, U. K., Sharma, A., Singh, S. K., Tomar, P. S., Dixit, K., & Upreti, K. (2023). Security and privacy aspect of cyber physical systems. In *Cyber Physical Systems* (pp. 141–164). Chapman and Hall/CRC.

Sloan, J. A., Halyard, M., El Naqa, I., & Mayo, C. (2016). Lessons from large-scale collection of patient-reported outcomes: Implications for Big data aggregation and analytics. International Journal of Radiation Oncology* Biology* Physics, 95(3), 922-929.

Smith, J., & Brown, A. (2023). Challenges and solutions in healthcare data storage and management. *Health IT Journal*, 45(2), 123–135. DOI: 10.1016/j.healthit.2023.01.003

Smith, J., & Johnson, M. (2021). Real-time data warehouses and stream processing in healthcare: Technologies and applications. *Journal of Biomedical Informatics*, 113, 103651. DOI: 10.1016/j.jbi.2021.103651

Soares, F. L., Campelo, D. R., Yan, Y., Ruepp, S., Dittmann, L., & Ellegard, L. (2015, March). Reliability in automotive ethernet networks. In 2015 11th International Conference on the Design of Reliable Communication Networks (DRCN) (pp. 85-86). IEEE.

Sobb, T., Turnbull, B., & Moustafa, N. (2023). A holistic review of cyber–physical–social systems: New directions and opportunities. *Sensors (Basel)*, 23(17), 7391. DOI: 10.3390/s23177391 PMID: 37687846

Sokolsky, O., Lee, I., & Heimdahl, M. P. E. (2011). Challenges in the regulatory approval of medical cyber-physical systems. 2011 Proceedings of the Ninth ACM International Conference on Embedded Software (EMSOFT). DOI: 10.1145/2038642.2038677

Sontan, A. D., & Samuel, S. V. Adewale Daniel SontanSegun Victor Samuel. (2024). The intersection of Artificial Intelligence and cybersecurity: Challenges and opportunities. *World Journal of Advanced Research and Reviews*, 21(2), 1720–1736. DOI: 10.30574/wjarr.2024.21.2.0607

Sourin, A. (2017). *Case study: Shared virtual and augmented environments for creative applications*. SpringerBriefs in Computer Science.

Statti, A., & Torres, K M. (2020, January 1). Digital Literacy: The Need for Technology Integration and Its Impact on Learning and Engagement in Community School Environments. Taylor & Francis, 95(1), 90-100. DOI: 10.1080/0161956X.2019.1702426

Stefan, P. S., & Groza, B. (2020). *Efficient Physical Layer Key Agreement for Flex-Ray Networks*. Transactions on Vehicular Technology.

Strengholt, P. (2023). *Data management at scale*. O'Reilly Media, Inc.

Subhan, F., Mirza, A., Su'ud, M. B. M., Alam, M. M., Nisar, S., Habib, U., & Iqbal, M. Z. (2023). AI-enabled wearable medical internet of things in healthcare system: A survey. *Applied Sciences (Basel, Switzerland)*, 13(3), 1394. DOI: 10.3390/app13031394

Sukhramani, K., Kumre, H., Rasool, A., & Jadav, A. (2024, February). Binary Classification of News Articles using Deep Learning. In 2024 IEEE International Students' Conference on Electrical, Electronics and Computer Science (SCEECS) (pp. 1-9). IEEE. DOI: 10.1109/SCEECS61402.2024.10482129

Sumra, I. A., Ahmad, I., & Hasbullah, H., & bin Ab Manan, J. (2011). Classes of attacks in VANET, in 2011 Saudi International Electronics, Communications and Photonics Conference (SIECPC).

Surya, K. C. N., & Rajam, V. M. A. (2023, January 2). Novel Approaches for Resource Management Across Edge Servers. *Springer Nature*, 11(1), 20–30. DOI: 10.1007/s44227-022-00007-0

Sykownik, P. (2022) "Something Personal from the Metaverse: Goals, Topics, and Contextual Factors of Self-Disclosure in Commercial Social VR". Conference on Human Factors in Computing Systems – Proceedings. DOI: 10.1145/3491102.3502008

Synnestvedt, M. B., Chen, C., & Holmes, J. H. (2005). CiteSpace II: Visualization and Knowledge Discovery in Bibliographic Databases. *AMIA ... Annual Symposium Proceedings - AMIA Symposium. AMIA Symposium*, 2005, 724–728. PMID: 16779135

Tao, P., Liu, N., & Dong, C. (2024). Research progress of MIoT and digital healthcare in the new era. *Clinical eHealth, 7*, 1–4. https://doi.org/DOI: 10.1016/j.ceh.2023.11.004

TechTarget. (n.d.). *Cybersecurity skills gap: Why it exists and how to address it.* https://www.techtarget.com/searchsecurity/tip/Cybersecurity-skills-gap-Why-it-exists-and-how-to-address-

Tevaarwerk, A. J., Karam, D., Gatten, C. A., Harlos, E. S., Maurer, M. J., Giridhar, K. V., Haddad, T. C., Alberts, S. R., Holton, S. J., Stockham, A., Leventakos, K., Hubbard, J. M., Mansfield, A. S., Halfdanarson, T. R., Chen, R., Jochum, J. A., Schwecke, A. S., Eiring, R. A., Carroll, J. L., & Mandrekar, S. J. (2024). Transforming the oncology data paradigm by creating, capturing, and retrieving structured cancer data at the point of care: A Mayo Clinic pilot. *Cancer*, cncr.35304. DOI: 10.1002/cncr.35304 PMID: 38662502

The Intact One. (2023, April 8). *Cyber Regulation Appellate Tribunal.* https://theintactone.com/2023/04/08/cyber-regulation-appellate-tribunal/

The Personal Data Protection Bill. (2019). https://prsindia.org/billtrack/the-personal-data-protection-bill-2019

Theodorou, A., Panno, A., Carrus, G., Carbone, G. A., Massullo, C., & Imperatori, C. (2021). Stay home, stay safe, stay green: The role of gardening activities on mental health during the Covid-19 home confinement. *Urban Forestry & Urban Greening*, 61, 127091. DOI: 10.1016/j.ufug.2021.127091 PMID: 35702591

Thompson, L., & Garcia, S. (2020). The double-edged sword of healthcare data volume. *International Journal of Medical Informatics*, 136(1), 102–114. DOI: 10.1016/j.ijmedinf.2020.104094

Törngren, M., & Sellgren, U. (2018). Complexity challenges in development of Cyber-Physical systems. In Lecture Notes in Computer Science (pp. 478–503). DOI: 10.1007/978-3-319-95246-8_27

Tozzi, C. (2022). Will the Metaverse Help or Hinder Sustainability. ITPro Today, 10.

Tresp, V., Overhage, J. M., Bundschus, M., Rabizadeh, S., Fasching, P. A., & Yu, S. (2016). Going digital: A survey on digitalization and large-scale data analytics in healthcare. *Proceedings of the IEEE*, 104(11), 2180–2206. DOI: 10.1109/JPROC.2016.2615052

Tripathi, S., & Ghatak, C. (2018). Artificial Intelligence and Intellectual Property Law, 8, *Christ University Law Journal*, http://journals.christuniversity.in/index.php

Tudosi, A., Graur, A., Balan, D. G., & Potorac, A. D. (2023). Research on Security Weakness Using Penetration Testing in a Distributed Firewall. *Sensors (Basel)*, 23(5), 2683. DOI: 10.3390/s23052683 PMID: 36904890

Tuovila, A. (2024, May 12). Relative valuation model: Definition, steps, and types of models. Investopedia. https://www.investopedia.com/terms/r/relative-valuation-model.asp

Tyagi, A. K., & Sreenath, N. (2021). Cyber Physical Systems: Analyses, challenges and possible solutions. *Internet of Things and Cyber-Physical Systems*, 1, 22–33. DOI: 10.1016/j.iotcps.2021.12.002

U.S. Congress. (2021). *S.1353 – International Cybercrime Prevention Act.* https://www.congress.gov/bill/117th-congress/senate-bill/1353/text

U.S. Senate Committee on Armed Services. (2024). *FY25 National Defense Authorization Act executive summary.* https://www.armed-services.senate.gov/imo/media/doc/fy25_ndaa_executive_summary.pdf

United Nations. (2022). *Sustainable Development Goals*. United Nations.

Upreti, K., Syed, M. H., Alam, M. S., Alhudhaif, A., Shuaib, M., & Sharma, A. K. (2021). Generative adversarial networks based cognitive feedback analytics system for integrated cyber-physical system and industrial iot networks.

Upreti, K., Arora, S., Sharma, A. K., Pandey, A. K., Sharma, K. K., & Dayal, M. (2024). Wave Height Forecasting Over Ocean of Things Based on Machine Learning Techniques: An Application for Ocean Renewable Energy Generation. *IEEE Journal of Oceanic Engineering*, 49(2), 1–16. DOI: 10.1109/JOE.2023.3314090

Upreti, K., Peng, S., Kshirsagar, P. R., Chakrabarti, P., Al-Alshaikh, H. A., Sharma, A. K., & Poonia, R. C. (2023). A multi-model unified disease diagnosis framework for cyber healthcare using IoMT- cloud computing networks. *Journal of Discrete Mathematical Sciences and Cryptography*, 26(6), 1819–1834. DOI: 10.47974/JDMSC-1831

Upreti, K., Syed, M. H., Khan, M. A., Fatima, H., Alam, M. S., & Sharma, A. (2022). Enhanced algorithmic modelling and architecture in deep reinforcement learning based on wireless communication Fintech technology. *Optik (Stuttgart)*, 272, 170309. DOI: 10.1016/j.ijleo.2022.170309

Upreti, K., Vats, P., Srinivasan, A., Sagar, K. V. D., Mahaveerakannan, R., & Babu, G. C. (2023). Detection of Banking Financial Frauds Using Hyper-Parameter Tuning of DL in Cloud Computing Environment. *International Journal of Cooperative Information Systems*, 2350024. Advance online publication. DOI: 10.1142/S0218843023500247

van Eck, N. J., & Waltman, L. (2010). Software survey: VOSviewer, a computer program for bibliometric mapping. *Scientometrics*, 84(2), 523–538. DOI: 10.1007/s11192-009-0146-3 PMID: 20585380

van Eck, N. J., & Waltman, L. (2017). Citation-based clustering of publications using CitNetExplorer and VOSviewer. *Scientometrics*, 111(2), 1053–1070. DOI: 10.1007/s11192-017-2300-7 PMID: 28490825

Viswas, A., Dabla, P. K., Gupta, S., Yadav, M., Tanwar, A., Upreti, K., & Koner, B. C. (2023). SCN1A Genetic Alterations and Oxidative Stress in Idiopathic Generalized Epilepsy Patients: A Causative Analysis in Refractory Cases. *Indian Journal of Clinical Biochemistry*. Advance online publication. DOI: 10.1007/s12291-023-01164-x

Wang, K., Belt, M C D., Heath, G., Walzberg, J., Curtis, T L., Berrie, J., Schroeder, P., Lazer, L., & Altamirano, J. (2022, November 1). Circular economy as a climate strategy: current knowledge and calls-to-action. DOI: 10.2172/1897625

Wang, B., Zheng, P., Yin, Y., Shih, A., & Wang, L. (2022). Toward human-centric smart manufacturing: A human-cyber-physical systems (HCPS) perspective. *Journal of Manufacturing Systems*, 63, 471–490. DOI: 10.1016/j.jmsy.2022.05.005

Wang, J., & Xu, Z. (2015). *Crowd anomaly detection for automated video surveillance*. IEEE Xplore., DOI: 10.1049/ic.2015.0102

Wang, L., & Liu, X. (2018). NOTSA: Novel OBU With Three-Level Security Architecture for Internet of Vehicles. *IEEE Internet of Things Journal*, 5(5), 3548–3558. DOI: 10.1109/JIOT.2018.2800281

Wang, Q., Kung, L., & Byrd, T. A. (2018). Levels of data quality in healthcare settings. *Journal of Health Informatics*, 24(2), 189–201. DOI: 10.1177/1460458218784611

Wang, X., Jerome, Z., Wang, Z., Zhang, C., Shen, S., Kumar, V., Bai, F., Krajewski, P. E., Deneau, D., Ahmad, J., Jones, R., Piotrowicz, G., & Liu, H. (2024, February 20). Traffic light optimization with low penetration rate vehicle trajectory data. *Nature Communications*, 15(1), 1306. Advance online publication. DOI: 10.1038/s41467-024-45427-4 PMID: 38378680

Wati, K. M. (2024). Cyber Physical System For Automated Weather Station And Agriculture Node In Smart Farming. Globe: Publikasi Ilmu Teknik, Teknologi Kebumian. *Ilmu Perkapalan*, 2(1), 13–27.

Wickramasinghe, C. S., Marino, D. L., Amarasinghe, K., & Manic, M. (2018). Generalization of deep learning for cyber-physical system security: A survey. IECON 2018 - 44th Annual Conference of the IEEE Industrial Electronics Society, Washington, DC, USA, 745-751. DOI: 10.1109/IECON.2018.8591773

Wiederhold, B. K., & Riva, G. (2022). Metaverse creates new opportunities in healthcare. *Annual Review of Cybertherapy and Telemedicine*, 20, 3–7.

Williams, P., & Davis, R. (2021). Innovative approaches to healthcare data storage and management. *BMC Medical Informatics and Decision Making*, 21(3), 567–579. DOI: 10.1186/s12911-021-01524-5

Williams, R., & Taylor, M. (2022). Merging heterogeneous data sources in healthcare: Techniques and challenges. *International Journal of Medical Informatics*, 160, 104692. DOI: 10.1016/j.ijmedinf.2022.104692

Wilson, R., & Garcia, L. (2022). Leveraging in-memory computing for real-time healthcare data processing. *International Journal of Medical Informatics*, 160, 104693. DOI: 10.1016/j.ijmedinf.2022.104693

Wong, E. K. S., Ting, H. Y., & Atanda, A. F. (2024). Enhancing Supply Chain Traceability through Blockchain and IoT Integration: A Comprehensive Review. *Green Intelligent Systems and Applications*, 4(1), 11–28. DOI: 10.53623/gisa.v4i1.355

Woo, H., & Yi, J., (2008). Design and development methodology for resilient cyber-physical systems. In *Proceedings of The 28th International Conference on Distributed Computing Systems Workshop*, 525-528. DOI: 10.1109/ICDCS.Workshops.2008.62

Wu, F., Gao, J., Kang, J., Wang, X., Niu, Q., Liu, J., & Zhang, L. (2022). Knowledge Mapping of Exosomes in Autoimmune Diseases: A Bibliometric Analysis (2002–2021). *Frontiers in Immunology*, 13, 939433. https://www.frontiersin.org/journals/immunology/articles/10.3389/fimmu.2022.939433. DOI: 10.3389/fimmu.2022.939433 PMID: 35935932

Wu, H., Cheng, K., Guo, Q., Yang, W., Tong, L., Wang, Y., & Sun, Z. (2021). Mapping Knowledge Structure and Themes Trends of Osteoporosis in Rheumatoid Arthritis: A Bibliometric Analysis. *Frontiers in Medicine*, 8, 787228. DOI: 10.3389/fmed.2021.787228 PMID: 34888333

Wu, Z., Cheng, K., Shen, Z., Lu, Y., Wang, H., Wang, G., Wang, Y., Yang, W., Sun, Z., Guo, Q., & Wu, H. (2023). Mapping knowledge landscapes and emerging trends of sonodynamic therapy: A bibliometric and visualized study. *Frontiers in Pharmacology*, 13, 1048211. https://www.frontiersin.org/journals/pharmacology/articles/10.3389/fphar.2022.1048211. DOI: 10.3389/fphar.2022.1048211 PMID: 36699067

Xiang, J., & Liu, N. (2022). Crowd density estimation method using deep learning for passenger flow detection system in exhibition center. *Scientific Programming*, 2022(1), 1990951.

Xiong, G., Li, Z., Wu, H., Chen, S., Dong, X., Zhu, F., & Lv, Y. (2021, January 26). Building Urban Public Traffic Dynamic Network Based on CPSS: An Integrated Approach of Big Data and AI. *Applied Sciences (Basel, Switzerland)*, 11(3), 1109–1109. DOI: 10.3390/app11031109

Xu, B., Da Xu, L., Cai, H., Xie, C., Hu, J., & Bu, F. (2014). Ubiquitous Data Accessing Method in IoT-Based Information System for Emergency Medical Services. *IEEE Transactions on Industrial Informatics*, 10(2), 1578–1586. DOI: 10.1109/TII.2014.2306382

Yaacoub, J. P. A., Salman, O., Noura, H. N., Kaaniche, N., Chehab, A., & Malli, M. (2020). Cyber-physical systems security: Limitations, issues and future trends. *Microprocessors and Microsystems*, 77, 103201. DOI: 10.1016/j.micpro.2020.103201 PMID: 32834204

Yadav, V. S., Singh, A. R., Raut, R. D., Mangla, S. K., Luthra, S., & Kumar, A. (2022). Exploring the application of Industry 4.0 technologies in the agricultural food supply chain: A systematic literature review. *Computers & Industrial Engineering*, 169, 108304. DOI: 10.1016/j.cie.2022.108304

Yang, T., & Lv, C. (2021). A secure sensor fusion framework for connected and automated vehicles under sensor attacks. *IEEE Internet of Things Journal*, 9(22), 22357–22365. DOI: 10.1109/JIOT.2021.3101502

Yashi., (2023). Artificial Intelligence and Laws in India, *Legal Service India*, https://legalserviceindia.com/legal/article-8171-artificial-intelligence-and-laws-in-india.html

Yeung, A. W. K., & Mozos, I. (2020). The Innovative and Sustainable Use of Dental Panoramic Radiographs for the Detection of Osteoporosis. *International Journal of Environmental Research and Public Health*, 17(7), 7. Advance online publication. DOI: 10.3390/ijerph17072449 PMID: 32260243

Yu, B., Zhou, J., & Hu, S. (2020). Cyber-physical systems: An overview. In Big data analytics for cyber-physical systems (pp. 1-11)

Yuan, C., Ding, S., Wang, Y., Feng, J., & Ma, N. (2023, January 1). Emergency Resource Layout with Multiple Objectives under Complex Disaster Scenarios. Cornell University. https://doi.org/DOI: 10.48550/arXiv.2304

Yu, T., & Wang, X. (2022). Real-time data analytics in Internet of Things systems. In *Handbook of real-time computing* (pp. 541–568). Springer Nature Singapore. DOI: 10.1007/978-981-287-251-7_38

Zaidan, E., Ghofrani, A., Abulibdeh, A., & Jafari, M. A. (2022, March 2). Accelerating the Change to Smart Societies- a Strategic Knowledge-Based Framework for Smart Energy Transition of Urban Communities. *Frontiers in Energy Research*, 10, 852092. Advance online publication. DOI: 10.3389/fenrg.2022.852092

Zaidi, K. (2021). Artificial Intelligence and Cyber Law, *Ipleaders*, https://blog.ipleaders.in/artificial-intelligence-cyber-law

Zainurin, S. N., Ismail, W. Z. W., Mahamud, S. N. I., Ismail, I., Jamaludin, J., Ariffin, K. N. Z., & Kamil, W. M. W. A. (2022, October 28). Advancements in Monitoring Water Quality Based on Various Sensing Methods: A Systematic Review. *International Journal of Environmental Research and Public Health*, 19(21), 14080–14080. DOI: 10.3390/ijerph192114080 PMID: 36360992

Zeng, L., Wang, W., Feng, D., Zhang, X., & Chen, X. (2023, January 1). A3D: Adaptive, Accurate, and Autonomous Navigation for Edge-Assisted Drones. Cornell University. https://doi.org/DOI: 10.48550/arXiv.2307

Zhang, L., Sridhar, K., Liu, M., Lu, P., Chen, X., Kong, F., . . . Lee, I. (2023). Real-time data-predictive attack-recovery for complex cyber-physical systems. In 2023 IEEE 29th Real-Time and Embedded Technology and Applications Symposium (RTAS) (pp. 209-222). IEEE DOI: 10.1109/RTAS58335.2023.00024

Zhang, X.-L., Zheng, Y., Xia, M.-L., Wu, Y.-N., Liu, X.-J., Xie, S.-K., Wu, Y.-F., & Wang, M. (2020). Knowledge Domain and Emerging Trends in Vinegar Research: A Bibliometric Review of the Literature from WoSCC. *Foods*, 9(2), 2. Advance online publication. DOI: 10.3390/foods9020166 PMID: 32050682

Zhang, Y., & Yen, I., (2009). Optimal adaptive system health monitoring and diagnosis for resource constrained cyber-physical systems. In *Proceedings of the 20th International Symposium on Software Reliability Engineering*, 51-60. DOI: 10.1109/ISSRE.2009.21

Zhang, Y., Zhou, D., Chen, S., Gao, S., & Ma, Y. (2016). Single-image crowd counting via multi-column convolutional neural network. *2016 IEEE Conference on Computer Vision and Pattern Recognition (CVPR)*, Las Vegas, NV, USA, 589-597. DOI: 10.1109/CVPR.2016.70

Zhao, W., Jiang, C., Gao, H., Yang, S., & Luo, X. (2021). Blockchain-Enabled Cyber–Physical Systems: A review. *IEEE Internet of Things Journal*, 8(6), 4023–4034. DOI: 10.1109/JIOT.2020.3014864

Zheng, X. (2022, July 27). Application of CPS Under the Background of Intelligent Construction. *Application of CPS Under the Background of Intelligent Construction.*, 6(4), 22–33. DOI: 10.26689/jwa.v6i4.4190

Zhou, K., Liu, T., & Liang, L. (2017). Security in cyber-physical systems: Challenges and solutions. *International Journal of Autonomous and Adaptive Communications Systems*, 10(4), 391–408. DOI: 10.1504/IJAACS.2017.088775

Zhou, Z.-H. (2021). *Machine Learning*. Springer Nature. DOI: 10.1007/978-981-15-1967-3

Zhu, H. J., You, Z. H., Zhu, Z. X., Shi, W. L., Chen, X., & Cheng, L. (2018). *Effective and robust detection of android malware using static analysis along with rotation forest model*. Semantic Scholar. DOI: 10.1016/j.neucom.2017.07.030

Ziehn, A., Grulich, P. M., Zeuch, S., & Markl, V. (2024). Bridging the gap: Complex event processing on stream processing systems. In EDBT (pp. 447-460)

Ziro, A., Gnatyuk, S., & Toibayeva, S. (2023). Improved Method for Penetration Testing of Web Applications. Intel*ITSIS'2023: 4th International Workshop on Intelligent Information Technologies and Systems of Information Security*, March 22–24, Khmelnytskyi, Ukraine.

Zitouni, M. S., Dias, J., Al-Mualla, M., & Bhaskar, H. (2015). *Hierarchical crowd detection and representation for big data analytics in visual surveillance.* IEEE Xplore., DOI: 10.1109/SMC.2015.320

About the Contributors

Ramesh Chandra Poonia is a Professor at the Department of Computer Science, CHRIST (Deemed to be University), India. He holds a double Postdoctoral Fellowship, first from the Cyber-Physical Systems Laboratory Lab, Department of ICT and Natural Sciences, Norwegian University of Science and Technology - NTNU, Norway and second fellowship on the Collaborative Project of Oakland University USA, and Imam University, Saudi Arabia. He obtained his PhD in Computer Science from Banasthali University, Banasthali, India, in July 2013. He received his MTech in DSE from Birla Institute of Technology and Science, Pilani, India. He has authored over 126 journal and conference articles, edited 11 books/conference proceedings. He is the main guest editor of more than 30 special issues of journals including Springer, Taylor and Francis, Inderscience, IGI Global and Elsevier. He is an associate editor of the Journal of Sustainable Computing: Informatics and Systems, Elsevier. He is serving as an editorial board member in several international journals/conferences. He was endorsed with the prestigious 'Faculty Appreciation Award' in 2013 for commendable services. His research interests include Sustainable Technologies, Cyber-Physical Systems, Computational Intelligence, and Network Protocol Evaluation.

Kamal Upreti is currently working as an Associate Professor in Department of Computer Science, CHRIST (Deemed to be University), Delhi NCR, Ghaziabad, India. He completed is B. Tech (Hons) Degree from UPTU, M. Tech (Gold Medalist), PGDM(Executive) from IMT Ghaziabad and PhD in Department of Computer Science & Engineering.He has completed Postdoc from National Taipei University of Business, TAIWAN funded by MHRD. He has published 50+ Patents, 32+Magazine issues and 110+ Research papers in in various reputed Journals and international Conferences. His areas of Interest such as Modern Physics, Data Analytics, Cyber Security, Machine Learning, Health Care, Embedded System and Cloud Computing. He has published more than 45+ authored and edited books under CRC Press, IGI Global, Oxford Press and Arihant Publication. He is having enriched years' experience in corporate and teaching experience in Engineering Colleges. He worked with HCL, NECHCL, Hindustan Times, Dehradun Institute of Technology and Delhi Institute of Advanced Studies, with more than 15+ years of enrich experience in research, Academics and Corporate . He also worked in NECHCL in Japan having project – "Hydrastore " funded by joint collaboration between HCL and NECHCL Company. Dr. Upreti worked on Government project – "Integrated Power Development Scheme (IPDS)" was launched by Ministry of Power, Government of India with the objectives of Strengthening of sub-transmission and distribution network in the urban areas. Currently, he has completed work with Joint collaboration with GB PANT & AIIMS Delhi, under funded project of ICMR Scheme on Cardiovascular diseases prediction strokes using Machine Learning Techniques from year 2017-2020 of having fund of 80 Lakhs .He got 5 Lakhs fund from DST SERB for conducting International Conference, ICSCPS-2024, 13-14 Sept 2024. Recently, he got 10 Lakhs fund from AICTE - Inter-Institutional Biomedical Innovations and Entrepreneurship Program (AICTE-IBIP) for 2024-2026. He has attended as a Session Chair Person in National, International conference and key note speaker in various platforms such as Skill based training, Corporate Trainer, Guest faculty and faculty development Programme. He awarded as best teacher, best researcher, extra academic performer and Gold Medalist in M. Tech programme.

***

Susanta Das is a seasoned educator with a proven record of teaching and mentoring diverse students, working at multiple institutions, and providing service to the various initiatives of organizations. He received his Ph.D. and M.A. degrees from Western Michigan University (WMU), USA, and M.Sc. degree from Banaras Hindu University (BHU), India all in Physics. He continued his research as a Marie-Curie post-doctoral fellow at Stockholm University, Stockholm, Sweden on beam diagnostics for the DESIREE (Double ElectroStatic Ion Ring ExpEriment) in the project DITANET (Diagnostic Techniques for particle Accelerators – a Marie-Curie initial training NETwork), at the Indian Institute of Science Education and Research-Kolkata, India on high-pressure physics as a Project Scientist-B, and at the University of Electro-Communications, Tokyo, Japan on ion-surface interactions as a Post-doctoral fellow. Throughout his career, Dr. Das worked and collaborated with many researchers, Ph.D., master, and visiting students from different countries. He further visited many countries to discuss research and an international conference participant (UK, Italy, Germany, Greece, Belgium, Bulgaria, Romania, Brazil etc.). He co-authored several research articles in WoS/Scopus indexed international journals, conference proceedings, and scholarly book chapters published by renowned international publishing houses. Beyond his research endeavors, Dr. Das has a storied history of service to academic and administrative committees, exhibiting his commitment to the growth and development of educational institutions. His experience includes tenure at Central University South Bihar, Sri Sri University, and P.K. University before assuming his current role at Ajeenkya DY Patil University. He received the Marie-Curie post-doctoral fellowship in Sweden, Science Academies' Summer Research Fellowship in India, Gwen Frostic Doctoral Fellowship, Department Graduate Research and Creative Scholar Award by WMU, and Leo R. Parpart Doctoral Fellowship by Dept. of Physics, WMU, among many others, throughout his academic journey. At present, Dr. Das continues to delve into cutting-edge fields, with a keen interest in nanotechnology, quantum computing, and data science. His multifaceted contributions, spanning teaching, research, and administrative leadership, showcase a dedicated professional who is instrumental in advancing the vision and mission of the institutions he serves.

Shubham Ekatpure Expertise in new product management, FABs and OSAT customer Management, Overseas ODM/OEM Coordination, and supplier tech development in Semiconductor. I leverage my skillsets in product management, supply chain network design, Python coding, and manufacturing to drive efficient operations, time to market product releases, reduce costs, and enhance customer satisfaction.

G. Sowmya is an Assistant Professor in the Department of Computer Science and Engineering at MLRIT, Hyderabad, with over a decade of teaching experience. Currently pursuing a Ph.D. in CSE from JNTUH, her research focuses on the formal verification of smart contracts. Sowmya holds an M.Tech and B.Tech in CSE and has received the prestigious Prathibha Award for academic excellence. She has contributed to numerous international journals and conferences, authored several patents, and actively engages in various FDPs and certifications, particularly in Blockchain, AI, and IoT. She has also conducted workshops and webinars on cutting-edge technologies such as blockchain and machine learning

Kushal Gaddamwar is an emerging professional in the field of Computer Science and Engineering, with a deep passion for cutting-edge technologies such as Artificial Intelligence and Machine Learning. A graduate of PDPM IIIT Jabalpur, Kushal has consistently demonstrated his ability to merge innovation with practicality, working on various impactful projects that aim to solve real-world challenges. His work spans the development of intelligent systems and personalized solutions, focusing on improving user experiences across multiple domains. As a proactive leader, Kushal has also contributed to mentoring and guiding fellow students, fostering a collaborative environment that encourages learning and growth. With a commitment to advancing technology for the betterment of society, Kushal continues to explore new avenues in AI, web development, and beyond

Himanshu Gupta is a seasoned professional with over 15 years of experience in the IT industry. He has expertise in software development, project management, and team leadership. Himanshu has worked with various technologies.

Rituraj Jain is currently working in the Department of Information Technology, Marwadi University, Rajkot, GUjarat, India. He has published many papers in International Journals and Conferences. He has published 4 books. He is an Oracle Database SQL Certified Expert. He is also associated with professional bodies like the Indian Society of Technical Education (ISTE), Computer Science Teachers Association (CSTA), Academy & Industry Research Collaboration Center (AIRCC), International Association of Computer Science and Information Technology (IACSIT), International Association of Engineers (IAENG). His research interests include Cloud Computing, Machine Learning, Deep Learning, Fuzzy Logic, IoT, Genetic Algorithms, Big Data and Software Engineering.

Sheetal Kalra did her Ph.D. in Computer Science & Engineering from Guru Nanak Dev University, Amritsar. She completed her MCA from the same University. She is currently working as an Assistant Professor in department of Computer Science and Engineering, G.N.D.U. Regional Campus, Jalandhar. She has 9 years of teaching experience and has more than 20 research publications in highly reputed journals to her credit. Her research areas are network & information security, cloud computing, big data and Internet of Things.

Parminder Kaur is working as an Associate Professor in the Department of Computer Science, at Guru Nanak Dev University Amritsar, India. She has done her post-graduation in Mathematics as well as System Software and doctorate in the field of Computer Science. She has more than 60 publications in International/National Journals/Conferences. Her research interests include Component-based Software Engineering, Web Engineering, Software Security, Semantic Web, Digital Image Processing and Service-Oriented Architecture.

Amna Kausar is a final year B.Tech. Computer Science Engineering student at Ajeenkya DY Patil University with specialising in AI and a minor in robotics. Her journey into technology started at a young age, fixing broken electronics, which evolved into winning academic and inter-school competitions, including awards in robotics and business challenges. She is particularly interested in the intersection of her two biggest strengths, creativity and logical thinking, finding a deep interest in the concept of biologically inspired systems like AI and robotics. Her professional experience includes internships in security, research, development, and as an AI solution engineer, she also contributes to developing software for NGOs. Outside of studies, she enjoys playing chess, reading, exploring different art mediums, and practising mindfulness through walks. Currently, she is moving forward with one of her startup ideas, participating in the IUCAA summer school programme in Astronomy and Astrophysics, while working on a personal project aimed at developing solutions for individuals with hearing impairments. By aligning her technical skills with her personal goals, she reflects her commitment to innovation and societal impact.

Rupesh Kumar M I specialize in marketing with 21 years of combined expertise in academic and industry roles. My doctoral research is focused on the retail sector. I've contributed 21 articles to national and international referred journals. Over the past decade, I have dedicated my research efforts to multi-brand outlets, publishing 12 articles in this domain. My primary research interests revolve around retail marketing, consumer behaviour, and advertising.

Rajeevan M S is a passionate library professional and information scientist based in Kerala, India. With a background in Physics and a postgraduate degree in Library and Information Science from the University of Kerala, Rajeevan is dedicated to fostering access to knowledge and supporting academic research. Currently serving as a Library professional at the Knowledge Resource Centre, Indian Institute of Technology, Hyderabad, Rajeevan's commitment to lifelong learning and knowledge dissemination shines through in his work.

Katelyn Medows B.Tech in Computer Science and Engineering Graduate with Minor in Psychology, passionate about using technology to solve complex problems with a strong fondation in machine learning, data analysis, and software development

Michael Moses T is a dedicated and highly accomplished Assistant Professor in the Department of Computer Science and Engineering at CHRIST University since 2013. He holds a Ph.D. from CHRIST University, specializing in Affective Computing for Improving Classroom Engagement through Deep Learning methodologies. Dr. Moses earned his M.Tech. in Software Engineering from Periyar Maniammai University and a B.Tech in Computer Science and Engineering from Anna University. In addition to his academic role, Dr. Moses plays a significant role as an NCC Officer for the Airwing, contributing to cadet training, leadership development, and aviation-related skills within the NCC program. His research contributions are widely recognized, with numerous journal articles and indexed book chapters in prestigious publications. Dr. Moses' expertise includes areas such as secure data sharing in cloud environments, multimodal emotional analysis using video summarization, IoT applications in smart mirror structures, and engagement detection through facial emotional recognition. He has presented his work at esteemed conferences organized by IEEE, and Springer and has published extensively in leading journals that have expertise in Multimedia Tools, Image processing, Deep learning, and Information Technology. Dr. Moses' research extends to areas like brain tumor detection, diabetic retinopathy diagnosis, emotional detection, Affective Computing, Student Engagement Detection, and stock market prediction using advanced deep-learning techniques.

Arul Kumar Natarajan currently serves as an Assistant Professor in the Department of Computer Science at the Samarkand International University of Technology in Uzbekistan. He earned his Doctor of Philosophy degree from Bharathidasan University, India, in 2017. Concurrently, he is engaged in postdoctoral research in Generative AI for Cybersecurity at the Singapore Institute of Technology, Singapore. Throughout his 14-year teaching career, Dr. Arul has held esteemed positions at various institutions, including Christ University, Bishop Heber College in India, and Debre Berhan University in Ethiopia. Dr. Arul has made significant contributions to academia, specializing in cybersecurity and artificial intelligence, evidenced by his portfolio of scholarly works. He has authored 32 international publications indexed in Scopus and 03 international publications and has delivered 34 conference presentations. Additionally, he has edited and published three books with IGI Global Publisher, USA, focusing on Python data structures, algorithms, and geospatial application development. In addition to his academic pursuits, Dr. Arul is a prolific innovator, holding 17 patents in India and 01 granted patent in the United Kingdom across diverse domains, including communication and computer science. He demonstrates proficiency in networking and cybersecurity, having completed the CCNA Routing and Switching Exam from CISCO and the Networking Fundamentals exam from Microsoft. He maintains a sincere interest in GenAI for Cybersecurity.

Damodharan Palaniappan works as an Associate Professor & Head in the Department of Information Technology, Marwadi University, Rajkot, Gujarat. He completed his Ph.D. at Anna University, Chennai, M.E. in Computer Science and Engineering at C.I.E.T, Coimbatore and B.E. in Computer Science and Engineering at CSI College of Engineering, Ketty, Nilgiri. He has 19 years' experience of teaching and research. Currently, he guides 6 Ph.D. research scholars. He has authored 8 book chapters, and he has published research papers in reputed journals. His interested areas of research are Data Mining, Image Processing, Software, Machine Learning and Deep Learning.

Vishakha Kuwar is an Assistant Professor and founder of a startup company. She completed Bachelor of Engineering in Computer Science Engineering, Masters in Business Administration and qualified UGC-NET in Management. With 9.5 years of diverse experience, including 5 years in teaching and 4.5 years in industrial roles, she has worked across both service and product-based industries. Her expertise spans International Business Operations, Marketing, Production & Operations Management, and IT Management. Vishakha has published research in three national and five international conferences and is skilled in business development, instructional design, and analytics. Vishakha's analytical and strategic approach positions her to drive informed decision-making and contribute significantly to research and organizational success.

Atharva Saraf is pursuing his B.Tech. in Computer Science and Engineering at School of Engineering, Ajeenkya DY Patil University, Pune, India. His research interests are cybersecurity, software designing, and artificial intelligence.

Ravindra Babu Sathyanarayana Has more than 25 years of Industry and Teaching (Academic) experience and more than eight years of research experience. Has been teaching and researching in financial modeling, financial analytics, derivatives, SAPM, International finance, and strategic financial management in reputed B-schools. Is currently working as a Professor at the School of Business and Management (SBM), CHRIST University, Kengeri, Bangalore. Has presented and published several research articles and case studies in various reputed journals (ABDC, WOS, Scopus, UGC-peer). Has published eight books so far in Financial Management, Security Analysis and Portfolio Management, International Financial Management, Banking & Insurance, and Stock and Commodity Markets for MBA students of Bangalore and VTU. Has been on Academic Council, BOS, and BOE for prestigious autonomous colleges and Deemed to be a University. Has been part of the CORE Team for NBA, AACSB and NACC accreditations.

Aditya Sharma is pursuing his B.Tech in Computer Science and Engineering at School of Engineering, Ajeenkya DY Patil University. His research interests are cybersecurity and artificial intelligence.

Jitendra Nath Shrivastava is a Professor in the Department of Computer Science & Engineering. He has been the Vice-Chancellor, Officer on Special Duty (OSD), and Advisor to the proctorial Board in the past. He is having more than 24 years of experience in teaching, research & administration. He has more than 24 research papers in international & national journals of repute to his credit. He has also delivered many expert lectures in various reputed academic institutions. His research area of expertise involves data mining, artificial intelligence, and database management system. He always attempts to provide a student-friendly environment in the campus.

Anoop V S is a dedicated and accomplished professional with a proven track record in Applied Artificial Intelligence Research and Teaching. Possessing a solid foundation in Natural Language Processing and Text Mining, Anoop has consistently delivered impactful results in both academic and practical settings. He is committed to advancing the field of Applied Artificial Intelligence, specifically Natural Language Processing and Computational Social Science, through innovative research and fostering a dynamic learning environment for students. Dr. Anoop has led multiple research projects in AI, resulting in published papers in reputable international journals and conferences. His proven ability to communicate complex AI concepts in an accessible manner consistently generated positive student feedback for engaging and effective teaching methods. He is highly skilled in creating curricula, delivering lectures, handling governmental and corporate projects, and mentoring students. Dr. Anoop has worked in various roles, such as AI Research Scientist, Assistant Professor, Data Scientist, and Visiting Faculty for reputed organizations and institutions in India and abroad. Dr. Anoop has completed his Ph.D. and M.Phil. in Artificial Intelligence from the Faculty of Technology, Cochin University of Science and Technology, and his Postdoctoral Fellowship at the Smith School of Business at Queen's University, Canada. Anoop's research interest is in Natural Language Processing and Computational Social Science, and he has several publications, including edited books, book chapters, articles in international journals, and conference proceedings. He is a recipient of the Mozilla Responsible Computing Research Grant from the Mozilla Foundation and the United States Agency for International Development.

Srinidhi Vasan is an accomplished finance professional with over five years of experience in finance, data analysis, and project management. He has worked at the London Stock Exchange Group. He has received several awards, including the latest, Best Paper Award for his research on crisis management in the real estate sector. He has also excelled in international Taekwondo championships, securing gold medals and earning a place in the International Martial Arts Hall of Fame.

Index

A

Advanced Sensors 215, 222, 224
Artificial intelligence 33, 38, 47, 50, 57, 59, 120, 121, 127, 149, 155, 156, 157, 158, 159, 161, 163, 170, 172, 173, 179, 180, 181, 182, 183, 227, 228, 235, 236, 238, 239, 242, 245, 248, 250, 252, 262, 271, 278, 280, 281, 282, 285, 302, 304, 325, 359, 362, 366, 374, 376, 382, 387, 389
Attack Surface 5, 195, 200, 202

B

Bibliometric Analysis 343, 344, 346, 348, 365, 366, 371, 388
Big Data 41, 82, 88, 89, 116, 121, 153, 223, 229, 231, 261, 284, 287, 288, 289, 290, 292, 293, 294, 297, 298, 300, 301, 302, 303, 304, 305, 306, 344, 361, 362, 364, 368, 389
Blockchain 1, 2, 3, 6, 7, 8, 9, 10, 12, 16, 17, 18, 19, 20, 21, 22, 23, 24, 25, 26, 27, 28, 29, 30, 31, 32, 33, 34, 35, 149, 219, 285, 325, 361, 362, 366, 368, 369, 374, 375, 383, 385, 386, 387, 388

C

CAN 2, 5, 6, 8, 10, 11, 12, 13, 14, 15, 16, 17, 18, 19, 20, 21, 22, 23, 24, 25, 26, 27, 28, 29, 30, 31, 32, 37, 39, 40, 42, 43, 49, 50, 51, 52, 53, 58, 59, 60, 61, 64, 67, 69, 70, 78, 79, 80, 83, 85, 86, 88, 89, 91, 93, 96, 99, 103, 110, 113, 118, 119, 124, 125, 126, 127, 128, 129, 130, 131, 132, 133, 134, 135, 136, 137, 139, 141, 142, 143, 144, 146, 147, 155, 156, 157, 158, 159, 160, 161, 162, 163, 164, 166, 167, 168, 170, 171, 173, 174, 175, 176, 177, 178, 179, 185, 186, 187, 188, 189, 190, 192, 193, 194, 195, 196, 198, 201, 202, 203, 204, 205, 210, 211, 217, 218, 219, 220, 222, 223, 224, 225, 237, 239, 240, 241, 242, 243, 245, 246, 247, 248, 249, 252, 253, 254, 260, 261, 262, 263, 264, 265, 266, 267, 268, 269, 270, 271, 272, 273, 274, 276, 277, 278, 279, 280, 288, 296, 298, 299, 300, 302, 304, 311, 312, 313, 314, 315, 316, 317, 318, 319, 320, 321, 322, 323, 326, 328, 329, 330, 331, 332, 333, 334, 335, 336, 338, 339, 340, 344, 366, 374, 376, 377, 378, 380, 381, 383
CERT 166, 167, 168, 181
Challenges 2, 3, 4, 6, 8, 9, 10, 11, 16, 20, 21, 22, 23, 24, 25, 26, 27, 28, 29, 30, 31, 32, 34, 35, 39, 42, 43, 45, 47, 48, 49, 50, 58, 59, 60, 61, 64, 70, 79, 80, 85, 86, 87, 89, 90, 116, 117, 118, 123, 124, 126, 130, 136, 137, 139, 140, 141, 142, 143, 144, 145, 146, 148, 149, 152, 156, 158, 164, 171, 173, 176, 178, 180, 183, 205, 206, 207, 212, 218, 219, 221, 222, 223, 225, 229, 230, 232, 233, 238, 246, 249, 253, 255, 258, 261, 262, 263, 264, 269, 272, 276, 283, 284, 285, 289, 290, 291, 292, 302, 303, 304, 305, 306, 307, 309, 310, 311, 314, 316, 318, 320, 323, 324, 325, 327, 343, 344, 345, 346, 368, 369, 370, 374, 378, 379, 383, 384, 385, 386, 387
Communication Protocols 10, 14, 25, 26, 27, 185, 187, 222, 236, 238, 262, 274, 275, 378, 379, 385
Computer Vision 83, 90, 98, 118, 119, 120, 121, 135, 157, 159, 197, 239
Confidential data 158, 159, 161, 162, 165, 167, 177, 328
consensus mechanisms 2, 7, 10, 17, 18, 19, 22, 23, 29, 33
Convolutional Neural Network 85, 86, 87, 89, 95, 117, 118, 121, 150
CRAT 165, 166
CrowdNet 87, 88, 89, 90, 95, 96, 97, 98, 117
cryptographic protocols 1, 30, 32

443

Cryptographic Techniques 7, 8, 13, 207, 211
Cyberattack 162, 261, 329
cyber-physical systems 1, 2, 3, 4, 8, 9, 10, 11, 12, 13, 14, 15, 16, 17, 18, 19, 20, 21, 23, 24, 25, 26, 27, 28, 29, 30, 32, 33, 34, 35, 57, 58, 59, 64, 69, 70, 78, 79, 80, 81, 83, 85, 123, 124, 125, 126, 127, 128, 130, 132, 133, 134, 135, 136, 137, 138, 139, 140, 141, 142, 143, 144, 145, 146, 148, 150, 151, 153, 186, 215, 216, 218, 220, 221, 222, 223, 224, 225, 226, 227, 228, 229, 230, 231, 232, 233, 235, 238, 245, 248, 250, 252, 253, 255, 257, 258, 259, 266, 270, 274, 278, 280, 281, 282, 283, 284, 285, 369, 373, 374, 375, 379, 381, 382, 383, 384, 385, 387, 388, 389
Cybersecurity 10, 13, 14, 24, 57, 58, 59, 60, 61, 63, 64, 70, 77, 79, 80, 81, 82, 136, 148, 155, 156, 157, 158, 159, 160, 161, 163, 164, 165, 166, 167, 168, 169, 171, 173, 174, 177, 178, 179, 180, 181, 182, 183, 211, 222, 230, 258, 261, 278, 281, 283, 327, 332, 375

D

Data-Driven 37, 38, 39, 40, 41, 125, 135, 181, 241, 253, 289, 300, 303, 374, 387, 389
data tampering 1, 2, 9, 32
decentralized architecture 1, 2, 7, 8, 11, 12, 19, 32
Decision-Making 4, 5, 12, 15, 16, 18, 19, 20, 35, 37, 38, 39, 40, 41, 42, 43, 52, 59, 82, 89, 123, 141, 156, 170, 176, 177, 179, 181, 219, 220, 223, 228, 237, 239, 240, 241, 244, 245, 246, 248, 249, 251, 252, 257, 259, 261, 262, 263, 264, 265, 270, 271, 272, 273, 274, 275, 276, 277, 278, 279, 280, 284, 288, 289, 290, 291, 292, 293, 295, 298, 299, 300, 301, 302, 303, 304, 305, 373, 374, 376, 377, 379, 380, 381, 382, 384, 385, 389

Deep Learning 58, 60, 82, 83, 85, 86, 87, 89, 90, 94, 103, 117, 118, 120, 121, 156, 159, 161, 163, 164, 178, 222, 231, 239, 278, 283, 362, 366
Density Map 94, 98, 100, 101, 102, 117
Detection 14, 17, 25, 58, 59, 60, 61, 63, 64, 65, 68, 69, 70, 71, 72, 78, 79, 80, 81, 82, 83, 84, 85, 86, 87, 88, 89, 90, 116, 118, 119, 120, 121, 134, 152, 156, 158, 159, 160, 161, 162, 163, 173, 177, 178, 181, 182, 183, 187, 189, 193, 194, 197, 198, 199, 203, 204, 205, 207, 208, 209, 210, 211, 212, 217, 222, 227, 230, 231, 239, 244, 248, 252, 262, 265, 270, 277, 283, 285, 286, 292, 300, 305, 319, 346, 372, 373, 382
Digital Divide 136, 137, 151, 243, 319, 322, 383
D. Narayana 343

E

Edge computing 15, 50, 219, 257, 258, 259, 262, 264, 267, 268, 269, 273, 278, 279, 280, 282, 319, 373, 376, 379, 380, 386, 388
efficiency 1, 2, 11, 12, 16, 19, 20, 21, 22, 23, 25, 28, 30, 32, 41, 44, 47, 58, 80, 85, 87, 88, 89, 90, 96, 116, 117, 118, 119, 124, 126, 128, 129, 131, 140, 141, 143, 146, 147, 162, 178, 205, 216, 218, 219, 220, 221, 223, 224, 226, 238, 240, 241, 246, 247, 248, 252, 253, 263, 265, 271, 273, 274, 277, 287, 291, 292, 293, 295, 298, 299, 300, 301, 303, 328, 346, 373, 374, 375, 379, 380, 381, 382, 383, 384, 385
ESG Integration 42, 46, 49

F

Financial Analysis 38, 46
FlexRay 185, 186, 187, 188, 189, 190, 191, 192, 193, 194, 195, 196, 197, 198, 199, 200, 201, 202, 203, 204, 205, 206, 207, 208, 209, 210, 211, 212, 213

G

Gaussian Filter 95, 117
Generative Artificial Intelligence 57, 59, 172

I

IDS 62, 69, 199, 211
Indian Penal Code 169, 181
Information Technology 33, 82, 123, 165, 167, 168, 170, 171, 181, 182, 215, 222, 283, 327, 337
Intelligent Automation 215, 216, 220, 221
Internet of Things 33, 34, 35, 118, 149, 158, 213, 220, 226, 230, 232, 245, 284, 285, 292, 297, 343, 344, 345, 346, 348, 349, 350, 352, 354, 355, 356, 357, 359, 361, 362, 364, 365, 366, 368, 369, 370, 371, 374, 379, 387, 388, 390

L

LIN 61, 82, 185, 186, 188, 212, 246, 248, 255, 304, 388

M

Machine learning 14, 19, 38, 41, 50, 60, 81, 94, 124, 129, 156, 157, 158, 159, 160, 161, 162, 163, 174, 180, 199, 207, 231, 235, 236, 238, 239, 242, 245, 247, 248, 250, 252, 255, 268, 270, 271, 272, 273, 278, 280, 285, 291, 298, 299, 301, 302, 305, 362, 373, 374, 376, 381, 382, 385, 386, 387, 389
Medical Internet of Things 343, 344, 345, 346, 348, 349, 350, 352, 354, 355, 356, 357, 359, 361, 364, 365, 366, 368, 369, 371
Metaverse 309, 310, 311, 312, 313, 314, 315, 316, 317, 318, 319, 320, 321, 322, 323, 325, 326
M. S. Rajeevan 343

P

Penetration Testing 210, 327, 328, 329, 330, 331, 332, 333, 334, 335, 336, 337, 338, 339, 340, 341, 342
Phishing Attacks 62, 161
Privacy 7, 9, 21, 24, 26, 27, 28, 29, 30, 35, 50, 57, 58, 59, 60, 78, 79, 80, 81, 118, 119, 135, 136, 138, 139, 140, 141, 145, 150, 152, 158, 162, 163, 164, 165, 167, 170, 172, 173, 174, 176, 177, 178, 179, 180, 193, 219, 223, 225, 229, 230, 232, 238, 242, 243, 244, 245, 249, 253, 259, 261, 265, 270, 279, 283, 289, 303, 310, 316, 320, 321, 322, 346, 362, 364, 365, 369, 373, 378, 380, 381, 383, 384

R

Real-time data 19, 20, 22, 23, 26, 29, 30, 89, 127, 220, 224, 232, 246, 252, 257, 259, 260, 261, 262, 263, 264, 265, 266, 267, 268, 269, 270, 274, 276, 280, 284, 291, 292, 298, 299, 306, 307, 373, 374, 375, 376, 377, 379, 380, 382, 383, 385, 389, 390
Real-time data analytics 268, 373, 374, 375, 376, 377, 382, 385, 389, 390
Real-Time Feedback 215, 216, 218, 220, 265, 382
Research Trends 236, 344, 345, 365
resilience 1, 6, 7, 8, 9, 10, 11, 12, 13, 14, 16, 18, 20, 30, 43, 48, 60, 79, 126, 144, 146, 150, 152, 160, 161, 192, 193, 194, 200, 219, 223, 227, 249, 252, 258, 278, 368

S

Scenario-Based 37, 42, 43, 45, 46, 47, 48, 51, 54
security 1, 2, 4, 5, 6, 7, 8, 9, 10, 11, 12, 13, 14, 16, 17, 19, 20, 21, 22, 23, 26, 30, 32, 33, 34, 35, 57, 58, 59, 60, 61, 64, 70, 78, 79, 80, 81, 82, 83, 84, 85, 88, 118, 121, 126, 132, 135, 136, 138,

139, 140, 145, 153, 157, 158, 159,
160, 161, 162, 163, 164, 165, 166,
167, 168, 169, 170, 172, 173, 174,
175, 177, 178, 179, 180, 181, 182,
185, 186, 187, 191, 192, 193, 194,
195, 196, 197, 198, 200, 203, 204,
205, 206, 208, 209, 210, 211, 212,
213, 218, 219, 221, 222, 223, 224,
225, 226, 227, 230, 232, 233, 238,
246, 249, 251, 253, 257, 258, 259,
261, 262, 263, 265, 266, 270, 275,
276, 277, 278, 279, 280, 281, 295,
303, 316, 320, 321, 327, 328, 329,
330, 331, 332, 333, 334, 335, 336,
337, 338, 339, 340, 341, 342, 346,
361, 362, 364, 365, 366, 368, 369,
373, 374, 375, 378, 380, 381, 383,
384, 385, 388
Security and privacy 26, 35, 57, 58, 59,
78, 79, 180, 232, 238, 259, 261, 279,
320, 369, 373, 378, 381
Smart Cities 2, 3, 16, 19, 20, 58, 128, 136,
138, 142, 143, 144, 148, 150, 152, 224,
227, 228, 238, 246, 248, 249, 264,
265, 266, 281, 282, 284, 325, 362,
366, 370, 375, 379, 383, 385
Social Impact 318
Standardization 25, 26, 27, 29, 30, 31, 42,
206, 209, 246, 278, 292, 293, 296, 305
Strategies 6, 11, 20, 25, 30, 37, 38, 39, 40,
41, 42, 46, 49, 58, 59, 60, 81, 84, 87,
111, 114, 119, 124, 127, 133, 137, 145,
146, 147, 152, 156, 171, 173, 174, 177,
219, 221, 223, 225, 229, 230, 237, 243,
246, 247, 252, 253, 280, 293, 299, 303,
305, 312, 328, 329, 381
Sustainability 38, 39, 40, 41, 42, 43, 46,
47, 48, 51, 52, 53, 54, 81, 124, 128,
134, 135, 139, 143, 144, 146, 147, 148,
150, 219, 225, 226, 287, 288, 294,
300, 303, 324, 325, 326, 341, 369, 384
Sustainable Development 53, 148, 280,
309, 310, 311, 313, 314, 316, 317,
318, 319, 320, 323, 326
Sustainable Investment 37, 38, 39, 41,
42, 49
System resilience 12, 258

T

Thematic Analysis 345
transparency 1, 2, 6, 8, 9, 10, 13, 17, 18,
19, 20, 24, 28, 29, 32, 84, 85, 87, 118,
119, 141, 146, 151, 164, 165, 170,
172, 176, 177, 178, 241, 242, 245,
249, 253, 271, 383, 385

U

unauthorized access 1, 2, 5, 8, 10, 13, 14,
27, 32, 58, 59, 135, 156, 175, 179,
193, 201, 242, 244, 279, 328, 329,
330, 332, 335, 337
Urban Infrastructure 126, 136, 238
Urban Living 123, 124, 143
Urban Mobility 152
User Experience 206, 216, 217, 220, 224,
225, 226, 227, 228, 231, 240, 241,
242, 318

V

Variability 86, 129, 287, 288, 290, 293,
294, 295, 299, 303
Variety 68, 93, 156, 178, 245, 287, 288,
290, 292, 293, 294, 295, 296, 299,
301, 303, 311, 312, 317, 318, 329,
333, 334, 336, 338, 374
VCPS 186, 187, 192
Velocity 129, 178, 190, 203, 245, 287, 288,
289, 290, 291, 293, 294, 295, 297, 298,
299, 301, 302, 303, 304
Veracity 287, 288, 290, 292, 294, 295, 298,
299, 302, 303
Virtual Reality 226, 310, 311, 312, 313,
314, 315, 316, 325, 326
Volume 11, 22, 129, 151, 163, 178, 203,
245, 263, 279, 287, 288, 289, 290,
293, 294, 295, 299, 301, 302, 303, 307
V. S. Anoop 343
Vulnerabilities 1, 2, 5, 6, 10, 18, 33, 51,
58, 59, 70, 79, 81, 152, 157, 160, 162,
164, 165, 168, 175, 177, 178, 195,
196, 198, 201, 206, 207, 222, 223,
225, 238, 251, 253, 262, 277, 327,

328, 329, 330, 331, 332, 333, 334, 335, 336, 337, 338